ENVIRONMENT MANAGEMENT TECHNOLOGY
環境管理科技詞匯

Environment Management Technology
環境管理科技詞匯

A Glossary of Modern Terminology
English-Chinese / Chinese-English

by Gottfried Eigenmann
Translated by Lu Fu-run

G・艾耕曼 著　　呂富潤 譯

The Chinese University Press
and
Novartis Country Organization in China

中文大學出版社
暨
諾華中國總部
出 版

ISBN: 962-201-856-4
      962-201-771-1 (Simplified Chinese characters edition)

The Chinese University Press
The Chinese University of Hong Kong
Sha Tin, N.T., Hong Kong
Fax: +852 2603 6692, +852 2603 7355
E-mail: cup@cuhk.edu.hk
Web-site: http://www.cuhk.edu.hk/cupress/w1.htm

Novartis Country Organization in China
16th Floor, Golden Land Building
No. 32, Liang Ma Bridge Road
Chao Yang District
Beijing 100016, China
Tel: +86-10-6464 1188
Fax: +86-10-6464 3747

國際統一書號(ISBN)：962-201-856-4
              962-201-771-1(中文簡體字版)

出版：中文大學出版社
      諾華中國總部

Printed in Hong Kong

Contents

目　錄

Preface

Why this glossary?

Environmental issues have become worldwide issues. Waste discharges have caused environment problems and hazards, which are often felt far from their sources in other regions and countries. Hence, efficient environmental protection activities are important, especially in countries under industrial development. China, the home of one-fifth of the world's population, is no exception. Her present rapid economic development must go hand in hand with effective protection of the environment.

Many reasons speak for supporting environmental protection in countries under development. Such activities are also in the interest of industrial nations who have accumulated a wealth of knowledge on environmental protection. Support may be given through technology transfer, exchange of information and sharing of knowledge, according to the needs of the receiving parties.

A common technical language will make clear and efficient communication possible, and will also facilitate mutual understanding. Global enterprises doing business in developing countries are in constant contact with local managers and engineers. Unfortunately, efficient communication by using a common language is sometimes difficult to establish. This glossary tries to fill this gap in Chinese-speaking areas.

The idea of this book came up at a discussion between the author and Mr. Guy Clayton, the Head of Ciba-Geigy (Hong Kong) Ltd., at a conference in Shanghai which focused on development issues. Guy Clayton's strong personal identification with the ideas of sustainable development (of which the solution of environmental problems is an important part) made this book possible. The book also received full support from Ciba-Geigy in Switzerland.

Ciba-Geigy, a worldwide manufacturer of a wide range of chemicals based in Switzerland, has been active and successful in the environmental field for many years. The know-how and skills developed by Ciba-Geigy over a long period of time form the main basis of this glossary.

The recent merger of Ciba-Geigy and Sandoz has led to two new enterprises: Novartis Ltd. and Ciba Specialty Chemicals Ltd. Both will carry on with the long tradition of Ciba-Geigy in environmental management and achievements. Both will also be active in China and continue the long-standing cooperation with the Chinese business concerns. With this book, which will be a contribution to technology transfer, Novartis Ltd. and Ciba Specialty Chemicals (China) Ltd. will continue their active support in solving environmental problems in China.

The objective of this bilingual English-Chinese glossary is to facilitate communication in environmental management. It will be a useful reference for specialists and should be of help and interest to non-professionals, to persons engaged in policy-making and environmental management, and to executives in public administration or private enterprises.

Solutions to environmental problems are multidisciplinary and require much technical and management skills. In this glossary, over one thousand entries of terms are included, providing basic information in all fields related to ecology and the environment, especially those in connection with manufacturing of goods and with general human activities. The very important concept of solving environmental problems at the source and by improving manufacturing processes has been emphasized. Analysis is treated as a fundamental tool for measuring and quantifying the level of pollution and the efficiency of counter measures. The various manage-ment aspects are related to an integral environmental management concept for eco-efficient solutions. Terms in the field of legislation include both traditional environmental legislation as well as explanations of the new market-oriented regulatory instruments. Entries on pollution control technology reflect modern methods used today in industrialized countries. They cover the fields of water, air and waste management. This book is designed in such a way that any reader who knows either English or Chinese is able to use it. The explanation of each term is given in both English and Chinese, and plain language is used to explain the technical terms.

The author and all sponsors and supporters of this book hope it will fulfill its purpose: to facilitate and improve communication on environmental matters in Chinese speaking areas.

Acknowledgments

No single person is able to have complete knowledge in a field as broad and as interdisciplinary as that covered in this book. The author has relied on help from various sources when compiling this book. Without these numerous contributions this glossary would not exist. The author acknowledges most gratefully the help and the support of all contributors.

Apart from information from published sources such as reference works, publications and technical documents, the book contains also the knowledge, skills and experience of the many managers, scientists and engineers engaged actively in environmental protection at Ciba-Geigy in Switzerland.

Mr. Lu Fu-run of the Environmental Protection Research Institute under the Ministry of Chemical Industry in Beijing, China, translated the English texts into Chinese. Special thanks go to him for his careful and never tiring help and important contribution.

Bridging the gap between the Swiss and Chinese thinking and understanding was a delicate work. Thanks go to Ms. Miranda Szeto of Ciba-Geigy (Hong Kong) Ltd., who in many respects served as a link between Switzerland, Hong Kong and China.

Ms. Desiree Allenspach and Ms. Monica del Bondio at the Ciba-Geigy Offices in Switzerland provided valuable administrative support, for which the author is very grateful.

Contributors

The author is also grateful to the following managers, specialists, scientists, engineers of Ciba-Geigy who had provided valuable technical information, suggestions and comments during the compilation of this book:

Constantine Amblianitis studied business administration at the University of South Africa where he graduated with a B.Comm. He is now with the Finance and Management Information Systems Group at Ciba, Horsham, England.

Rolf Bentz graduated with a doctorate in chemistry at the University of Basel. In the context of general environmental management, he specializes in remediation management.

Erich Buchta carried out studies in environmental technology and physics at the Technical University of Berlin and at Wuppertal, Germany. He is a specialist and expert in noise pollution control.

Michel Buser studied chemistry and graduated with a diploma in chemistry at the Institute of Engineering in Muttenz, Basel. He has also completed a post-graduate course in environmental protection. His present activities include effluent treatment technology and remediation projects.

Stephan Buser studied mechanical engineering at the Federal Institute of Technology in Zurich, Switzerland and graduated with a doctorate in technical sciences. He is now working on air pollution control problems.

Pierpaolo Cedraschi studied chemical process engineering at the Federal Institute of Technology in Zurich, Switzerland where he graduated with a diploma in chemical engineering. He is presently the Head of the Off-gas and Incineration Technology Group.

Kaspar Eigenmann studied chemistry at the Federal Institute of Technology in Zurich, Switzerland and graduated with a doctorate in physical chemistry. He is the Head of Corporate Safety and Environment.

Urs Gujer studied technical chemistry at the Federal Institute of Technology in Zurich, Switzerland where he graduated with a doctorate in technical sciences. He is the Head of Corporate Environmental Protection.

Hans Gysin graduated with a doctorate in organic chemistry from the University of Basel. He was the Head of the first Staff-Office of Environmental Protection.

Heinz U. Huper studied chemistry at the Institute of Technology in Winterthur, Switzerland and public health science at the University of North Carolina, USA. He graduated with a master degree in public health. He is a specialist in the design, operation and management of effluent plants.

Michael Reisinger studied organic chemistry at the Karl Franzens University in Graz, Austria where he graduated with a doctorate in chemistry. His current activities focus on effluent treatment technology and remediation problems.

Sigrid Rembold (Ms.) studied chemistry at the Technical University in Stuttgart, Germany. She graduated with a doctorate in natural sciences and is now engaged in effluent treatment and remediation projects.

Franz Schmalz studied airplane- and jet-propulsion systems at the Technical University of Berlin, Germany and graduated with a doctorate in engineering. He is with the Safety Management Group, engaged in explosion prevention.

Gerhard Stucki studied biochemistry and microbiology at the Federal Institute of Technology in Zurich, Switzerland and graduated with a doctorate in natural sciences. He is now a consultant on environmental biotechnology.

Markus Thuer studied process engineering at the Federal Institute of Technology in Zurich, Switzerland and graduated with a doctorate in technical sciences. He now heads the Effluent Management and Remediation Group.

Joachim Voigt studied technical chemistry at the University of Hannover, Germany, where he graduated with a doctorate in natural sciences. He is the Head of the Membrane and Filtration Technology Group.

Martin Wenk studied microbiology at the University of Zurich, Switzerland and graduated with a Ph.D. In the scope of environmental technology, he is a specialist in waste water technology and remediation questions.

Kurt Wyss studied chemistry at the University of Basel, Switzerland and graduated with a doctorate in chemistry. He is a specialist in environmental analysis.

Chun Xiao Yu-Jiang (Ms.) studied environmental chemistry at Hangzhou University (China) and at the University of Bern, Switzerland. She graduated with a Ph.D. Her present activities focus on effluent technology.

Gottfried Eigenmann
December 1996

序　言

編寫本詞典的緣由

環境問題已成為全世界共同關注的問題。廢物排放已引起許多環境問題和危害，而這些問題又常常在遠離它們產生的地方，甚至在其他地區或國家才顯現出來。因此，有效的環境保護是非常重要的，尤其是在工業發展中的國家。就中國這個佔世界人口五分之一的國家而言，也不例外。目前中國經濟的迅速發展，必須與保護環境協調一致。

有許多理由表明需要支持發展中國家的環境保護工作，而這對具有豐富環境保護經驗的工業化國家也是有利的。支持包括技術轉讓、信息交流和知識分享，視接受方的需要而定。

在技術上具有共同的語言將會促進交流及相互理解。許多跨國公司積極在發展中國家開展業務，經常要與當地的經理人員和工程師接觸交往，但是有時卻很難以共同的語言進行溝通。這本詞匯就是為了彌補這一缺陷而編寫的。

編寫這本詞匯的想法，始於作者與汽巴-嘉基(香港)有限公司董事長G.克禮頓先生在上海召開的一次研討有關發展問題的會議上相互交談討論產生的。克禮頓先生以可持續發展的觀點高瞻遠矚 —— 解決環境問題是可持續發展的一個重要組成部分 —— 使得這本詞匯得以問世。這一想法也得到了瑞士汽巴-嘉基公司總部的充分肯定。

總部設在瑞士的汽巴-嘉基公司，是生產多類化學品的全球性化工公司。多年來，它積極並成功地參與環境保護的活動。它長期以來研究開發的專門技術和知識，是構成本詞匯的重要基礎。

最近汽巴-嘉基公司和山道士公司的合併，導致形成了兩間新的公司，即諾華有限公司和汽巴精化有限公司。它們將繼承汽巴-嘉基公司環境管理工作的傳統，珍視在這方面所取得的業績，並且將在中國積極發展業務，繼續保持與中國工商界長期、持久的合作。以這本詞匯作為對技術合作的貢獻，諾華有限公司及汽巴精化(中國)有限公司將繼續積極支持中國解決環境問題。

　　這本英漢對照環保詞匯的目的，在於促進在環境管理方面的交流。它將是專家們有用的參考書，而對於關注環境問題的非專業技術人員、政府部門或私人企業的決策者和管理人員也將會有所幫助。

　　解決環境問題涉及多種學科，需要許多技術和管理的技能。本詞匯包括了一千多條術語，提供了與生態和環境有關的基本知識，特別是與產品生產和一般人類活動有關的環境管理與技術方面的知識。還強調了通過改進生產工藝及在源頭消除或減輕污染這一解決環境問題十分重要的觀念。分析技術是測定和定量評估污染程度及污染防治措施效果的基本工具。環境管理，尤其是在企業中樹立綜合環境管理概念，是保持生態平衡和解決污染問題的重要保證。本詞匯中法律法規方面的術語，包括了傳統的環境法律法規和新近的以市場經濟為導向的法制手段。污染控制技術方面的術語，則反映了現今工業發達國家所採用的現代技術，包括廢水、廢氣和廢物管理的控制技術。為了使英文讀者和中文讀者都能使用這本詞匯，對每一名詞都以中英兩種文字作了簡要解釋，對專門技術術語都盡量用簡單淺白的語句解釋。

　　作者及所有支持及參與編寫本詞匯的人都希望：這本詞匯能實現它原來設想的意圖，即在講華語的地區中能幫助和促進對環境事務的溝通和交流。

致　謝

　　本詞匯涉及的學科領域如此寬廣，以致沒有那一個人能具有全部所有的知識。在編寫本詞匯的過程中，作者參考了許多資料。沒有這些幫助，本詞匯是不可能編寫完成的。在此，作者對所有提供幫助和支持的先生和女士們表示衷心的感謝。這些參考資料包括參考書、出版物和技術文件。除了這些公開發表的資料外，本詞匯還包括了積極參與環境保護工作的瑞士汽巴-嘉基公司的許多經理、專家、科技人員和工程師的知識和經驗。

　　中國化學工業部環境保護研究所的呂富潤先生，將本詞匯的內容翻譯成為中文。對於他認真不倦的努力和所給與的重要幫助，特予致謝。

　　在瑞士和中國之間駕起一座溝通思想及理解的橋樑，是一項需予以格外關注的工作。在此作者要感謝汽巴－嘉基(香港)公司的司徒少貞女士,她在瑞士、香港、北京之間做了大量聯絡工作。

另外,作者也感謝瑞士汽巴－嘉基公司總部辦公室的D.艾倫斯佩克女士和M.邦迪歐女士,她們在行政工作上給與了大力的支持。

對本詞典作出貢獻的同仁

下面所列的汽巴－嘉基公司的經理、專家、科技人員和工程師,為本詞匯提供了有關的技術資料、建議和意見,並對許多內容進行了校核,特此致謝。

C.安布利安尼蒂斯,曾在南非大學修讀工商管理,獲商業學士學位。現在在英國霍爾舍姆汽巴公司財務和管理信息系統處工作。

R.本茨,畢業於瑞士巴塞爾大學化學系,獲博士學位。在環境管理方面,專長於污染場地的恢復補救管理。

E.布赫塔,曾在德國烏帕塔爾和柏林工業大學修讀環境工程和物理學,是噪聲污染控制專家。

M.布謝爾,畢業於瑞士巴塞爾的穆坦茲工程學院,獲化學學位證書,並且修讀了環境保護研究生課程。現在從事廢水處理技術和污染場地恢復補救工程方面的工作。

S.布謝爾,畢業於瑞士蘇黎士的聯邦工業大學機械工程系,獲技術科學博士學位。現在從事空氣污染控制方面的工作。

P.塞德拉西,畢業於瑞士蘇黎士的聯邦工業大學化學工程系,獲化學工程學位證書。現在是廢氣和焚燒技術小組的負責人。

K.艾格曼,畢業於瑞士蘇黎士的聯邦工業大學化學系,獲物理化學博士學位,現在是汽巴－嘉基公司安全與環境部的負責人。

U.古傑爾,曾在瑞士蘇黎士的聯邦工業大學修讀工藝化學,獲技術科學博士學位,現在是汽巴－嘉基公司環境保護處的負責人。

H.傑欣,畢業於瑞士巴塞爾大學,獲有機化學博士學位。曾任汽巴－嘉基公司最初的環境保護辦公室的負責人。

H.胡佩爾,畢業於溫特圖爾工業大學化學系,後在美國北卡羅來納大學獲公共衛生碩士學位,是廢水處理廠設計、運行和管理方面的專家。

M. 雷欣格，曾在奧地利格拉茨的卡爾弗蘭津斯大學修讀有機化學，獲化學博士學位。現在主要從事廢水處理和污染場地恢復補救工程方面的工作。

S. 雷姆博爾德，曾在德國斯圖加特工業大學修讀化學，獲自然科學博士學位。她現在從事廢水處理和污染場地恢復補救工程方面的工作。

F. 施馬爾茲，曾在德國柏林工業大學修讀飛機和噴氣推進系統，獲工程博士學位。現在在汽巴-嘉基公司安全管理處從事防爆方面的工作。

G. 斯圖基，曾在瑞士蘇黎士的聯邦工業大學修讀生物化學和微生物學，獲自然科學博士學位。現在從事環境生物技術諮詢工作。

M. 圖爾，曾在瑞士蘇黎士的聯邦工業大學修讀工藝工程，獲技術科學博士學位。現在是廢水管理和污染場地恢復補救小組的負責人。

J. 瓦格特，曾在德國漢諾威大學修讀工藝化學，獲自然科學博士學位。現在是膜和過濾技術小組的負責人。

M. 溫克，曾在瑞士蘇黎士大學修讀微生物學，獲哲學博士學位，是廢水處理技術和污染場地恢復補救工程的專家。

K. 偉斯，畢業於瑞士巴塞爾大學化學系，獲化學博士學位，是環境分析專家。

江春曉，曾在中國杭州大學和瑞士伯爾尼大學攻讀環境化學，獲博士學位。她現在主要從事廢水處理技術方面的工作。

G. 艾耕曼

1996年12月

About the author

Gottfried Eigenmann is Swiss and studied chemistry at the Institute of Technology in Burgdorf, Switzerland and at the University of Missouri, USA, where he graduated in 1957 with a Ph.D. in Organic Chemistry. At Ciba-Geigy, he initially carried out applied research in physical organic chemistry. Since 1973, his assignments included the environmental management for the whole Ciba-Geigy group, worldwide, and finally he became the Head of the Environmental Audit Group. He has been a private consultant since 1991, specializing on environmental auditing.

He has rich experience in various fields of environmental management. From projects in China, he is well aware of the situation there. He has given numerous lectures and has organized training programs in environmental protection and auditing.

About the translator

Lu Furun is Chinese and graduated from the Department of Meteorology, Nanjing University, the People's Republic of China in 1959. Since 1971, he has been involved in the development and application of environmental engineering technology at the Environmental Protection Research Institute of the Ministry of Chemistry Industry, the People's Republic of China. He was promoted to a senior engineer with professor title in 1988 and later assigned as Deputy Director and Chief Engineer of the Institute.

He has assisted in Ciba-Geigy's cooperative projects in environmental issues with the Ministry of Chemical Industry of the People's Republic of China, including Dr. G. Eigenmann's lectures and training programs in environmental protection and auditing in China.

作者簡介

　　G. 艾耕曼，瑞士人，曾在瑞士布格多夫工業大學修讀化學，1957年畢業於美國密蘇里大學，獲有機化學博士學位。然後一直在汽巴-嘉基公司工作，最初從事物理有機化學應用研究，從1973年起參與汽巴-嘉基公司在世界各地的環境管理工作，最後被任命為該公司環境審計處的負責人。從1991年開始，他轉為私人顧問，主要從事環境審計諮詢工作。

　　作者在環境管理的各個領域具有豐富經驗。通過參與在中國的項目，作者對中國的情況有一定的了解，並曾在中國就環境保護和審計進行過多次演講和培訓工作。

譯者簡介

　　呂富潤，中國人，1959年畢業於中國南京大學氣象系。自1971年起，一直在中國化學工業部環境保護研究所工作，從事環境工程技術的開發和應用研究。1988年晉升為教授級高級工程師，並先後擔任該所副所長和總工程師等職務。

　　他曾協助汽巴-嘉基公司與中國化學工業部在中國開展環境保護方面的合作項目，包括G. 艾耕曼博士就環境管理和審計在中國進行的講課和培訓。

ENGLISH-CHINESE GLOSSARY

The English term in each entry is followed first by its translation in Chinese and then by the pinyin pronunciation of its Chinese translation in italics. Its explanation is given first in Chinese and then in English. Terms in the explanation which appear also as or have meaning similar to entries in this glossary are also in italics. Readers may refer to these entries for further explanations.

此英中詞匯中跟著每一英文名詞的是其中譯名詞，隨著是用斜體字表示的中譯名詞的漢語拼音。另段則是此名詞的解釋，先是中文後是英文。在解釋中出現的名詞，若是包括在此詞匯中，或與其意思相近的，也用斜體字表示，以便讀者參閱。

A

AA analysis 原子吸收分析 *yuanzi xishou fenxi*

見：atomic absorption analysis。

See *atomic absorption analysis.*

abiotic 非生物的 *feishengwude*

無生命的，與生物的、有生命的含義相反。

Lifeless, without life, in contrast to biotic, living, with life.

abiotic components of an ecosystem 生態系統的非生物組成部分 *shengtai xitongde feishengwu zucheng bufen*

生態系統的無生命組成部分，如土壤中的礦物質、營養物質、濕度和水、空氣等。

Lifeless components of an ecosystem, like minerals in ground and soil, nutrients, humidity and water, atmospheric components and others.

abiotic ecological factors 非生物的生態因子 *feishengwude shengtai yinzi*

表示生態系統特徵的一類因子。非生物因子包括溫度、濕度、水、空氣、土壤、能量投入和光。

A group of factors characterizing an ecosystem. Abiotic factors include the temperature, humidity, water, air, ground/soil and also energy input and light.

absolute filter 純粹過濾器 *chuncui guolüqi*
一種對氣流中的物質具有很高截留效率和很低透過率的過濾器。它用合成纖維織物作為過濾材料，或者在某些情況下也可用燒結多孔材料。這種過濾器一般只在氣流預處理後作為最終淨化段使用。例如，它在"絕對清潔室"保持無菌條件的通風設備中和在核技術方面的應用，就是最典型的實例。而被濾除的物質，則可以加以回收。

A filter for gas streams with a very high retention efficiency and a very low rate of passage of matter. Absolute filters use mats of synthetic fibers as filter elements, or, in certain cases also sintered porous materials. Absolute filters are only used as a last step after pretreatment of a gas stream. Typical applications are in ventilation for 'clean rooms' to maintain sterile conditions or in nuclear technology. Materials, which are filtered off, may be recovered.

absorption 吸收(作用) *xishou (zuoyong)*
某一化合物被液體吸收的過程。在此過程中，化合物可發生化學變化，或者也可以不發生化學變化。

The process by which a chemical compound is taken up (absorbed) by a liquid. The compound may or may not undergo chemical changes during the process.

acceptable daily intake (ADI) 容許日攝入量 *rongxu risheruliang*
可被人每天吸收而不會對健康產生任何長期不良影響的某種化學物質的量。在考慮與農藥的使用有關的環境問題時，要應用容許日攝入量來計算食物中污染物的容許殘留濃度。容許日攝入量值與物質的特性有關，但也取決於與接觸該物質的人有關的一些其他因素。如果討論的問題涉及到土壤污染、建築物污染或整個場地污染，那麼對容許日攝入量還必須考慮場地的使用情況以及可能接觸的人群等因素。

The quantity of a chemical substance which can be absorbed daily by man without showing any long term negative health effects. In environmental considerations related to the use of pesticides, the ADI value is used in the calculation of permissible residual concentrations of contaminants in food. The ADI value is characteristic for the substance, but depends also on different factors related to the persons who are exposed to a substance.

If ADI is discussed in the context of the contamination of soil or ground, of buildings, or of a whole site, additional factors must be considered, such as the use of the site and the groups of the population which are potentially exposed.

accident control plan 事故控制計劃 *shigu kongzhi jihua*

見：containment measures; alarm plan。

See *containment measures*; *alarm plan*.

accidental releases 事故釋放 *shigu shifang*

由於事故向環境的釋放。

Releases to the environment as a result of an accident.

accumulation in the food chain 食物鏈中的積累 *shiwulianzhongde jilei*

化合物從一個生態組成部分向另一個生態組成部分轉移的過程，同時伴隨發生該化合物在*食物鏈*中的富集作用。它取決於化合物的性質，如污染物的*親脂性*和非生物降解性等。食物鏈中的積累會導致某些生物體和人的健康問題。

The process of transfer of chemical compounds from one ecological compartment to another which is coupled with a simultaneous concentration-increase of the compound in the *food chain*. The accumulation depends on the properties of the compound: the *lipophilic* properties, the non-biodegradable nature of a contaminant and others. Accumulation in the food chain can lead to health problems of certain organisms, also in man.

accuracy of a test method 試驗方法的準確度 *shiyan fangfade zhunquedu*

由分析測試方法測定得到的某組分的量與實際存在量之間相同一致的程度，它可用百分比數值表示。（參見：試驗方法的精密度）

Accuracy refers to the agreement between the amount of a component measured by the analytical test method and the amount actually present. Accuracy may be expressed numerically as a percentage value. (See also *precision of test method*)

acid rain 酸雨 *suanyu*

比正常降水具有較高酸度（即pH值較低）的降水、雨或雪。由於大氣中存在二氧化碳的緣故，所以正常雨水通常呈微弱酸性，其pH值在6.5−7範圍內。因為吸收了硫氧化物和氮氧化物等酸性污染物，故酸雨的pH較低。這些酸性氧化物來自不同的人為污染源，如熱電廠、車輛、工廠和家庭。據報導，在污染地區酸雨的pH值可低於4。

酸雨對生態系統的影響，取決於土壤對酸的緩衝能力和雨的酸度。曾報導，在土壤和地下層對酸不具緩衝能力的（花崗岩）地區，酸雨會引起湖泊酸化，造成魚類死亡的問題。另外，酸雨還會造成森林逐漸乾枯死亡的問題。（參見：森林破壞）

Precipitation, rain or snow, with an acidity higher (i.e. pH lower) than normal. Normal rain generally is faintly acidic with pH values in the range of 6.5−7 as a result of carbon dioxide in the atmosphere. The pH of acid rain is lower because of the absorption of acidic pollutants, like oxides of sulfur and oxides of nitrogen. These acidic oxides are resulting from different man-made sources: thermal power plants, traffic, industry and households. pH values below 4 have been reported for acid rain in polluted areas.

The effect of acid rain in ecosystems depends on the acid buffering capacity of the soil system and on the acidity of rain. Problems with fish have been reported in acidified lakes in areas with unbuffered (granite) soil and underground. Problems with slowly dying forest trees have also been attributed to acid rain. (See also *forest damages*)

activated carbon 活性碳 *huoxingtan*

用特殊方法製備的一種具有微細孔徑和高比面積的碳，它可吸附某些化合物，例如廢水和廢氣中的污染物。吸附過程是可逆的。

A specially prepared type of finely dispersed carbon with a high specific surface and with the capacity to adsorb certain compounds, e.g. pollutants in effluent and/or off-gas streams. The adsorption process is reversible.

activated carbon buffer system 活性碳緩衝系統 *huoxingtan huanchong xitong*

在空氣污染控制系統中用活性碳 *吸附塔* 作為處理過程的第一段。在高負荷操作期，活性碳吸附污染物；然後在低負荷操作期，活性碳進行 *解吸*。因此，減少了實際處理系統進口處濃度的波動。

The application of an activated carbon *adsorption tower* as a first process step in an air pollution control system. The activated carbon serves to adsorb contaminants during peak load operating periods, followed by *desorption* when the loads coming from processing are low. The concentration fluctuation at the input side of the actual treatment is thus reduced.

activated sludge 活性污泥 *huoxing wuni*

基於 *活性污泥技術* 運行的廢水生物處理場，其污泥 *混合液* 中具有生物活性的微生物。活性污泥含有能分解廢水中組分的各種好氧微生物。在好氧系統中，必須有空氣或氧存在，以保證好氧生物的生存。

The biologically active mixture of microorganism in the *mixed liquor* of a biological effluent treatment station, based on the *activated sludge technology*. The activated sludge contains different aerobic microorganisms which are capable to break down waste components of the effluent. In an aerobic system, air or oxygen must be present to ascertain survival of the aerobic organisms.

activated sludge basin 活性污泥池 *huoxingwunichi*

廢水生物處理場中持有污泥 *混合液* 及裝備有曝氣器的池子。生物降解在該池中進行。它可以建造在地上（如 *塔式生物反應器*）、地表面或地下（如 *深井曝氣技術*）。

The basin of a biological effluent treatment plant holding the *mixed liquor* and being equipped with aerators. Biodegradation takes place in this basin. Activated sludge basins can be constructed above ground (*tower bioreactor*), at the ground level, or below ground, e.g. as in the *deep shaft technology*.

activated sludge on supports 載體上的活性污泥 *zaitishangde huoxing wuni*

見：fixed film biological systems。

See *fixed film biological systems*.

activated sludge technology 活性污泥技術 *huoxingwuni jishu*

　　基於活性污泥微生物進行生物降解的廢水生物處理技術。其工藝過程包括以下基本單元：中和、一次沉降、生物降解、二次沉降、活性污泥回流和剩餘污泥處理。各單元的設計可以有很大的變化。該技術在19世紀末最早應用於英國和德國。

The biological effluent treatment technology based on the biodegradation properties of microorganisms of activated sludge. The process units include the basic steps: neutralization; primary clarification; biodegradation; secondary clarification; activated sludge recycling; handling of excess sludge. The design of the different steps may vary considerably. The technology was first used around the end of the 19th century in England and in Germany.

activation basin (tower) 活化池(塔) *huohuachi (ta)*

　　見：aeration basin, tall, tower。

See *aeration basin, tall, tower*.

active substance 活性物質 *huoxingwuzhi*

　　工業產品中賦予產品特定活性的化學物質。一種出售的產品可能含有一種或幾種不同的活性物質。

The chemical substance in a commercial product which is responsible for the specific activity of the product. One sales product may contain one or more different active substance.

active surface 活性表面 *huoxing biaomian*

　　含有分子相互作用活性部位的任何固體物質的表面，它對活性碳吸附具有重要作用。另外，它在痕量分析上也是非常重要的，因為在痕量分析中實際上可通過從樣品瓶表面無控制釋放出物質，或者無控制釋放物質到樣品瓶表面，而使樣品受到污染或變化。

A surface of any solid material which contains active sites for molecular interactions. An important aspect of activated carbon adsorption. Also important in trace analysis, where a sample flask may actually contaminate or change the sample through uncontrolled release of substances from or to the surface.

acute effect 急性效應 *jixing xiaoying*

突然開始且持續時間短的效應。

An effect with a clearly defined beginning and which manifests itself within a short period of time.

adaptation of activated sludge 活性污泥的適應(馴化) *huoxing wunide shiyin (xunhua)*

廢水生物處理活性污泥中的微生物逐漸適應於被處理的物質和廢水環境的過程。這一事實具有兩方面的意義：首先，可用來解釋生物降解的實驗結果，表明*接種菌種*的選擇對實驗室的實驗結果會有很大的影響。其次，在廢水*活性污泥*處理裝置的操作上，隨著活性污泥的馴化，會使得生物降解去除率逐漸提高。

Microorganisms in the biologically active sludge of an effluent treatment plant adapt themselves to some extent to the substances to be treated and the environment of the effluent. This fact has consequences in two areas: First, in the interpretation of laboratory results on biodegradation, the choice of the *inoculum* has an effect on the results. Second, in the operation of an *activated sludge* effluent treatment plant, the adaptation process may gradually lead to an improvement of an initially low rate of biodegradation.

add-on corrective measures 末端改善措施 *moduan gaishan cuoshi*

作為最終步驟，添加在生產過程末端的改善措施，如污染控制措施。(參見：末端工藝過程處理)

Corrective measures, e.g. pollution control measures which are added at the end, as a last step, of a manufacturing process. (See also *end-of-process treatment*)

ADI 容許日攝入量 *rongxu risheruliang*

見：acceptable daily intake。

See *acceptable daily intake*.

adsorbable organic halogen compounds 可吸附的有機鹵素化合物 *kexifude youji lusu huahewu*

見：AOX。

See *AOX*.

adsorbent 吸附劑 *xifuji*

能通過吸附作用束縛化合物的固體物質。活性碳常常可用作為吸附劑。

The solid substrate binding a compound by adsorption. Often activated carbon is used as an adsorbent.

adsorption 吸附(作用) *xifu (zuoyong)*

將化學物質束縛在*吸附劑*固體表面的作用。決定吸附過程的因子有：物質和活性表面之間的結合力的性質和強度、内比活性表面的大小、結合點佔的百分比。吸附可以發生自氣相或發生自液相。通常，吸附過程不發生化學變化。

Binding of a chemical substance on the solid surface of an *adsorbent*. Factors determining the adsorption process are: the nature and strength of the binding forces between the substance and the active surface, the magnitude of the specific internal active surface and the fraction of binding sites which are occupied. Adsorption may take place from the gaseous phase or from a solution. Generally, no chemical changes take place during the process.

adsorption columns 吸附塔 *xifuta*

吸附過程使用的裝填有吸附劑(如活性碳)的塔式裝置。吸附塔的有效容積為其總容積的60%到90%。

Processing units in the form of towers, containing the adsorptive material, e.g. activated carbon, used in an adsorption process. The active volume of an adsorption tower is in the order of 60 % to 90% of the total volume.

adsorption on activated carbon 活性碳吸附 *huoxingtan xifu*

淨化廢氣或廢水的一種方法。在活性碳吸附過程中，污染物被結合在活性碳上而得到去除。可用粒狀碳(*如填料塔*),也可用粉末碳進行吸附。

A method for the purification of off-gases or aqueous effluents. In this adsorption process the pollutant is removed by binding on activated carbon. The application is either in granular form (e.g. *packed column*) or as a powder.

aeration 曝氣 *baoqi*

把空氣中的氧或純氧引入好氧微生物系統中的過程,如引入水中或生物處理的活性污泥*混合液*中。

The introduction of oxygen from air or pure oxygen into an aerobic microbial process, in the present context into water or *mixed liquor* in the activated sludge process.

aeration basin, flat 曝氣池(淺平式) *baoqichi (qianpingshi)*

含有活性污泥*混合液*的廢水生物處理池。傳統的淺平式曝氣池是敞開式的,由*表面曝氣機*或*擴散器*引入空氣中的氧。增加曝氣池的深度是最近的發展趨勢,其目的在於提高充氧效率和減少惡臭污染,這類曝氣池可能是加蓋完全密閉的。

The basin of a biological effluent treatment station containing the biologically active *mixed liquor*. In the traditional flat form, the basin is open and the oxygen from air is introduced by *surface aerators* or *diffusors*. Newer developments include deeper tanks to increase oxygenation efficiency and to reduce air pollution from odorous components; these basins may be completely covered.

aeration basin, tall/tower 曝氣池(塔式) *baoqichi (tashi)*

傳統淺平式曝氣池的變型。活性污泥*混合液*在較高的塔式構築物中通過安裝在塔底的*曝氣器*引入空氣或純氧進行曝氣。它引入每公斤氧所消耗的能量比淺平式曝氣池小。

A variation of the traditional flat form of aeration basins. The biologically active *mixed liquor* is contained in a relatively tall, tower-like construction and the air or oxygen is introduced by *aerators* mounted at the bottom of the tower. The energy needed per kg of oxygen introduced into the mixed liquor is lower than in flat basins.

aeration equipment 曝氣設備 *baoqi shebei*

將氧引入污泥混合液或曝氣池中的專門設備。實際使用的曝氣設備有：*表面曝氣機、擴散器、射流曝氣系統和生物轉盤*。

The specific equipment used to introduce oxygen into the mixed liquor or an aeration basin. Several possibilities are used in practical equipment: *surface aerators, diffusors, jet aeration systems, rotating biological contactor.*

aeration residence time 曝氣停留時間 *baoqi tingliu shijian*

廢水在生物處理曝氣池中的平均停留時間。其最佳停留時間取決於廢水的生物降解性和比負荷。對於生活污水，典型的停留時間為2小時；對於高濃度的工業廢水，停留時間可能為10小時至24小時以上。在*延時曝氣系統*，停留時間可長達1–2天，這會使得剩餘活性污泥產生量較少，但能耗較高。

The average residence time of the effluent to be treated in the biological basin of the effluent plant. The optimal residence time depends on the biodegradability and the specific loading of the effluent. For domestic sewage, a residence time of 2 hours is typical, for highly loaded industrial effluents, the time may be 10 to more than 24 hours. In *extended aeration systems*, residence times as high as 1–2 days may be used. Such systems may lead to lower quantities of excess activated sludge but also to higher energy consumption.

aeration systems with oxygen 純氧曝氣系統 *chunyang baoqi xitong*

廢水生物處理靠引入純氧而不是空氣至污泥混合液中的曝氣系統，其優點是需處理的尾氣量較小，缺點是費用較高和需有純氧。工業規模的純氧曝氣系統有：*優諾克斯*(UNOX)系統和*林多克斯*(LINDOX)系統。

Aeration systems in biological effluent treatment which are based on the introduction of technically pure oxygen into the mixed liquor instead of air. The advantage of such systems lies in the smaller volumes of off-gas to be handled, the disadvantage in the cost and availability of pure oxygen. Several systems are available commercially: *UNOX, LINDOX*.

aerators 曝氣器 *baoqiqi*

見：aeration equipment。

See *aeration equipment*.

aerobic biodegradation 好氧生物降解 *haoyang shengwu jiangjie*

見：biodegradation, aerobic/anaerobic。

See *biodegradation, aerobic/anaerobic*.

aerobic process 好氧過程 *haoyang guocheng*

在空氣存在下發生的過程，更具體地說，這是基於好氧微生物（在有氧存在時才能生活的微生物）新陳代謝作用進行的生物降解過程。

A process taking place in the presence of air. More specifically, the biological degradation process which is based on the metabolism of aerobic microorganisms, which are microorganisms living in the presence of oxygen.

aerobic sludge digestion 污泥好氧消化 *wuni haoyang xiaohua*

見：sludge digestion, aerobic。

See *sludge digestion, aerobic*.

aerosol 氣溶膠 *qirongjiao*

氣流中充分分散的微滴。氣溶膠粒子的大小約為50–100毫微米。氣溶膠微滴可能含有生產過程產生的污染物，如試劑、原料、成品、副產品、溶劑、氯化銨等。氣溶膠微滴保持懸浮狀態，一般它們不能用常規方法分離，而需用更有效的方法如一次*凝集*、*撞擊和文丘里系統*加以去除。

Finely dispersed droplets in an air stream. The aerosol particles range in size up to about 50–100 nm. The aerosol droplets may contain other contaminants from a production process, like reagents, raw materials, finished products, by-products, solvents, ammonium chloride, etc. Aerosol droplets are kept in suspension and they generally do not separate by ordinary processes. Their removal requires more efficient methods like primary *agglomeration, impingers, venturi systems.*

aerosol separation 氣溶膠分離 *qirongjiao fenli*

將氣溶膠從污染的氣流中除去。如果將氣溶膠與微滴先進行*凝集*，然後在第二段加以分離，則能從廢氣中高效地將氣溶膠除去。

The elimination of aerosols from a polluted gas stream. Aerosols can be eliminated from off-gas streams with high efficiency if the polluting aerosols and fine droplets are first *agglomerated* and then separated in a second step.

agglomeration 凝集 *ningji*

許多小的粒子形成較大粒子的過程。廢氣中的氣溶膠通過凝集作用可形成較大的微滴。在水溶液中也能發生凝集過程，形成可以分離的沉澱物。(參見：混凝、絮凝)

The process of formation of larger particles from a number of smaller ones. The agglomeration of aerosols in an off-gas produces larger droplets. In aqueous solution the process of agglomeration can also take place and lead to the formation of precipitates which can then be separated. (See also *coagulation; flocculation*)

aggregate parameter 綜合指標(參數) *zonghe zhibiao (canshu)*

見：group parameter。

See *group parameter.*

air 空氣 *kongqi*

在自然未污染的地球大氣中氣體的混合物。空氣的組成(體積比)大約為：氧20.9%、氮78.1%、二氧化碳360 ppm(1990年)、水蒸汽(含量不定)、氬0.9%以及微量的其他稀有氣體如氦、氖、氪等。空氣中存在的臭氧濃度很低，而且隨高度而不同。此外，還存在有微量二氧化硫、氮氧化物、甲烷、塵粒和許多其他的有機和無機污染物。這些微量氣體有些是源於自然的，其餘是人為產生的。而要明確地區分哪些是自然存在的微量氣體，哪些是人為產生的污染氣體，需對具體情況加以仔細考慮。(參見：全球大氣的二氧化碳)

The physical mixture of gases in the natural unpolluted atmosphere of the earth. Air is a gas mixture containing approximately by volume: 20.9% of oxygen, 78.1% of nitrogen, ca. 360 ppm of carbon dioxide (1990), variable amounts of water vapor, 0.9% of argon, trace quantities of other rare gases like helium, neon, krypton. Ozone is present in trace concentrations depending on the altitude. In addition, traces of SO_2, oxides of nitrogen, methane, dust particles and many other organic and inorganic contaminants are present. Some of these trace-gases are of natural origin, others are man-made. A clear separation into natural trace gases and man-made pollution requires careful consideration of the specific circumstances. (See also *global atmospheric carbon dioxide*)

air and oxygen injection systems 空氣和氧噴射系統 *kongqiheyang penshe xitong*

見：aeration equipment。

See *aeration equipment*.

air pollution 空氣污染 *kongqi wuran*

通過污染物排放使空氣或大氣的自然組成發生變化的一切過程。污染物的排放主要是由人類的活動引起的。空氣污染會對人和生態系統中的生物產生影響，也會由於腐蝕作用對文化古迹和建築造成影響。有明顯的徵兆表明，空氣污染對地球氣候具有一定的影響。

The overall process of changing the natural composition of the air or atomsphere through emissions of pollutants. Emissions are caused mainly by *anthropogenic activities*. Air pollution has an impact on man, on the *biotic components of ecosystems*, and also on cultural monuments and technical structures as a result of increased corrosion. There are strong indications that air pollution has an impact on the overall climate of the earth.

air pollution bioindicators 空氣污染的指示生物 *kongqi wurande zhishi shengwu*

見：biomonitoring。

See *biomonitoring*.

air pollution control (APC) 空氣污染控制 *kongqi wuran kongzhi*

將空氣污染置於控制之下的一切活動，包括補救或預防技術、管理措施和改變人們的行為。

All activities to keep the pollution of the air under control. Such activities may include corrective or preventive technologies, administrative measures, and also changes of the behavior of the population.

air pollution control (APC) equipment 空氣污染控制設備 *kongqi wuran kongzhi shebei*

用於從廢氣中去除或分離污染物的設備。根據廢氣類型的不同，污染物可能是粉塵、顆粒物或各種氣體，還可能存在反應產物或副產物。

空氣污染控制設備主要有乾式和濕式兩類空氣污染控制系統。每一類還包括一些不同的單元裝置。（參見：濕式空氣污染控制系統、乾式空氣污染控制系統、吸收系統）

Equipment used to eliminate or separate pollutants from an off-gas stream. Depending on the type of the off-gas stream, pollutants may be dusts, particulate matter or gases. Reaction products or byproducts may be present.

The main groups of APC equipment are dry APC-systems and wet APC-systems. Each group includes several different unit types. (See also *APC-systems, wet*; *APC-systems, dry*; *absorption systems*)

air pollution control (APC) technology 空氣污染控制技術 *kongqi wuran kongzhi jishu*

空氣污染控制技術可用來分離或消除廢氣中的污染物，達到保護環境所要求的水平。然而，與水和固體污染控制一樣，*源頭削減*對空氣污染控制是非常重要的。對於不可避免產生的廢氣排放物，根據污染物的類型和所要求的處理效率，通常包括若干處理步驟。

例如，去除粉塵和氣溶膠，可採用以下具有不同作用的步驟：對於易於分離的較大顆粒，可直接進行分離；對於微細顆粒，則需調節顆粒表面以使它們凝集，並從氣相轉移到洗滌液中；然後將洗滌液小液滴加以分離。

對於溶劑蒸氣(*揮發性有機化合物*)和無機氣體的分離,可以採用*吸收、液化、填料洗滌塔*(*濕式空氣污染控制系統*)或*廢氣燃燒*等方法。

為了達到所要求的*排放限值*,可以設計一個包括乾式和濕式空氣污染控制過程單元的聯合處理系統。

APC technology serves to separate or eliminate pollutants from off-gas streams to levels required by environmental considerations. Here, as in other areas, *source reduction* is important. Technology to treat unavoidable emissions usually includes several steps, depending on the type of pollutant and the required efficiency.

For the removal of dusts and aerosols, the following steps serve different functions: direct separation of relatively coarse particles, which can easily be separated; conditioning of fine particle surfaces to allow for their subsequent agglomeration and transfer from the gas phase to the scrubbing liquor; separation of droplets of scrubbing liquid.

For the separation of solvent vapors (*VOC*'s) and inorganic gases, a number of other methods, such as *sorption processes*, *liquefaction*, *packed column scrubbers* (*wet APC-systems*), or *off-gas combustion* are also available.

The design of a combined system may include dry and wet APC-process units in order to reach the required *emission limits*.

air quality data types 數據類別,空氣質量 *shuju leibie, kongqi zhiliang*

見:data types, air quality。

See *data types, air quality*.

air quality goals 空氣質量目標 *kongqi zhiliang mubiao*

改善環境空氣質量的長期目標。它規定了一般污染物如粉塵、二氧化硫、氮氧化物、臭氧和顆粒物的環境空氣濃度。這些目標應在一定的時限內達到。如果需要,還可包括一些其他的參數。

因為*排放和環境空氣濃度*之間的關係是複雜的,這些目標只能作為中期或長期目標來實現。此外,由於來自鄰近地區或國家的排放對實現目標會有一定的影響,所以需要相互進行磋商與合作。空氣質量目標雖不可能直接實施,但其實是個排放控制的問題。

Long range goals for an amelioration of the quality of ambient air. Such goals may specify ambient air concentrations for common pollutants such as dust, sulfur dioxide, oxides of nitrogen, ozone and particulate matter. These goals should be reached within a specified time period. Additional parameters may be included if needed.

As the relationship between *emissions* and *ambient air concentrations* is complex, such goals can only be reached in the medium or long term. Since emissions from neighboring regions or countries may play a role in meeting these goals, consultations and cooperation between countries are necessary. Air quality goals cannot be mandated directly, but reaching them is a question of the control of emissions.

air stripping 空氣氣提 *kongqiqiti*

從水溶液中去除如溶劑等揮發性組分的過程。操作時,在氣提塔中鼓入小的空氣氣泡,使溶液中的揮發性組分轉入氣相而被去除。在空氣氣提中,化合物的蒸氣壓和它在水中的溶解度是決定因素。此技術被應用於工藝過程與分析上。

A process for removing volatile components, e.g. solvents, from an aqueous solution. By blowing small air bubbles through the solution in a column, components are transferred to the gas phase and are removed. The vapor pressure of the compound and its solubility in water are determining factors. The process is used on a technical scale as well as in analysis.

alarm plan 警報計劃 *jingbao jihua*

說明事故發生時必須執行的行動及其負責人的有關指令。這些行動包括事故遏制和消滅的措施,以及向企業職工、政府主管部門和可能受影響的公眾進行通報。(參見:緊急響應管理)

Instructions which list the activities to be executed in case of an accident. The list also specifies the persons responsible for these actions. The actions include containment and accident combat measures, as well as informing the employees of the enterprise, the authorities, and the public who may be affected. (See also *emergency response management*)

algae 藻類 *zaolei*

大多是水生植物，範圍包括原生生物到藻絲體和大的藻類。它總共有2萬多個不同的種類。綠藻是如高等植物那樣含葉綠素的生物體，它們能*同化*二氧化碳供自身生長。在此過程中，綠藻還能在水體中產生氧。除了綠藻之外，還有不同顏色的藻類，它們不含葉綠素，但含有產生其顏色的其他物質。蘭藻是由單細胞組成的像原生細菌那樣的生物體。因此，這些藻類的生物化學過程很不相同。藻類是*浮游植物*的一個重要組成部分。

Mostly aquatic plants, ranging from primitive organisms to filaments and large algae. There are over 20,000 different species. Green algae are higher plant-like organisms that contain chlorophyll. They are able to *assimilate* carbon dioxide for growth. In this process, green algae also produce oxygen in the waterbody. In addition to the green types, there are algae of different colors which contain no chlorophyll but other substances responsible for their colors. Blue algae are primitive bacteria-like organisms made up of a single cell. As a consequence, the biochemistry of these groups differs considerably. Algae constitute an important fraction of *phytoplankton*.

allocation of pollution control costs 污染控制費用的分攤 *wuran kongzhi feiyongde fentan*

見：cost allocation。

See *cost allocation*.

ambient air 環境空氣 *huanjing kongqi*

人、生物或物體周圍的空氣。

The atmospheric air around a person, an organism or an object.

ambient air concentrations 環境空氣濃度 *huanjing kongqi nongdu*

環境空氣或大氣中污染物的濃度。環境空氣濃度與空氣質量及其目標有關，但它是不能直接實施的。（參見：數據類別、環境空氣）

Concentrations of pollutants in the ambient air or the atmosphere. Ambient air concentrations are relevant in the context of air quality discussions and of quality goals. Ambient air concentrations cannot be enforced directly. (See also *data types, ambient air*)

ambient air goals 環境空氣目標 *huanjing kongqi mubiao*

　　某地區或國家空氣污染控制法規所確定的質量目標，它指的是環境空氣中污染物的濃度，而不是法定限制的排放濃度或排放負荷。當污染是由不同來源引起時，環境空氣目標與排放濃度或排放負荷之間的關係是複雜的。

Quality goals defined in air pollution control legislation for a certain region or country. Ambient air goals refer to concentrations of pollutants in the ambient air, in contrast to legislation limiting emission concentrations or emission loads. The relationship between ambient air goals and emissions concentrations or loads is complicated when a number of sources contribute to the pollution.

ambient noise 環境噪聲 *huanjing zaosheng*

　　見：noise, ambient。

See *noise, ambient.*

ambient noise (immission) limits 環境噪聲 (散發) 限值 *huanjing zaosheng (sanfa) xianzhi*

　　見：noise, ambient (immission) limits。

See *noise, ambient (immission) limits.*

anaerobic biodegradation 厭氧生物降解 *yanyang shengwu jiangjie*

　　見：biodegradation, aerobic/anaerobic。

See *biodegradation, aerobic/anaerobic.*

anaerobic biological processes 厭氧生物過程 *yanyang shengwu guocheng*

　　通常在無空氣存在或無氧條件下發生的過程。在環境保護中，這是在無氧條件下發生的*厭氧微生物*的新陳代謝過程。厭氧過程是還原過程，發酵和發穢臭就是厭氧過程，某些廢水處理過程也是通過厭氧微生

物進行的。厭氧過程常常導致生成小分子有機物或具惡臭的無機還原產物。(參見：廢水厭氧處理)

In general, a process which takes place in the absence of air or oxygen. In environmental protection, it refers to a group of biological metabolic processes of *anaerobic microorganisms* which take place in the absence of oxygen. Anaerobic processes are reductive processes. Fermentation and fouling processes belong to this group. Certain effluent treatment processes are based on anaerobic micro-organisms. Anaerobic processes often lead to small organic molecules or inorganic reduction products with a strong odor. (See also *anaerobic effluent treatment*)

anaerobic effluent treatment 厭氧廢水處理 *yanyang feishui chuli*

應用*厭氧生物過程*將廢水中污染物轉化為甲烷和二氧化碳。此外，也會生成一些還原化合物。厭氧過程是很有效的，與好氧過程相比，它產生的剩餘污泥量要少得多。由於厭氧過程常常產生成有臭味的物質，在其後通常應有好氧後處理過程。

The application of *anaerobic biological processes* for the conversion of pollutants in effluents to methane and carbon dioxide. In addition, some other reduced compounds are formed. The process can be very efficient and a considerably lower quantity of excess sludge is produced than in aerobic processes. Since smelling components are often formed, an anaerobic step is generally followed by an aerobic aftertreatment.

anaerobic microorganisms 厭氧微生物 *yanyang weishengwu*

生活在無氧環境中的微生物，它們能使含氧物質還原並放出分子氧。例如，硝酸鹽、硫酸鹽或有機化合物都可作為供氧體。

Microorganisms that live in oxygen-free environments. They reduce substances with bound oxygen in the place of molecular oxygen. For example, nitrate, sulfate or organic compounds may serve as oxygen donors.

anaerobic sludge digestion 污泥厭氧消化 *wuni yanyang xiaohua*

見：sludge digestion, anaerobic。

See *sludge digestion, anaerobic*.

anaerobic waterbody 厭氧水體 *yanyang shuiti*

　　缺少充足的氧來維持湖泊或河流中生物正常生活的水體。*厭氧狀態*是由於供氧不足以及死去的有機體生物在降解過程中大量耗氧而造成的。通常，湖泊的深處呈現缺氧狀態，但在表層仍存在有一定濃度的氧。在這樣的水體中，在底部會發生*厭氧生物過程*。厭氧水體的*恢復*，可以通過降低其*生物需氧量*（BOD）和補充供氧，如人工曝氣充氧來進行。（參見：富營養化）

A waterbody which lacks sufficient oxygen to support normal life in a lake or a river. The *anaerobic* state is a result of insufficient oxygen supply, coinciding with a high demand of oxygen for biodegradation processes, e.g. of dead organic matter. Often, the deeper regions of a lake lack oxygen, while there is still a reasonable concentration in the surface layers. In such a waterbody, *anaerobic biological processes* take place at the bottom. The *remediation* of anaerobic waterbodies may be promoted by lowering the *BOD* demand and by additional oxygen supply, e.g. through artificial aeration or oxygenation. (See also *eutrophication*)

analog recorder 模擬記錄器 *moni jiluqi*

　　見：recorder, analog。

See *recorder, analog.*

analog signal 模擬信號 *moni xinhao*

　　由某種作用產生的信號，它與作用的強度成正比。例如，由電子儀器測量流量、液面、壓力或pH等物理性質產生的電壓或電流信號。模擬信號可以表現為儀器（如電流表或電壓表）指針的位置，或者壓力計中指示液體的液面。為進一步處理，需把模擬信號轉換成*數字信號*。

A signal derived from a certain action. The signal is proportional to the intensity of the action. Examples are: the voltage or current signal derived from an electronic measurement of physical properties, e.g. flow, level, pressure or pH. An analog signal may be given as the position of a pointer of an instrument (e.g. a mA or voltmeter) or the level of an indicator fluid in a manometer. For further processing, analog signals need to be transformed into *digital signals.*

analysis, continuous 連續分析 *lianxu fenxi*

　　對某項指標（參數）連續進行分析或測定。例如，在廢水處理場排放口處進行的pH的連續測定。測定結果可以用圖示或數字記錄在*數據記錄儀上*，而不直接反饋到控制環路上。

　　An analysis or measurement of a parameter which is carried out continuously. The continuous measurement of pH at the outflow of an effluent treatment plant may serve as an example. The result may be recorded graphically or digitally on a *data-logger*. There is no direct feedback to a control loop.

analysis, environmental 環境分析 *huanjing fenxi*

　　對環境的各個組成部分（空氣、水和土壤）中的污染和污染物進行的分析，它是控制排放、評價環境狀況、研究環境各組成部分之間的相互作用和質量傳遞的基礎。它通常採用化學分析的標準分析方法，但必須適合於痕量分析。由於污染物呈混合狀態而且為痕量含量，因此在取樣、樣品保存、分析方法和分析結果的解釋上必須小心謹慎。

　　The analysis of contamination and pollutants in all compartments of the environment (air, water and ground). Environmental analysis serves to control emissions, assess the state of the environment, and to support investigations of interactions and mass transfers between the environmental compartments. Standard analytical methods of chemical analysis are generally applied, though these methods must be adapted to trace concentrations. In view of the complexity of mixtures and the trace levels of contamination, care must be taken with sampling, sample pre-servation, methodology and the interpretation of results.

analysis, on-line 在線分析 *zaixian fenxi*

　　在連續分析的體系中將所獲得的信號直接輸入到控制系統，經過加工處理後來調節閥門或控制參數。（參見：連續分析、模擬信號、數字信號）

　　A continuous analysis system in which the resulting signal is used directly as input into a control system, where it is processed to set certain valves or control parameters. (See also *analysis, continuous*; *analog signal*; *digital signal*)

analysis of special wastes 特殊廢物分析 *teshu feiwu fenxi*

　　為滿足廢物處置的需要對特殊廢物進行的分析，它隨廢物管理和許可規定的要求而不同。由於全面分析十分複雜，因此最好採用分步分析方法（參見：層次分析計劃）。對於擬在*受控填埋場處置*的廢物，*浸出液*的分析是一個重要的項目；而對於進行焚燒處理廢物，焚燒特性和*煙氣*中排放污染物的分析是很重要的。

Special wastes need to be analyzed to satisfy the requirements before their disposal. Details of the analysis depend on the management and permissible procedure for the wastes. In view of the complexity of a full analysis, a stepwise procedure is advantageous (See also *analysis plan in tiers*). For wastes to be deposited in a *controlled landfill*, the analysis of the *leachate* is an important element. For wastes to be incinerated, incineration characteristics as well as the emission of pollutants in the *flue gas* are important.

analysis plan in tiers 層次分析計劃 *cengci fenxi jihua*

　　作為決策的依據，在需要進行許多不同的測定與測試時，為了減少總的分析工作量而採用的一種工作體系。在層次分析計劃中，要對測試參數排出不同的優先次序，然後根據這一次序進行分析。如果第一級分析的結果表明需要進行第二級分析，則才進行第二級分析。例如，在特殊廢物的特性鑒別與管理中，情況就是這樣。

A system for economizing the overall analytical efforts in cases where a number of different measurements and tests are needed as a decision base. In the tier plan, parameters are given different priorities and the analyses are carried out according to these priorities. Second level analyses are only carried out if first level results warrant this. Analyses in tiers are carried out for instance in the characterization and management of special wastes.

anemometer 風速計 *fengsuji*

　　測定風速或廢氣管道中氣流速度的儀器。

An instrument for measuring the velocity of wind or an off-gas stream in an off-gas duct.

anion exchange 陰離子交換 *yinlizi jiaohuan*

通過陰離子交換樹脂的吸收作用，去除水溶液中的陰離子。

The removal of anions from an aqueous solution by absorption on anion exchange resins.

anoxic biodegradation 缺氧生物降解 *queyang shengwu jiangjie*

生物氧化降解的一種特殊情況，在此過程中，分子氧不是取自空氣，而是取自廢水中的物質，如硝酸鹽可以作為供氧體而同時被還原。缺氧生物降解常應用於生物反硝化過程。(參見：硝化、反硝化)

A special case of an oxidative biodegradation in which the molecular oxygen is not taken from air but from substances in the effluent. In this process, nitrates may serve oxygen donors; and they are thereby reduced. Anoxic biodegradation is used in the process of biological denitrification. (See also *nitrification*; *denitrification*)

anthropogenic activities 人(類)的活動 *ren (lei) de huodong*

人們的活動。

Activities of human beings.

anthropogenic emissions 人為排放 *renwei paifang*

由人們的活動，如食物生產、商品製造、運輸、維持正常的生活條件所產生和排放的各種廢物。人為排放與自然現象引起的排放，如火山噴發、生態系統中的生物過程和風蝕引起的排放是完全不同的。

Emissions and wastes of all kinds resulting from the activities of human beings, such as production of food, manufacturing of goods, transport, maintaining proper living conditions etc. Anthropogenic emissions are in contrast to emissions resulting from natural phenomena, like e.g. emissions from volcanoes, from biological processes in the ecosystems, and from wind erosion.

anti-pollution equipment 污染防治設備 *wuran fangzhi shebei*

減少廢物量或排放量或減少它們影響的技術設備。污染防治設備可能是生產過程一個組成部分，也可以應用於工藝過程的末端，或作為獨立的污染控制設備。(參見：水污染控制、空氣污染控制、廢物處置)

Technical equipment to reduce the quantity of waste streams or emissions, or to reduce their effect. Anti-pollution equipment can be part of a manufacturing process, or it can be applied at the end of the process, or as separate pollution control equipment. (See also *water pollution control*; *air pollution control*; *waste disposal*)

antifoaming agent 消泡劑 *xiaopaoji*

為了抑制廢水中表面活性物質產生的泡沫而投加到廢水處理場*活性污泥池*中的物質。消泡劑會增大液體的表面張力。

A substance added to the *activated sludge basin* of an effluent treatment plant in order to subdue the foam resulting from surface active substances in the effluent. Antifoaming agents increase the surface tension of the liquid.

AOS 可吸附的有機硫 *kexifude youjiliu*

Adsorbable organic sulfur的縮寫。它是用於度量和評價水溶性含硫化合物的一項*綜合指標*。目前(1996年)還沒有標準化的測定方法,因此對其測定結果必須謹慎加以說明。

Adsorbable organic sulfur, a *group parameter* intended for the measurement and assessment of water soluble compounds containing sulfur. The method is at present (1996) not yet standardized. Results must therefore be interpreted with caution.

AOX 可吸附的有機鹵素化合物 *kexifude youji lusu huahewu*

Adsorbable organic halogen compounds的縮寫。它是水污染控制中使用的一項*綜合指標*,指那些可被活性碳吸附的有機鹵素化合物。其含量的測定是將活性碳全部燃燒,其氣體用水溶液加以吸收,然後用銀鹽滴定法或庫倫分析法測定鹵化物。

Adsorbable organic halogen compounds, a *group parameter* used in water pollution control. It includes those organic halogen compounds which can be adsorbed on activated carbon. The determination of the quantity is carried out by total combustion of the activated carbon. The gases are absorbed in an aqueous solution. Halide is determined by argentometry or coulometry.

APC 空氣污染控制 *kongqi wuran kongzhi*

　　見：air pollution control。

　　See *air pollution control*.

APC equipment 空氣污染控制設備 *kongqi wuran kongzhi shebei*

　　見：air pollution control equipment。

　　See *air pollution control equipment*.

APC technology 空氣污染控制技術 *kongqi wuran kongzhi jishu*

　　見：air pollution control technology。

　　See *air pollution control technology*.

APC-concept 空氣污染控制觀念 *kongqi wuran kongzhi guannian*

　　生產場地空氣污染控制與管理的總體觀念。它以空氣污染物 *排放*（包括 *點源排放* 和 *擴散排放*）清單為基礎。對於不同類型的 *廢氣*，可以採用不同的空氣污染控制技術。所選用的技術要符合法律標準和公司內部目標與標準。這種總體觀念為空氣污染控制設施和措施的管理奠定了良好的基礎。

　　An overall concept for managing the air pollution situation of a manufacturing site. The concept is based on an *emissions inventory* including *point source emission* and *diffuse emissions*. Air pollution control technology takes into consideration the different *off-gas* types. Technology is selected to meet legal standards and company internal objectives and standards. Such a concept forms one base for a well managed implementation of air pollution control facilities and measures.

APC-systems, biological 生物法空氣污染控制系統 *shengwufa kongqi wuran kongzhi xitong*

　　基於生物降解過程的濕式廢氣處理系統。（參見：生物濾池、生物滴濾池、生物洗滌器）

　　Wet off-gas treatment systems based on biodegradation processes. (See also *biofilter*; *bio trickling filter*; *bio scrubber*)

APC-systems, combined 組合空氣污染控制系統 *zuhe kongqi wuran kongzhi xitong*

　　由若干段濕式和／或乾式空氣污染過程組合設計而成的空氣污染控制系統，其每一段都有它特定的作用。它可根據污染物的類別和所要求的去除效率，將不同的空氣污染控制過程組合成一個整體。一個完整的組合空氣污染控制系統還必須包括用*洗滌液*去除污染物後對含在洗滌液中的污染物進行處理所需的後續處理單元。（參見：空氣污染控制技術、濕式空氣污染控制系統、乾式空氣污染控制系統）

A custom-designed air pollution control system which combines several wet and/or dry APC process steps, each having a specific function. Depending on the type of a pollutant and the required efficiency, different APC process steps are integrated. A complete APC-system must also include the units necessary for subsequent treatment of the pollutants which are removed with, and contained in, the aqueous *scrubbing liquor*. (See also *APC technology*; *APC-systems, wet*; *APC-systems, dry*)

APC-systems, dry 乾式空氣污染控制系統 *ganshi kongqi wuran kongzhi xitong*

　　主要用於分離粉塵和顆粒物的空氣污染控制設備。該系統不使用洗滌液，當廢氣通過系統時，由於重力、慣性、離心力、靜電吸引作用和尺度區分等不同的作用原理，使污染物得到分離。（參見：重力沉降箱、旋風分離器、廢氣過濾器、靜電分離器）

Air pollution control equipment used mainly for the separation of dust and particulate matter. No scrubbing liquor is used. The off-gas is passed through the equipment and different principles lead to the separation of the pollutants: gravity force, inertia, centrifugal forces, electrostatic attraction, and size discrimination. (See also *gravity settling box*; *cyclone separator*; *filters, off-gas*; *electrostatic separator*)

APC-systems, wet 濕式空氣污染控制系統 *shishi kongqi wuran kongzhi xitong*

　　基於污染物從氣相傳遞到洗滌液液相的原理而設計的一組空氣污染控制設備，大多數系統用*洗滌液*洗滌，但有些系統用有機溶劑工作。為了進行有效的質量傳遞，必須強化氣體與液體之間的接觸，並盡可能增大接觸表面(傳遞表面)。

其技術體系包括*填料洗滌塔*、*噴淋塔*、*泡罩洗滌塔*、*噴射洗滌器*和*文丘里型洗滌器*。根據*生物降解*原理，有些系統利用*生物濾池*進行污染物的生物降解處理。(參見：生物洗滌器(廢氣))

A group of air pollution control devices which are based on the principle of transferring the pollutant from the gas phase to the liquid phase of a scrubbing liquor. Most systems work with aqueous *scrubbing liquors*, but some applications make use of organic solvents. To provide for an efficient mass transfer, the contact between the gas and the liquid must be intense and the contact surface (transfer surface) must be as large as possible.

Technical systems include *packed column scrubbers*; *spray towers*; *bubble column scrubber*; *jet scrubbers*; *Venturi type scrubbers*. Some systems are based on the *biodegradation* of pollutants in a *biofilter*. (See also *bio scrubber, off gas*)

apparatus off-air 裝置廢氣 *zhuangzhi feiqi*

直接在某工藝設備收集的廢氣，通常其污染物濃度很高，流量在1000–15000米³/小時(或更大)範圍內。例如，室內空氣進入反應器人孔被溶劑蒸氣污染後由通風系統排出的廢氣。(參見：其他類型廢氣)

An off-gas stream collected directly at a processing unit. The pollutant concentration is usually high. Flow volumes of apparatus off-air range between 1,000 and 15,000 m³/h and higher. Example: air that enters the manhole of a reactor as room air becomes contaminated with solvent vapor and is then withdrawn as off-air by a ventilation system. (See also *off-gas classes for other types*)

aquatic toxicity of single substances 單一物質的水生生物毒性 *danyi wuzhide shuisheng shengwu duxing*

見：toxicity, aquatic, of single substances。

See *toxicity, aquatic, of single substances*.

aquatic toxicology 水生生物毒理學 *shuisheng shengwu dulixue*

見：toxicology, aquatic。

See *toxicology, aquatic*.

assimilation 同化作用 *tonghua zuoyong*

生物（主要是植物）吸收營養物和其他化合物到其體內合成細胞物質而生長的能力。同化作用建立在一系列生物化學過程的基礎上。例如，利用葉綠素和太陽能，二氧化碳通過光化學同化作用，可生成碳水化合物和細胞物質。

The capacity of organisms, mainly plants, to integrate nutrients and other compounds into their system in order to grow and build up cell mass. Assimilation is based on biochemical processes. An example is the photochemical assimilation of CO_2 to produce carbohydrates and cell mass, using chlorophyll and solar energy.

ASTM Standards 美國材料試驗學會標準 *meiguo cailiao shiyan xuehui biaozhun*

由設在首都華盛頓特區的美國材料試驗學會制定的*環境分析*的標準方法。在環境分析中，美國材料試驗學會標準方法已為全世界所採用。

Standards, including analytical methods for *environmental analysis*, issued by the American Society for Testing Materials (ASTM) in Washington, DC. In environmental analysis, ASTM standard methods are applied worldwide.

atmosphere 大氣 *daqi*

泛指包圍行星的氣體層，在特殊意義上指地球周圍的*空氣層*。大氣圈與*陸圈*和*水圈*是地球*生態圈*的三個基本組成部分。大氣可延伸至500公里以上的高度，那裏氣體的濃度很低。關於大氣的組成大約100公里高度範圍內。（參見：空氣）

In the general sense, the layer of gas enveloping a planet. In the more specific use of the word, the layer of *air* around the earth. Together with the *lithosphere* and the *hydrosphere*, the atmosphere is one of the three basic compartments of the global *ecosphere*. Atmospheric gases extend in ever diminishing concentrations up to an altitude of more than 500 km. For the composition of the atmosphere up to a level of about 100 km. (See also *air*)

atomic absorption (AA) analysis 原子吸收分析 *yuanzi xishou fenxi*

測定水樣中存在的多種微量金屬元素的分析方法。用這種方法不需在分析前進行*富集*處理。在進行適當準備後，如在某些情況下用硝酸進行消解後，將樣品注入*火焰等離子體*使原子電離，然後，在每一元素特定的波長處通過紫外光的吸收來測定離子化的金屬原子。在環境分析中，原子吸收分析主要用來測定鋅、汞、鎘和錫等金屬。

An analytical method to measure a wide range of metallic elements in trace quantities, e.g. in water samples. No *enrichment* is needed before the analysis. After suitable preparation, in some cases digestion with nitric acid, the sample is introduced into a *flame plasma* which ionizes the atoms. The ionized metal atoms are measured by the absorption of UV light at a specific wavelength for each element. In environmental analysis, atomic absorption analysis is mainly used to determine metals like zinc, mercury, cadmium and tin.

atomic absorption flame spectrometry 原子吸收火焰光譜法 *yuanzi xishou huoyan guangpufa*

見：atomic absorption analysis。

See *atomic absorption analysis*.

audit 審計 *shenji*

由獨立審計人或審計小組對實際情況進行系統、全面和綜合的評價。審計過程就是將實際情況與確認的目標相比較，並確定可能需要採取的行動與措施。目標可來自法規、國際準則或公司內部標準。通常，審計由企業領導提出申請，審計的建議也應提交給企業領導。

A systematic, comprehensive and overall assessment and evaluation of a situation by an independent auditor or audit team. In the audit process, the actual situation is compared to valid objectives and goals, and possible needs for action are identified. Objectives and goals may be derived from legislation, from international norms or from company internal standards. Generally, an audit is requested by top management and the audit recommendations go to top management.

audit findings 審計結果 *shenji jieguo*

工廠的實際情況與*審計標準*之間的偏離。審計結果是工廠採取改進措施的依據。

Deviations of the actual situation at a site from *audit standards*. Audit findings form the base of corrective actions to be carried out at the site.

audit procedure 審計程序 *shenji chengxu*

由*審計小組*在工廠對其狀況進行評價與分析的系統管理程序。評價是在查閱文件材料以及在現場會見有關人員的基礎上作出的。分析則是將實際情況與*審計標準*進行比較。審計程序的最終步驟是針對工廠的狀況編寫出審計報告，其內容還應包括審計者提出的改進建議。

A systematic management procedure for the assessment and analysis of the situation at a plant by the *audit team*. The assessment is based on written evidence (documents) and interviews with personnel at the site. The analysis includes a comparison of the situation with *audit standards*. The audit procedure ends with the editing of a report on the situation. The report may also include recommendations by the auditors for corrective actions.

audit standards 審計標準 *shenji biaozhun*

涉及管理、工作效能和技術的標準，它包括適用的法律、公司內部的規定和*現代工藝技術水平*。審計過程就是將工廠的實際情況與這些標準相比較。審計報告中要反映與審計標準相偏離的情況，並提出需要採取的改正措施。

Standards referring to management, performance and technology. Audit standards may include elements of applicable legislation, company internal directives, as well as the *state-of-the-art*. In the audit process, the actual situation at a plant is compared with such standards. Deviations from audit standards are reported in the audit report as items which need corrective actions.

audit team 審計小組 *shenji xiaozu*

　　負責進行狀況審計的工作小組。它由該審計領域的不同專家所組成，並在審計小組組長的領導下進行工作。

The team charged with carrying out a situational audit. The team is generally led by a team leader who may be supported by different specialists in the fields to be audited.

auditor 審計者 *shenjizhe*

　　代表高層管理人員對工廠的實際情況獨立地進行分析和評價的人員。通常，審計者不是工廠所屬的人員，因此是公正的。審計者由一審計小組包括該領域的專家所支持。（參見：審計）

An independent person, analyzing and assessing the situation at a plant on behalf of senior management. The auditor generally is not part of the site organization and is therefore impartial. He may be supported by an audit team including specialists in the fields to be analyzed. (See also *audit*)

autecology 個體生態學 *geti shengtaixue*

　　著重研究環境條件及其對個別生物體或物種影響的生態學。（參見：生態學、群落生態學）

Ecology focusing on the environmental conditions and influences upon individual organisms or species. (See also *ecology*; *synecology*)

auto regeneration of a waterbody 水體的自動再生 *shuitide zidong zaisheng*

　　水生生態系統消化和礦化污染物的能力。自淨是微生物和高等生物所進行的生物化學過程的結果。在此過程中，細菌消化了有機污染物，使它們變為無機物，同時生成了生物質，而生物質又進一步被*原生動物*降解。水體的自動再生類似廢水處理場中的*生物降解*過程。自淨的能力取決於存在有上述生物體、有足夠的溶解氧以及其他因子。如果不能滿足連續再生的條件，水體就會迅速轉入另一種狀態，使得通過自淨機處理來恢復水體的水質大大延緩或不可能。

The ability of an aquatic ecosystem to digest and mineralize pollutants. Self-purification is the result of biochemical processes of microorganisms and higher life forms. Bacteria ingest the organic pollutants, mineralize them and produce biomass, which in turn is further degraded by *protozoa*. Auto regeneration is similar to the *biodegradation* process in an effluent plant. The capacity for self-purification depends on the presence of such organisms, sufficient dissolved oxygen and other factors. If the conditions for continuous regeneration are not fulfilled, a waterbody may rapidly flip over into a state from which recovery by self-purification mechanisms is greatly retarded or impossible.

automotive catalytic converters 汽車催化轉化器 *qiche cuihua zhuanhuaqi*

安裝在使用汽油的汽車發動機上的體積較小的催化轉化器。目前在大多數工業化國家中，法律規定必須採用這種轉化器。它具有多種作用：去除氮氧化合物、一氧化碳和殘餘碳氫化合物（痕量汽油）。為此，可使用"三效催化劑"（具有三種作用的混合催化系統）。它是將貴金屬催化劑混合物沉積在一整體陶瓷多孔體上而製成。汽車催化轉化器是在溫度350–500°C下工作，進行一系列複雜的氧化、還原反應。鉛會使該催化體系中毒，因此必須使用無鉛汽油。為了有效地運行，必須嚴格地控制尾氣中氧的含量。*蘭達傳感器*是這一系統的一個部件，由它發出的信號經過加工處理被用來調節發動機中空氣與汽油的比例。

Relatively small catalytic converters for automotive engines, using gasoline (OTTO motors). The use of such converters is nowadays prescribed by law in most industrialized countries. The converter has multiple functions: removal of nitrogen oxides, of carbon monoxide and of residual hydrocarbons (gasoline traces). To achieve this, 'three-way-catalysts' (mixed catalytic systems with three functions) are used. The noble metal catalyst mixture is deposited on a monolithic ceramic cellular body. A complex set of oxidation and reduction reactions takes place at the temperature of 350–500°C. This catalytic system can be poisoned by lead, therefore lead-free gasoline has to be used. For an efficient operation, the oxygen content of the exhaust gas must be carefully controlled. For this, a *lambda sensor* is part of the system. The signal from this sensor is processed and used to adjust the air to gasoline ratio in the engine.

autosampler 自動取樣器 *zidong quyangqi*

儀器分析中使用的一種取樣器，它能自動更送分析樣品，使樣品能在夜間自動進行分析。自動取樣器是與自動分析儀器一起使用的。

An instrument in instrumental analysis which automatically changes analytical samples and which allows samples to be analyzed unattended over night. Autosamplers are used in conjunction with automatic analytical instruments.

autotrophic organisms 自養生物 *ziyang shengwu*

像綠色植物或特殊微生物一類的生物，它們僅僅靠無機化合物如水、二氧化碳、氫、硫化氫（以及其他無機化合物和營養物）而生長及合成細胞物質。（參見：異養生物）

Organisms, like green plants or special microorganisms, which exclusively grow on inorganic compounds, like H_2O, CO_2, H_2, H_2S (and other inorganic compounds and nutrients) to synthesize cell material. (See also *heterotrophic organisms*)

auxiliary operations 輔助操作 *fuzhucaozuo*

與生產過程無直接關係的操作，但它們對生產過程具有重要的支持作用。在工廠中這些輔助操作包括：能源生產、貯運、廢物處置、維修等。在總體環境概念中，必須把輔助操作排出的"三廢"也包括在內。

Processing operations which are not a direct part of the manufacturing process but which have important support functions. In a factory such auxiliary operations are: energy production, transport and storage, waste disposal services, maintenance and repair, etc. In an overall environmental concept, emissions from such auxiliary operations must also be included.

B

background noise 背景噪聲 *beijing zaosheng*

見：noise, background。

See *noise, background.*

bacteria 細菌 *xijun*

一類單細胞生物體，它對人和動物的健康、發酵和發穢臭過程以及對生態系統中物質的降解等許多不同過程起著重要的作用。通過細菌的新陳代謝作用，可將有機化合物分解成較小的分子。代謝產物依細菌的類別和條件而不同。（參見：好氧、厭氧微生物）

A class of single cell organisms which play an important role in many different processes: human and animal health, fermentation and fouling processes, degradation of matter in ecosystems, etc. Through their metabolism, bacteria break down organic compounds into smaller molecules. The metabolic products depend on the type of bacteria and on the conditions. (See also *aerobic*; *anaerobic microorganisms*)

bag filter 袋式過濾器 *daishi guolüqi*

從氣流中去除塵粒的過濾設備。實際的過濾器被設計為袖式或袋式。工作時，污染空氣從外側通過布袋至內側，將塵粒收集在布袋外側。袋式過濾器可由若干過濾小區組合而成。現代的過濾器還裝備有自淨震蕩設備。不同的過濾材料具有不同的孔徑和應用範圍。（參見：濕式袖式過濾器）

A filtering device for removing dusts and particulates from an air stream. The actual filter is designed in the form of sleeves or bags. The polluted air passes from the outside to the inside, and the separated dust is collected outside the bag. Bag filters may incorporate a number of filter compartments. Modern versions are equipped with a self-cleaning shaking mechanism. Different filter materials provide for a range of pore sizes and applications. (See also *wet sleeve filter*)

baghouse 袋濾室 *dailüshi*

袖式或*袋式過濾器*系統中容納各單元的外套或容器。

The housing or container holding the individual elements of a system of sleeve filters or *bag filters*.

barometric BOD₅ measurement 壓力法五天生物需氧量測定 *yalifa wutian shengwu xuyangliang ceding*

見：BOD₅ measurement, barometric。

See *BOD₅ measurement, barometric*.

Basel Convention of 1989 1989年巴塞爾公約 *yijiubajiunian basai'er gongyue*

在*聯合國環境規劃署*主持下在瑞士巴塞爾制定的一個涉及廢物管理的國際公約（全稱為“控制危險廢物越境轉移及其處置的1989年巴塞爾公約”）。它的主要目的是防止非正當地將廢物從一國轉移到另一國所產生的環境問題。該條約對危險廢物的跨國轉運予以控制。受控制的危險廢物在*聯合國環境規劃署廢物分類系統*中作出了規定。對這些廢物的出口、過境和進口，必須取得所有直接當事國的同意。國家有權禁止某些廢物的進口（或出口）。該條約從1992年開始生效。除了一些主要工業國未簽字之外，已經幾乎有100個國家批准了該條約。

An international convention developed in Basel, Switzerland, under the auspices of *UNEP*, dealing with waste management (full name: 1989 Basel Convention on the Control of Transboundary Movements of Hazardous Wastes and Their Disposal). Its main objective is to prevent environmental problems resulting from improper waste movements from country to country. The treaty regulates the transnational transfer and movement of hazardous wastes. Hazardous wastes to be controlled are defined in the *UNEP waste classification system*. The export, transit and import of such wastes requires the consent of all nations directly concerned. Countries have the right to prohibit the import (or the export) of certain wastes. The treaty, in force since 1992, has been ratified by almost 100 parties, with the exception of some major industrial nations.

BAT 可獲得的最佳技術 *kehuodede zuijia jishu*
    見：best available technology。
    See *best available technology.*

BATEA 可獲得而經濟的最佳技術 *kehuode'er jingjide zuijia jishu*
    見：best available technology economically achievable。
    See *best available technology economically achievable.*

benefit-cost analysis 效益−費用分析 *xiaoyi-feiyong fenxi*
    見：cost-benefit analysis。
    See *cost-benefit analysis.*

benthnic ecosystem 底棲生態系統 *diqi shengtai xitong*
    水體底部如海底、湖底的生態系統。
    The ecosystem at the bottom of a waterbody, e.g. at the sea bottom or at the bottom of a lake.

benthon, benthos 底棲生物 *diqi shengwu*
    海底或湖底的動物和植物群落。
    Fauna and flora at the sea bottom or at the bottom of a lake.

best available technology (BAT) 可獲得的最佳技術 *kehuodede zuijia jishu*
    見：best available technology economically achievable。
    See *best available technology economically achievable.*

best available technology economically achievable (BATEA) 可獲得而經濟的最佳技術 *kehuode'er jingjide zuijia jishu*
    它是為完成某項工作任務，在目前的污染控制技術中可獲得的最佳技術。在某些法規中使用這一術語來定義和說明技術的水平。"經濟上可實現的"意味著所採用的技術具有經濟競爭能力，而不是尚處在研究與中試階段的技術。
    A definition and description of the best technology available to carry out a

certain task, in the present context of pollution control technology. The term is used in connection with certain legislation specifying technology. The qualifier 'economically achievable' implies technologies which can be applied in an economically competitive business situation, in contrast to technology which is yet in the research and piloting stage.

best practicable technology (BPT) 最佳實用技術 *zuijia shiyong jishu*

已在實際工業條件下應用並証明有效的技術，但並不包括最新發展的技術。在某些法規中使用這一術語來定義和説明技術的水平。

Technology which has been applied, and is proven efficient, under practical industrial conditions, but which may not yet include the latest developments. A term used in connection with certain legislation specifying technology levels.

binary reference gas mixer 二元參比氣體混合器 *eryuan canbi qiti hunheqi*

用來制備兩組分*校準混合氣*的儀器。這種混合氣可用於氣相色譜分析的線性校準。

A mechanical instrument which is used to produce *calibration gas mixtures* of two components. Such mixtures are used for calibration of the linearity of gas chromatographic analyses.

bio adaptation 生物適應(馴化) *shengwu shiying (xunhua)*

生物物種經過多代逐漸變化適應環境的過程。由於生物馴化作用的結果，會使生物物種的遺傳密碼逐漸變化，從而更好地適應它所在的*小生境*的條件。

因為微生物一代的壽命較短(細菌的壽命為幾小時)，因此它的馴化要比高等生物快得多。生物馴化對於*活性污泥適應*於廢水中的主要污染物具有重要作用，但它也會產生不良後果，例如當昆蟲適應殺蟲劑後就會產生抗性。(參見：活性污泥的適應或馴化)

A slow modification process of organisms which takes place over many generations of the species. As a result of bio adaptation, the genetic code of an organism is slowly changed, in order to adapt the organism better to the conditions in its *ecological niche*.

Because microorganisms have a relatively short life-span of one generation, (bacteria in the order of hours) adaptation is much more rapid than in higher organisms. Bio adaptation has positive consequences in the *adaptation of activated sludge* to critical pollutants in effluents. It can also have negative consequences, for instance when insects adapt themselves to insecticides and develop resistance. (See also *adaptation of activated sludge*)

bio diversity 生物多樣性 *shengwu duoyangxing*

見：diversity, biological。

See *diversity, biological.*

bio sciences 生物科學 *shengwu kexue*

見：life sciences。

See *life sciences.*

bio scrubber, off-gas 生物洗滌器（廢氣）*shengwu xidiqi (feiqi)*

與標準濕式洗滌器基本相似的廢氣處理系統，但其洗滌液與廢水處理場的污泥混合液相似。這些洗滌液大部分被保存在一個單獨的貯池中，從那裏循環回到洗滌塔，同時在洗滌塔進行曝氣。在洗滌塔中，污染物由廢氣轉移到洗滌液，然後在貯池中通過生物降解而被去除，但在貯池中可能需要進行補充曝氣。這種系統可應用於處理含水溶性、無毒、可生物降解物質的廢氣。

A system for treatment of off-gases which is basically similar to a normal *wet scrubber*. The composition of the scrubbing liquor, however, is similar to the *mixed liquor* of an effluent treatment plant. Its main volume is kept in a separate basin from where it is circulated through the scrubbing tower. Aeration takes place in the scrubbing tower. Pollutants are transferred from the off-gas stream to the scrubbing liquor and are destroyed in a separate tank by biodegradation. There, additional aeration may be needed. The system is applicable to off-gas streams containing substances which are water soluble, non-toxic and biodegradable.

bio trickling filter for off-gas 廢氣處理生物滴濾池 *feiqi chuli shengwu dilüchi*

含可生物降解、水溶性、無毒物質的廢氣的處理系統。在該系統中，廢氣以並流方式通過裝填*有規填料*的填料塔。填料係由惰性物質製成，它起到作為活性生物膜的*載體*的作用，並且依靠載體上的生物膜將廢氣中的污染物進行生物降解。為使生物膜保持潮濕狀態，需將含有微生物所需營養物的水在系統中循環。

生物滴濾池比一般*生物濾池*易於控制，它適用於去除生物降解的代謝產物 (如鹽酸或硫酸) 的情況。

A system for treatment of off-gas streams containing substances which are biodegradable, water-soluble and non-toxic. The off-gas is passed in co-current through a column with a *structural filling*. This is made of an inert material and acts as a support for a coating of biologically active material which serves to biodegrade contaminants in the off-gas. To stabilize the humidity of the biomass, water, containing nutrients for the microorganisms, is circulating.

The bio trickling filter is more controllable than the simple *biofilter*, and it can be applied in cases where the metabolites of biodegradation (like e.g. hydrochloric acid or sulfuric acid) need to be removed.

bio-assay 生物測定 *shengwu ceding*

基於應用生物反應和指示物的分析方法或評價方法。(參見：生物監測、免疫測定法、生態毒性)

An analytical or assessment method based on the use of biological reactions and indicators. (See also *biomonitoring*; *immuno-assay*; *ecotoxicity*)

bio-compost 生物堆肥 *shengwu duifei*

見：biofilter。

See *biofilter*.

bio-remediation 生物法恢復補救 *shengwufa huifu bujiu*

用生物方法通過真菌、細菌和高等植物的作用，對已污染的地下水和土壤進行恢復補救。

對於地下水的恢復補救，常常採用*固定膜生物反應器*，即用泵將已

39

污染的地下水抽取出來進行生物處理，然後再用泵打回到土壤中。處理後的廢水在重新滲濾之前，有時需進行充氧（氧源為空氣、過氧化氫或硝酸鹽）和添加就地補充處理所需的營養物。

對於污染土壤的恢復補救，在有些情況下採用類似堆肥處理的方法。（參見：恢復補救）

Remediation of contaminated groundwater and/or contaminated soil and ground using biological processes with fungi, bacteria and also higher plants.

For this type of groundwater remediation, often *fixed film bioreactors* are used. Contaminated groundwater is extracted by pumping, treated biologically, and is then pumped back into the ground. Before re-infiltration, treated water is sometimes enriched with oxygen (or sources of oxygen like air, hydrogen peroxide or nitrate) and nutrients for additional in-situ treatment.

For remediation of contaminated soils, a process similar to composting has been applied in some cases. (See also *remediation*)

biocenose 生物群落 *shengwu qunluo*

彼此相互作用並且生活在同一棲息地的植物和動物群體。它們是適應於棲息地特殊條件的生物群落。（參見：群落生境、生態系統、生態因素，有生命的，非生物的）

The populations of plants (flora) and animals (fauna) interacting with each other and occupying the same habitat or biotope. The biocenose is adapted to the particular conditions of the habitat. (See also *biotope*; *ecosystem*; *ecological factors, biotic, abiotic*)

biocides 生物殺傷劑 *shengwu shashangji*

能殺死生物體的物質。它既可以殺死特定的目標生物體，也可以殺死範圍廣泛的目標生物體。生物殺傷劑的作用潛力變化很大，取決於物質的化學結構、接觸的濃度和類型、目標生物體的敏感度。

Substances which kill organisms. Biocides act either specifically or towards a broad range of target organisms. The biocidal potential varies to a great extent and depends on the chemical structure of the substance, the concentration and mode of exposure and the specific sensitivity of the target organism.

bioconcentration 生物富集 *shengwu fuji*

親油化合物在經由食物鏈的每一不同營養級時，在生物體脂肪組織中濃縮富集的現象。（參見：食物鏈、營養級）

The phenomenon of increasing the concentration of a lipophilic compound in the fatty tissue of the organisms as the compound is passed along each subsequent trophic level of the food chain. (See also *food chain*; *trophic level*)

biodegradable 可生物降解的 *keshengwu jiangjiede*

某種有機化合物的屬性，即可通過生物體系中微生物的作用，使該有機化合物進行生物降解。（參見：礦質化）

The property of an organic chemical substance which allows for its biological degradation by microorganisms in a biological system. (See also *mineralization*)

biodegradation, aerobic/anaerobic 生物降解（好氧，厭氧）*shengwu jiangjie (haoyang, yanyang)*

有機物通過生物化學過程由大分子降解為小分子的過程，生物降解是由微生物進行的新陳代謝過程，好氧生物降解由好氧微生物進行，而厭氧生物降解則是在無氧條件下由厭氧微生物進行的。生物降解是生態系統中物質轉化和分解的一個重要而普遍的原理，在技術上已應用於水污染控制中。其應用包括好氧過程和厭氧過程。（參見：活性污泥技術、厭氧廢水處理、厭氧微生物、厭氧生物過程）

The process of degradation of organic material by biochemical processes whereby smaller molecules are formed from larger ones. Biodegradation is a metabolic process, carried out by microorganisms. Aerobic biodegradation takes place with oxygen consuming microorganisms, the anaerobic biodegradation takes place in the absence of oxygen, by anaerobic organisms. Biodegradation is an important general principle in the transformation and breakdown of matter in the ecosystems. This principle is applied technically in water pollution control. Both aerobic as well as anaeroble processes are used. (See also *activated sludge technology*; *anaerobic effluent treatment*; *anaerobic microorganisms*; *anaerobic biological processes*)

biodegradation rate 生物降解速率 *shengwu jiangjie sulü*

　　某物質通過生物降解被去除的速率（反應速度）。

　　The rate (reaction velocity) at which a substance is removed by biodegradation.

biofilm 生物膜 *shengwumo*

　　生長和生活在載體表面的呈薄膜狀的生物污泥。（參見：載體上的生物）

　　The thin layer of biomass which grows and lives on the surface of a support. (See also *biologies on supports*)

biofilter 生物濾池 *shengwu lüchi*

　　用生物法從廢氣中去除某些污染物的工藝設備。它利用生長在多孔固體載體上的好氧微生物進行生化氧化。可以將一定量的泥炭沼與生物堆肥混合用作過濾層，其厚度為0.5至1.5米左右。處理時，氣體的平均接觸時間應至少在60秒左右，可依污染物的性質而不同。生物濾池必須在其上噴水以始終保持濕潤狀態。所去除的物質必須可溶於水和可生物降解。生物濾池具有不同的結構，包括簡單的淺平活性材料層到複雜的樓式建築。

　　A process unit to remove certain pollutants from an off-gas stream by a biological system. A biochemical oxidation takes place in the aerobic biological growth on a solid, porous support. Certain qualities of peat moss, or heather twigs, mixed with e.g. bio-compost can be used. The thickness of the layer varies from 0.5 to ca. 1.5 m; the average contact time of the gas should be at least around 60 seconds, depending on the properties of the pollutant. The biofilter must always be kept in a moist condition by spraying water onto it. Substances to be removed must be somewhat soluble in water and must be biodegradable. The construction of a biofilter may vary from a simple flat layer of active material to elaborate systems housed in a separate building.

biogas 生物氣（沼氣）*shengwuqi (zhaoqi)*

　　廢水生物處理場剩餘活性污泥或其他有機廢物厭氧發酵過程產生的

42

以甲烷為主並含少量二氧化碳的混合氣，它可以通過許多方式被利用作為能源。（參見：污泥發酵、填埋氣體）

A mixture of mainly methane and a smaller amount of carbon dioxide which is produced by *anaerobic processes* by fermentation of excess activated sludge from a biological effluent plant, or from other organic wastes. Biogas may be used in various ways to supply energy from organic wastes. (See also *sludge fermentation*; *landfill gas*)

bioindicators 指示生物 *zhishi shengwu*

生態系統（如河流生態系統）中對污染有明顯和不同敏感性的特定生物體（重要的生物體）。評價這些指示生物體的狀況，可以定性說明所研究的生態系統的狀態。

Specific groups of organisms (key organisms) of ecosystems, such as a river ecosystem, with distinct and different sensitivities to pollution. An assessment of the condition of these indicator organisms may serve to qualify the state of the ecosystem under investigation.

bioindicators, air pollution 空氣污染的指示生物 *kongqi wurande zhishi shengwu*

見：biomonitoring。

See *biomonitoring*.

bioindicators in river ecosystems 河流生態系統中的指示生物 *heliu shengtai xitongzhongde zhishi shengwu*

生活在河流生境中對水和河床沉積物中的污染物具有不同程度敏感性的生物。非常敏感的生物（如某些蠅種的幼蟲）將只能存活在很乾淨的河流中；而要求比較不高的生物（如某些纖毛蟲）將可存活在輕度污染的環境中，並可能取代較敏感的生物。因此，估測不同主要生物群體的數量，就可表示河流的質量。所以，指示生物可起到補充化學分析的作用。已利用這種指示生物以及其他標準確定了一些河流如萊茵河的污染程度，並以此為依據對這些河流進行了分類。

Organisms living in a river biotope, which are sensitive in varying degrees to pollutants in the water and in the sediments of the river bed. Very sensitive organisms (for instance larvae of several species of flies) will only survive in a very clean river. Less demanding organisms (for instance several *ciliata*) will also survive in a somewhat polluted environment and may replace the more sensitive organisms. An assessment of the magnitude of different key populations will therefore give an indication of the river quality. Bioindicators may thus complement chemical analysis. Rivers, e.g. the Rhine river, have been classified according to their degree of pollution using, amongst other criteria, such organisms.

biological activated carbon adsorber 生物活性碳吸附裝置 *shengwu huoxingtan xifu zhuangzhi*

　　一種廢水處理的生物反應器，其中裝填活性碳作為微生物生長的載體。

A bioreactor for waste water in which the biomass is grown on the packing of activated carbon which acts as a random support.

biological APC-systems 生物法空氣污染控制系統 *shengwufa kongqi wuran kongzhi xitong*

　　見：APC-systems, biological。

　　See *APC-systems, biological.*

biological diversity 生物多樣性 *shengwu duoyangxing*

　　見：diversity, biological。

　　See *diversity, biological.*

biological effluent plant 廢水生物處理場 *feishui shengwu chulichang*

　　見：activated sludge technology。

　　See *activated sludge technology.*

biological off-gas treatment 廢氣生物處理 *feiqi shengwu chuli*

　　基於廢氣中污染物生物降解原理而設計的廢氣處理系統。當污染物為水溶性、可生物降解並對微生物無毒時,可應用此方法進行處理。廢氣生物處理系統包括*生物濾池、生物濕式洗滌系統和生物滴濾池*。

Off-gas treatment systems which are based on biodegradation of pollutants contained in an off-gas stream. These systems are applicable in cases where the pollutants are water soluble, biodegradable and non-toxic to the microorganisms. Applications include *biofilters, biological wet scrubbing systems* and *bio trickling filters*.

biological oxygen demand (BOD) 生物需氧量 *shengwu xuyangliang*

用於定量表示廢水中可生物降解物質進行氧化所需要的氧量的一項指標。它可用標準測試法進行測定。其下標"5"指在標準條件下進行為期5天的生物降解。但氨氮和某些無機化合物也可能被氧化。

A parameter used for quantification of oxygen required for the oxidation of biodegradable substances in sewage or an effluent. It is measured by a standardized test. The subscript '5' refers to a biodegradation time of 5 days under standard conditions. Ammonia-N as well as some inorganic compounds may also be oxidized.

biological treatment 生物處理 *shengwu chuli*

見:activated sludge technology。

See *activated sludge technology*.

biologies on supports 載體上的生物 *zaitishangde shengwu*

見:support biologies。

See *support biologies*.

biology 生物學 *shengwuxue*

從事生物體研究的廣大領域,它是生命科學的一部分。生物學可以進一步劃分為大生物學(研究大的生物體,如動物學研究動物,植物學研究植物);微生物學(研究微生物,如細菌、真菌);生物化學(研究生物體系中的化學反應及相互作用);生態學(研究生物群體在其群落中和

生態系統中的相互作用）。除了這些一般分類外，還有其他一些專門的
研究領域。

The large study field that investigates living organisms. It is thus one aspect of the life sciences. The field of biology can be subdivided into: macrobiology that investigates large organisms in zoology (the study of animals) and in phytology (the study of plants); microbiology or the study of microorganisms (bacteria, fungi) and biochemistry (study of the chemical reactions and interactions in biological systems); and ecology or the study of interactions of populations of organisms in their communities and in their ecosystems. In addition to these general groups, a number of specialized fields are being investigated.

biomagnification 生物放大 *shengwu fangda*

見：bioconcentration。

See *bioconcentration*.

biomass 生物量 *shengwuliang*

生態系統中有生命的有機物質（動物和植物群落）的總量。按其應用
場合的不同，可採用各自特定的測量單位。

The total of the mass of living organic material (flora and fauna) in an eco-system. Depending on the application of this term, specific units for measurement are used.

biomonitoring 生物監測 *shengwu jiance*

在實用上通過觀察和鑒定*指示生物*來評價痕量污染物的影響。一般
分析測定所提供的數據說明某一時刻的狀況，而生物監測所提供的指標
表示某一時期污染的總體情況。觀察位於工廠中重要地點的指示植物的
生長和損害情況，就可以對環境空氣中的污染物作出評價，但在鑒定其
損害時，必須認真考慮氣候因素。也可利用特定的指示生物來評價對水
體的影響。在任何情況下，為了作出解釋，必須了解各污染物的*生態毒
性*。（參見：指示生物）

A pragmatic possibility to assess the impact of trace pollutants by the observa-tion and evaluation of *bioindicators*. While normal analytical measurements provide data on the situation at a given moment, biomonitoring will give indications of the integral value of contamination over a certain period of time. Contaminants in ambient air can be assessed by observation of the growth and of possible damages to indicator plants located in strategic positions in a factory. For the evaluation of damages, a careful consideration of climatic factors must be included. Specific bioindicator organisms may serve to asses the impact of pollution on waterbodies. In all cases, for interpretation, the *ecotoxicity* of individual pollutants must be known. (See also *bioindicators*)

bioreactor, fluidized bed 生物反應器(流化床) *shengwu fanyingqi (liuhuachuang)*

用微生物法去除廢水中污染物的一種反應器。在流化床系統中，微生物生長在載體上形成生物膜。它可包括厭氧和好氧兩種過程。(參見：補救恢復、生物降解、生物法補救恢復、流化床系統)

A type of reactor for the removal of pollutants from polluted water by micro-biological methods. The microorganisms form a film on a support in a fluidized bed system. Anaerobic and/or aerobic processes are possible. (See also *remediation*; *biodegradation*; *bio-remediation*; *fluidized bed systems*)

biosludge 生物污泥 *shengwu wuni*

見：activated sludge。

See *activated sludge*.

biosludge/activated carbon systems 生物污泥/活性碳系統 *shengwu wuni/ huoxingtan xitong*

一種活性污泥處理技術。它係將少量粉末活性碳投加到曝氣池的混合液中以提高處理效率。活性碳有兩種作用：作為廢水中某些有毒有害化合物的吸附劑，以及作為活性污泥微生物生長的載體。(參見：粉末活性碳處理)

An activated sludge treatment technology in which powdered activated carbon is added in small amounts to the mixed liquor in the biology basin to increase its efficiency. The activated carbon has a double function: it may act as an adsorbent for certain noxious compounds in the effluent, and it may act as a support for the biological growth in the activated sludge. (See also *powdered activated carbon treatment*)

biosphere 生物圈 *shengwuquan*

不同生物體生活繁衍所在的地球各層的總稱，包括大氣圈、水圈和岩石圈。

The total of the different strata ('layers') of the earth which are populated by different organisms. The biosphere includes as main compartments: the *atmosphere*, the *hydrosphere* and the *lithosphere*.

biotechnology 生物技術 *shengwu jishu*

生物過程技術開發與應用的研究領域，其目的是應用生物技術來進行物質的轉化與生產。藻類、細菌和酵母等微生物可用於細胞培養、發酵過程和酶反應。但生物技術並不一定需要利用遺傳變異的生物體。

The field of investigation of the development and application of biological processes in technology. The objectives are to use biotechnological processes for conversion and production of substances. Microorganisms such as algae, bacteria and yeasts are used in cell cultures, fermentation processes and enzyme reactions. Biotechnology does not necessarily use genetically modified organisms.

biothane system 厭氧生物反應器系統 *yanyang shengwu fanyingqi xitong*

商業化的*升流式厭氧污泥床系統*。可應用它來處理高濃度的化工廢水，如廢發酵肉湯、濃度稀的醇類或脂肪酸。

A commercially available *upflow anaerobic sludge blanket system*. It can be applied to treat highly loaded waste waters from the chemical industry, like exhausted fermentation broths, dilute alcohols or fatty acids.

biotic components of ecosystems 生態系統中的生物組成部分 *shengtai xitongzhongde shengwu zucheng bufen*

生態系統組成中有生命的部分，包括植物群體和動物群體兩類。（參見：生物群落）

The living components of an ecosystem. The two groups are: populations of plants (flora) and populations of animals (fauna). (See also *biocenose*)

biotic ecological factors 生物生態因素 *shengwu shengtai yinsu*

影響和決定生態系統中群體相互作用的因素，這些因素源於生活在該生態系統中的*群體*。

Factors which influence and determine the interactions in an ecosystem and which are derived from the *populations* living in the ecosystems.

biotic equilibrium 生物平衡 *shengwu pingheng*

某群落生境中不同群體之間的*動態平衡*狀態，它會在某一平衡位置附近週期地變動，並取決於不同群落之間的相互作用，如由*捕食者與被捕食者之間關係產生*的自動調節機理。（參見：靜態平衡）

A *dynamic equilibrium* state of the different populations in a biotope. The position of the dynamic equilibrium fluctuates periodically around a certain equilibrium point. The position of the biotic equilibrium depends on self-regulating mechanisms of interactions between different populations, resulting for instance from *predator-prey relationships*. (See also *static equilibrium*)

biotope 群落生境 *qunluo shengjing*

由*非生命*的生態因素組合而成的一部分*生態系統*。它是生活在一起而且相互作用的某群植物、動物和微生物的自然生息地或生活空間，例如湖泊、深海、沙漠、雨林、牧場、人類皮膚、下水道的表面，等等。

Part of an *ecosystem* with a given combination of *abiotic* ecological factors. A biotope is the natural home or living space for a certain group of plants, animals and microorganisms, living and interacting together. Examples of biotopes are: a lake, the deep sea, a desert, a rain forest, a grassland prairie, the human skin, the surface of a sewer and many others.

BOD$_5$ 五天生物需氧量 *wutian shengwu xuyangliang*

在標準條件下在五天降解時間內水樣中污染物進行*好氧生物降解*和新陳代謝所需的氧量。

Biological oxygen demand, the quantity of oxygen which is required for the *aerobic biodegradation* and metabolism of pollutants in an aqueous sample under standard conditions in a degradation period of 5 days.

BOD$_5$ analysis 五天生物需氧量分析 *wutian shengwu xuyangliang fenxi*

　　測定廢水水樣中污染物因生物降解所需要的氧量。下標"5"指為期五天的降解時間。其測定結果在很大程度上取決於所採用方法的類別以及實驗室分析的質量。某些方法，測定的是實際利用和吸收的氧量；而另一些方法測定的是生物降解過程所去除的*總有機碳*；還有一些方法測定的是氧化過程中產生的二氧化碳的量。通常利用廢水處理場中同一的*微生物群體*作為生物降解的菌種。BOD$_5$正確測定的結果與廢水處理場污染物的生物降解有密切關係。（參見：傑欣－魏倫斯試驗、生物可降解性的連續測定（哈斯曼試驗）、經濟合作與發展組織－生物需氧量測定、BOD$_5$/COD比、壓力法五天生物需氧量測定）。

The measurement of the demand of oxygen by biological degradation of pollutants in an aqueous effluent sample. The subscript '5' refers to a degradation time of 5 days. The result of this measurement depends very much on the type of method used and on the quality of analysis in the laboratory. In some methods, the actual use and absorption of oxygen is measured. Other methods measure the removal of *TOC* by biodegradation processes. Still others measure the amount of carbon dioxide produced in the oxidation process. For biodegradation, the same *microcenose* is generally used as in a biological effluent plant. The result of a proper BOD$_5$ measurement relates well with the biodegradation of a pollutant in an effluent plant. (See also *Zahn-Wellens Test*; *continuous measurement of biodegradability (Husmann Test)*; *OECD-BOD Test*; *BOD$_5$ /COD ratio*; *BOD$_5$ measurement, barometric*)

BOD$_5$ loading of biosludge 活性污泥BOD$_5$負荷 *huoxingwuni BOD$_5$ fuhe*

　　見：food to biomass ratio。

　　See *food to biomass ratio*.

BOD$_5$ measurement, barometric 壓力法五天生物需氧量測定 *yalifa wutian shengwu xuyangliang ceding*

生物需氧量的靜態測定，即在標準條件下將水樣稀釋，並用*活性污泥接種*，然後將裝有樣品的燒瓶充以空氣，密蓋後於20℃恆溫下放置五天。用壓力測定法測量水樣上空氣的壓力，該期間的壓力降可用來衡量好氧過程中所消耗的氧量。在標準方法中對稀釋水的製備、*接種*和所需要的營養物作出了規定。

A static measurement of the biological oxygen demand. The sample is diluted and inoculated with *activated sludge* under standard conditions. The flask is filled with air, closed, and kept at a constant temperature of 20℃ for 5 days. The pressure of the air over the liquid is measured barometrically. The drop of the pressure over time is a measure of the consumption of oxygen in the aerobic process. Standard methods are prescribed for the preparation of the dilution water, the *inoculum* and the necessary nutrients.

BOD$_5$/COD ratio BOD$_5$/COD比 *BOD$_5$/COD bi*

表徵廢水*生物降解性*的一項指標。它表示廢水中可*生物降解*污染物所佔的大致比例。它是基於分別測定某一樣品的*BOD$_5$*值和*COD*值而求得的。如廢水BOD$_5$/COD比值大於0.6，則為易生物降解；如比值小於0.3，其污染物則是難生物降解的。

An indicator for the *biodegradability* of an effluent. The ratio gives the approximate fraction of the pollutant which is amenable to *biological degradation*. It is based on separate measurements of *BOD$_5$* and *COD* of a sample. Effluents with a BOD$_5$/COD ratio over 0.6 are easily biodegradable. If the ratio is below about 0.3, the pollutants are hardly biodegradable.

boundary layer 邊界層 *bianjieceng*

兩不混溶相（如有機溶劑相與含水母液）之間的分層。在理論上邊界層是很小與薄的，但在實際分離中，邊界層的界限常常是不明顯的，而且會被污染。這可能又會導致含水母液的污染。（參見：相界面、聚結器）

The layer between two immiscible phases, e.g. between an organic solvent phase and an aqueous mother liquor. In theory, the boundary layer is very minute and thin. In practical separations, the boundary layer is often not clearly defined and contaminated. This may lead to additional contamination of an aqueous mother liquor. (See also *phase boundary*; *coalescer unit*)

BPT 最佳實用技術 *zuijia shiyong jishu*
見：best practicable technology。
See *best practicable technology*.

branch contract 行業合同 *hangye hetong*
見：branch covenant。
See *branch covenant*.

branch covenant 行業契約 *hangye qiyue*
污染控制中的一種法律手段。行業契約是立法當局與同行業企業之間的協議，它可規定在一定期間內企業自願減少排放的數量，而不詳細涉及濃度和技術。行業契約是命令和控制型立法（管制法）的另一種選擇，它給企業留下更多的管理自由，因此使得污染控制更具生態效益。

An legal instrument in pollution control. Branch covenant includes an agreement between the legislating authority and a group of enterprises of the same industrial branch. The agreement may prescribe a voluntary reduction of emissions in a specified time period without going into details of concentrations and technology. Branch covenants are an alternative to command and control type legislation (*police laws*) and leave more management freedom to the enterprise. They may therefore lead to more *ecoefficient* pollution control.

breakthrough point 穿透點 *chuantoudian*
與活性碳吸附去除化合物（污染物）有關的術語。當運行達到穿透點時，吸附劑總吸附容量已耗竭，如果繼續使污染物通過吸附劑，則它將不再被吸附而穿透。

In relation to the adsorption of compounds (pollutants) by activated carbon: the point in operation at which the total adsorption capacity of the adsorbent is used up.

If the adsorber unit is charged with additional pollutant, this will not be adsorbed any more, it breaks through.

bubble column 泡罩塔 *paozhaota*

化工過程的一種單元裝置：塔或反應器，在其中氧氣或空氣氣泡通過液體而上升。因此，質量傳遞過程會伴隨著發生氣液之間的反應。泡罩塔具有某些內在結構的部件，以增加氣液之間的接觸時間。*空氣濕式氧化裝置中的反應器，就是一種泡罩塔的結構。*

A unit in chemical processing: a type of column or reactor in which gas bubbles, for instance of oxygen or air, rise through a liquid. Thereby, a mass transfer takes place with a subsequent reaction between gas and liquid. The bubble column may contain certain built-in structural elements to increase the contact time between gas and liquid. The reactor of a *wet air oxidation* unit is constructed as a bubble column.

bubble column scrubber 泡罩塔洗滌器 *paozhaota xidiqi*

用於空氣污染控制的一種簡單的濕式洗滌器，這種塔式設備內部具有盤狀結構，可持留住洗滌液；而氣體呈氣泡狀流經塔柱與洗滌液接觸，使氣態污染物被洗滌液吸收。

A simple wet scrubber for air pollution control. A processing column with built-in tray-like structures serves to hold a scrubbing liquid. The gas stream is bubbling through the column and the scrubbing liquid, and gaseous pollutants are absorbed in the liquid.

bubble principle 泡罩原理 *paozhao yuanli*

一種特殊類型的空氣污染控制法規的基礎。與傳統法規限制各煙囪的排放量相反，應用泡罩原理所限制的是在一假想的泡罩下工廠的總排放量，因此稱為"泡罩原理"。它使工廠在管理上有較大的自主性，以採取最經濟有效的辦法來控制空氣污染。（參見：經濟手段、排放權、行業合同）

The basis of a special type of air pollution control legislation. In the application of the bubble principle, the total emissions of a site under a hypothetical 'bubble' are limited, hence the name 'bubble principle', in contrast to traditional legislation which limits individual stack point emissions. The bubble principle leaves more freedom to the factory management to apply the most ecoefficient air pollution control solution. (See also *economic instruments*; *emission rights*; *branch contracts*)

buffer tanks 緩衝池 *huanchongchi*

見：equalization; equalization tanks。

See *equalization*; *equalization tanks*.

C

calcinated residue 煅燒殘渣 *duanshao canzha*

在600℃高溫下煅燒乾燥的殘渣後留下的無機殘灰。

The inorganic residue which remains after a total dry residue has been calcinated at an elevated temperature of 600°C.

calcium oxide equivalents 氧化鈣當量 *yanghuagai dangliang*

見：neutralization equivalents。

See *neutralization equivalents*.

calibration, gas chromatography 校準，氣相色譜法 *jiaozhun, qixiang sepufa*

見：calibration of analytical procedures; permeation tube in gas chromatography。

See *calibration of analytical procedures*; *permeation tube in gas chromatography*.

calibration gas mixtures 校準混合氣 *jiaozhun hunheqi*

微量氣體分析(如氮氧化物分析)中用來校準不同儀器的標準。它們可以從市場上購得，或者可以用二元參比氣體混合器來製備。

Standards in trace gas analysis, used for the calibration of different instruments, as e.g. in the analysis of nitrous oxides. Calibration gas mixtures can be obtained commercially, or be produced by a *binary reference gas mixer*.

calibration of analytical procedures 分析程序的校準 *fenxi chengxude jiaozhun*

微量雜質分析中的一個重要步驟。它是在標準分析條件下分析一定量的微量測試物質，以証實在分析信號和濃度之間存在線性關係。如果有偏差，則應對分析程序進行改進。(參見：校準標準、試驗方法的準確度、試驗方法的精密度、方法的檢測極限)

An important step in the analysis of trace impurities. Calibration is carried out by analyzing defined trace quantities of a test substance under standardized analytical conditions. This serves to verify the linearity of the relationship between the analytical signal and the concentration. In case of deviations, the analytical procedure should be improved. (See also *calibration standards*; *accuracy of a test method*; *precision of a test method*; *detection limit of a method*)

calibration standards 校準標準 *jiaozhun biaozhun*

　　檢驗分析程序用的含有確切已知量被分析化合物的樣品。有許多不同類別的校準標準，可供校準方法具體選用。對於氣體分析，市場上有標準混合氣供應。在實驗室，可以用二元參比氣體混合器或用擴散法（滲透管）來製備。對於水溶液的分析也有類似的標準。

Samples for testing analytical procedures, containing exactly known quantities of the compound under question. There are many different types which are specific for the methods being calibrated. For gas analysis, standard gas mixtures are available commercially. In the laboratory, they can be produced by mixing with a *binary reference gas mixer*, or by diffusion processes (*permeation tubes*). Similar standards are available for the analysis of aqueous solutions.

capillary column 毛細管柱 *maoxiguanzhu*

　　直徑小到毛細力起有效作用的柱子。它也可應用於氣相色譜法。（參見：毛細管氣相色譜法）

Any column whose diameter is so small that capillary forces become active; used also in gas chromatography. (See also *capillary gas chromatography*)

capillary gas chromatography 毛細管氣相色譜法 *maoxiguan qixiang sepufa*

　　見：gas chromatography, capillary。

See *gas chromatography, capillary*.

capital investments 基建投資 *jijian touzi*

　　見：investment costs。

See *investment costs*.

carbon cycle 碳循環 *tanxunhuan*

　　生態學中用來說明生態圈中元素碳自然循環的術語。簡單地說，碳

由大氣中的二氧化碳轉化為植物中的碳水化合物；動物利用碳水化合物
為食物而生成蛋白質；然後通過微生物的降解過程把碳水化合物和蛋白
質轉化為二氧化碳，再釋放到大氣中。由於碳以二氧化碳形式大量貯存
於大氣中，以二氧化碳形式溶解於海洋中，以碳酸鹽形式貯存於石灰石
中，以及貯存於礦石燃料（煤、石油和天然氣）中，因此碳的閉合循環是
很複雜的。

A term used in ecology to describe the natural cycle of the element carbon in
the ecosphere. In a very much simplified description, carbon is transformed from
carbon dioxide in the atmosphere to carbohydrates in plants; animals use
carbohydrates as food to produce proteins; microbial degradation processes of both
carbohydrates and proteins then yield carbon dioxide which is again released to the
atmosphere. The closed cycle is complicated by the fact that there are large
reservoirs of carbon: carbon dioxide in the atmosphere, CO_2 dissolved in the
oceans, carbon in form of carbonates in limestone, and carbon in form of fossil
fuels (coal, oil and natural gas).

carbon dioxide 二氧化碳 *eryanghuatan*

見：global atmospheric carbon dioxide。

See *global atmospheric carbon dioxide.*

carnivore 食肉動物 *shirou dongwu*

僅以肉類為營養的動物。（參見：捕食者、食草動物）

An animal species living exclusively on flesh and meat as nourishment. (See
also *predator; herbivore*)

carrier gas 載氣 *zaiqi*

氣相色譜儀中用來輸送揮發性有機化合物通過分離柱的氣體，通常
為氫氣或氦氣。

The gas, normally hydrogen or helium, which is used in gas chromatography to
transport the volatile organic compounds through the separation column.

catalyst bed 催化劑床 *cuihuajichuang*

　　具有催化劑的活性接觸層。催化劑被沉積在陶瓷載體基體上，而陶瓷基體可用氧化鋁來預塗漬以增加其活性表面。由於催化過程是在高溫下進行的，當在*汽車催化轉化器*上應用時，必須能抗熱衝擊和振動，所以陶瓷載體必須能抗熱應力和耐磨蝕。

　　The active contact zone holding the catalyst. Catalysts are deposited on a ceramic support matrix. The ceramic matrix may be precoated with alumina to increase its active surface. As catalytic processes take place at elevated temperatures and, in the case of *automotive catalytic converters*, have to withstand thermal shock and vibration, the ceramic support material must withstand thermal stress and be resistant to abrasion.

catalyst poisons 催化劑毒物 *cuihuaji duwu*

　　能全部或部分降低*催化劑*的催化性能的物質或元素。在催化燃燒系統中，鹵素、磷、硒、砷、硫、硅、鉛和其他重金屬化合物等物質會產生不良作用。研究開發*非敏感性的催化劑*是催化劑應用新的發展方向。

　　Substances or elements which may totally or partially reduce the catalytic properties of a *catalyst*. In catalytic combustion systems, substances like halogens, phosphorous, selenium, arsenic, sulfur, silicon, lead and other heavy metal compounds may exhibit such a negative effect. New developments in catalyst applications go in the direction of *non-sensitive catalysts*.

catalysts 催化劑 *cuihuaji*

　　在許多化學*催化過程*中使用的物質。通常，催化劑能提高反應速率，而本身不被消耗。催化劑的化學結構可有很大的變化，它可以是單一元素（金屬）、金屬氧化物，也可以是金屬有機化合物或有機化合物以及生化過程中的酶。

　　Substances which are used in many chemical *catalytic processes*. A catalyst will generally increase the reaction rate without being consumed. The chemical structure of catalysts may vary considerably and range from single elements (metals), oxides of metals to metal-organic or organic compounds and enzymes in biochemical processes.

catalytic combustion/incineration 催化燃燒/焚燒 *cuihua ranshao/fenshao*

在比高溫燃燒低得多的溫度下進行的燃燒或焚燒過程。（參見：催化氧化）

A combustion or incineration process operating at significantly lower temperature than high temperature combustion. (See also *catalytic oxidation*)

catalytic converters 催化轉化器 *cuihua zhuanhuaqi*

具有催化劑床層的工藝裝置。

The process unit holding the catalyst bed.

catalytic converters for diesel engines 柴油發動機的催化轉化器 *chaiyou fadongjide cuihua zhuanhuaqi*

柴油發動機排出的尾氣，除含有一氧化碳、殘餘碳氫化合物和煙灰外，還含有相當數量的氮氧化物。在大型固定柴油機和船舶柴油機上，都採用催化系統作為*空氣污染控制系統*。它們包括兩段獨自的催化過程：*氮氧化物的催化還原段*和*催化氧化段*。這兩個催化段組合在一雙效作用的轉化器中，其反應溫度為320-420℃，並用尿素來還原氮氧化物。

The exhaust emissions of diesel engines contain considerable amounts of oxides of nitrogen in addition to carbon monoxide, residual hydrocarbons and soot. *APC-systems* for large stationary diesels and ship diesels are based on catalytic systems. They make use of two separate catalytic stages, a *catalytic reduction of nitrogen oxides* and a *catalytic oxidation* stage. Both stages are combined in a double-action converter. The reaction temperature is between 320–420℃. Urea is used to reduce the oxides of nitrogen.

catalytic oxidation 催化氧化 *cuihua yanghua*

用特定催化劑來加速氧化反應的氧化過程。與高溫氧化相比較，催化氧化可在較溫和的反應條件（如在溫度300-450℃）下進行。它所使用的催化劑包括貴金屬（鉑、鈀和銠）和金屬氧化物（銅、錳）。在空氣污染控制中，利用催化氧化系統來氧化廢氣中的揮發性有機化合物。使用時必須小心防止*催化劑中毒*問題。（參見：催化劑床、催化轉化器）

An oxidation process which is promoted by a specific catalyst. A catalytic oxidation will take place under milder reaction conditions e.g. at a lower temperature between 300–450°C than a normal high temperature oxidation. For catalytic oxidation, catalysts from the group of noble metals (platinum, palladium and rhodium) and metal oxides (copper, manganese) are used. In air pollution control catalytic oxidation systems are used to oxidize volatile organic compounds (VOC) in off-gases. Care has to be taken to avoid problems due to *catalyst poisoning*. (See also *catalyst bed*; *catalytic converter*)

catalytic processes in APC 空氣污染控制中的催化過程 *kongqi wuran kongzhizhongde cuihua guocheng*

消除廢氣中污染物的過程。在此過程中氮氧化物(一氧化氮和二氧化氮)被催化還原為元素氮,而工業排放廢氣中所含的*揮發性有機化合物*被催化氧化。

由於汽車發動機中的燃燒過程進行得不徹底,其排放的尾氣也可能含有殘餘的碳氫化合物(揮發性有機化合物),這些污染物可在*汽車催化轉化器*中被氧化。催化系統在技術上的應用範圍很廣,從大型的煙道氣淨化裝置和含揮發性有機化合物廢氣處理裝置,到小型的汽車尾氣處理裝置。在某些情況下,催化系統可包括還原段和氧化段,而具有多效作用。(參見:選擇催化還原法)

Processes to eliminate pollutants in off-gases. Oxides of nitrogen (NO and NO$_2$) are catalytically reduced to elemental nitrogen, and volatile organic compounds (*VOC*) contained in industrial emissions are catalytically oxidized.

As a result of incomplete combustion in the engine, residual hydrocarbons (VOC) may also be contained in automotive emissions. These contaminants are oxidized in *automotive catalytic converters*. Technical applications of catalytic systems range from large scale units for flue gas purification and treatment of VOC emissions to small units for motor vehicles. In certain cases, catalytic systems with a multiple function, including reductive and oxidative stages are used. (See also *SCR process*)

catalytic processes in general 一般的催化過程 *yibande cuihua guocheng*

通過加入*催化劑*在不改變反應的平衡狀態下提高化學反應速率的過程。在此過程中催化劑參與反應並被再生，但本身不被消耗。催化過程開闢了一條節能和快速的反應途徑，因此在化工工藝過程中有著非常廣泛的應用。它可在均相或非均相中進行。在催化過程在生物化學酶變化中也起著重要的作用。在污染控制技術中，催化過程主要應用在空氣污染控制領域，但也應用於水污染控制上，如*濕式空氣氧化*或*廢水低壓氧化*中污染物的催化氧化。（參見：催化氧化過程、汽車催化轉化器）

Processes in which the rate of a chemical reaction is increased by the addition of a *catalyst* without changing the equilibrium position of the reaction. The catalyst enters into the reaction and is regenerated; it is not consumed during the process. It opens up a reaction pathway which generally is energetically favorable and faster.

Catalytic processes are used very widely in chemical processing technology. They may take place in a homogeneous or heterogeneous phase. Enzymatic catalytic processes also play an important role in biochemical enzymatic changes. In pollution control technology, catalytic processes are technically applied mainly in the field of air pollution control. Catalytic processes can also be found in water pollution control, e.g. the catalytic oxidation of pollutants in *wet air oxidation* or in *low pressure effluent oxidation*. (See also *catalytic oxidation process*; *automotive catalytic converters*)

catalytic reductive removal of NO$_X$ 催化還原去除氮氧化物 *cuihua huanyuan quchu danyanghuawu*

基於催化還原過程將廢氣中一氧化氮和二氧化氮進行還原的空氣污染控制技術，它大多應用於煙道氣的淨化。另外，也可以使用氮氧化物的*選擇催化還原法*這一術語。

在催化還原過程中，氮氧化物在*催化轉化器*中於200-340°C下與氨反應，被還原為氮氣和水；也可用尿素代替氨作為還原劑。在轉化器進料末端設有一熱交換器，以通過熱量傳遞與利用來達到節能的目的。

在催化還原過程中，使用含有特定金屬氧化物（如二氧化鈦、三氧化鎢和五氧化二釩）的*催化劑*，通常這些催化劑的活性組分被沉積在陶瓷

載體的基體上。（參見：催化劑床、催化劑毒物、氮氧化物的選擇性非催化還原）

An air pollution control technology to reduce nitrogen monoxide and nitrogen dioxide from off-gases which is based on catalytic reduction processes. The technology is mostly applied to the purification of flue gases. Alternatively, the terms *SCR process* or selective catalytic reduction of nitrogen oxides are also used.

In the catalytic reduction process, oxides of nitrogen (NOx) are reduced to nitrogen and water by reaction with ammonia, in a *catalytic converter* at 200–340°C. Alternatively, urea may be used as a source of reductive nitrogen instead of ammonia. A heat exchanger at the input-end of the converter renders the energy economy optimal by heat transfer and re-utilization.

In this process, *catalysts* based on specific metal oxides (e.g. TiO_2, WO_3 and V_2O_5) are used as a catalyst. Generally, the active catalyst is deposited on a ceramic support matrix. (See also *catalyst bed*; *catalyst poisons*; *selective non-catalytic NOx reduction*)

catch basin 集水池 *jishuichi*

見：retention basins。

See *retention basins*.

cation exchange 陽離子交換 *yanglizi jiaohuan*

通過陽離子交換樹脂的吸收作用，去除水溶液中的陽離子。

The removal of cations from an aqueous solution by absorption on a cation exchange resin.

cause-effect relationship 因果關係 *yinguo guanxi*

某一原因與某一結果之間的關係。了解這一關係，就可以通過排除產生的原因來避免造成某種不利的影響。在環境相互作用上，因果關係常常是很複雜的，因此必須在調查了解最後的細節和絕對肯定原因之前採取應對措施。（參見：相關系數）

A relationship between a certain cause and a certain effect. The knowledge of this relationship is basically necessary to avoid a certain negative effect by removing the cause. In environmental interactions, cause-effect relationships are very often complex and counter measures must be taken before the last detail is investigated and the cause is known with absolute certainty. (See also *correlation coefficient*)

CEFIC 歐洲化學工業聯合會中心 *ouzhou huaxue gongye lianhehui zhongxin*

歐洲各國國家化學協會的最高組織機構，辦公室設在比利時布魯塞爾。它頒佈了許多有關環境管理的指南。

The peak European organization of European National Chemical Societies, with offices in Brussels. CEFIC has published a number of guidelines on environmental management topics.

CERES 環境負責組織聯盟 *huanjing fuze zuzhi lianmeng*

見：Coalition for Environmentally Responsible Economies。

See *Coalition for Environmentally Responsible Economies.*

CERES Principles 環境負責組織聯盟原則 *huanjing fuze zuzhi lianmeng yuanze*

有關企業環境責任管理的10條基本規定。CERES原則即瓦爾德之原則是在1989年於阿拉斯加發生了"埃克森‧瓦爾德之"號油輪事故後，由環境負責組織聯盟（CERES）提出的。

其主要內容包括：生物圈保護、廢物減量、廢物正確處置、可持續利用資源和能源、安全產品、環境問題和風險的合理管理、損害的恢復與補救。該原則的一個重要方面是實施審計並以適當方式將審計結果公之於眾。公司可在自願基礎上採納瓦爾德之原則，並作出遵守這一原則的公開承諾。而公司的實際情況是否與該原則一致，則由環境負責組織聯盟進行檢查。在1996年年中，已大約有50個企業（主要是美國企業）贊同認可了瓦爾德之原則。（參見：責任關懷計劃、國際標準組織14000標準系列、環境管理與審計體系）

A set of ten basic rules for responsible environmental management of an enterprise. The Valdez Principles were developed by the *Coalition for Environmentally Responsible Economies (CERES)*, following the accident of the oil tanker 'EXXON VALDEZ' in 1989 in Alaska.

The key points encompass the protection of the biosphere, reduction of waste quantities and their proper disposal, sustainable use of resources and energy, safe products, sound management of environmental aspects and risks, as well as restoration and *remediation* of damages. An important aspect of the principles is the obligation to carry out audits and make public the audit results in a suitable way. Companies may adopt the Valdez Principles on a voluntary basis and pledge in public to abide by them. Compliance with the principles is monitored by CERES. In mid-1996, about 50 mainly American enterprises have endorsed the principles. (See also *Responsible Care Program*; *ISO 14000*; *EMAS*)

centrifugal separation forces 離心分離力 *lixin fenlili*

在旋風分離器中進行顆粒分離的物理作用。(參見：旋風分離器)

The physical forces responsible for the separation of particles in a cyclone separator. (See also *cyclone separator*)

certified auditor 合格審計者 *hege shenjizhe*

由國家許可機構正式批准從事審計工作的合格人員。資格證書中通常規定了審計者許可從事的審計工作的技術範圍。(參見：審計)

A person qualified officially through a state-operated approval system to carry out audits. The qualification scheme often specifies the technical areas for which an auditor is certified. (See also *audit*)

certified environmental laboratory 合格的環境實驗室 *hegede huanjing shiyanshi*

通過檢驗符合一切要求、包括成功地參與了實驗室間相互測試的實驗室。合格的環境實驗室通常需由專門的國家機構進行確認。

A laboratory which has passed all requirements for certification, including successful participation in *inter-laboratory tests*. Certification takes place usually through specific national organizations.

certified verifier 持有証書的檢查者 *chiyou zhengshude jianchazhe*

歐洲聯盟環境管理和審計體系指令中有關對工廠生產中的環境問題

進行檢查的持有合格證書的專家。其檢查是為證實該工廠的操作與其公諸於眾的環境報告中所陳述的內容的一致性。

In connection with the *EMAS Directive of the European Union*, a specialist certified officially to verify the conformity of the environmental aspects of the operation of a factory with the statements made in the Environmental Report to the public of the same factory.

CFC 氯氟烴 *lüfuting*

見：chloro fluoro carbons。

See *chloro fluoro carbons*.

chain of polluters 污染者之鏈 *wuranzhezhilian*

對生態系統某一部分的污染具有影響的不同但有相互聯繫的作用者（個人、單位或組織、生產者、消費者、立法者）。例如，汽車運輸產生的污染與以下因素有關：汽車的構造、車輛的數量、所供應汽油的類別與價格、車輛駕駛方式、公共交通發展的狀況、污染控制立法和交通運輸立法等。這些活動的所有作用者都可以被認為是污染者之鏈中的一環。因此，通常很難確定個別的污染者。

The different and interrelated actors (persons, institutions, manufacturers, consumers, legislators) having an influence on the pollution of a given part of an ecosystem. For example, pollution from motor vehicle transport is determined by: the construction of the motor vehicle; the availability of vehicles; the type and price of gasoline available; the way the vehicle is driven; the availability of public transport as an alternative; legislation dealing with pollution control; and legislation dealing with transport and traffic.

All actors of these actions may be regarded as being involved in the chain of polluters. It is therefore often difficult to identify a single polluter.

chain of value creation 價值創造鏈 *jiazhi chuangzaolian*

生產某種具有附加值產品所涉及的一系列過程，如從原料開採和一次能源的生產開始，到產品的銷售和使用及其最終處置為止。在產品的*生命周期分析*中，必須對所有這些階段進行評價。（參見：產品管理）

65

The sequence of steps involved in the manufacture a product with added value. The sequence starts with the mining of raw materials and primary energy production and ends with the distribution and use of the product and its final disposal. In the *life cycle analysis* of a product, all of these steps must be assessed. (See also *product stewardship*)

chemical oxygen demand 化學需氧量 *huaxue xuyangliang*

見：COD。

See *COD*.

chemical waste 化學廢物 *huaxue feiwu*

見：special waste。

See *special waste*.

chemical wet analysis 化學濕式分析 *huaxue shishi fenxi*

基於傳統方法的分析，即通常採用的容量分析方法和重量分析方法。傳統的化學分析與現代的分析不同，後者以*儀器分析方法*和計算機數據處理系統為基礎。

Analysis based on traditional general methods, e.g. analysis by volumetric and gravimetric methods. Traditional chemical wet analysis is in contrast to modern analysis which is based strongly on *instrumental methods* and computer data processing.

chemisorption 化學吸收 *huaxue xishou*

一種物質被某種液體吸收同時發生化學變化的過程。例如，二氧化硫被鹼洗滌液吸收，同時生成亞硫酸鹽。

Absorption of a substance in a liquid whereby the absorbed substance undergoes chemical changes. Example: the absorption of sulfur dioxide in an alkaline scrubbing liquid with simultaneous formation of sulfite.

chlorinated hydrocarbons 氯化烴 *lühuating*

具有一個或幾個氯原子的碳氫化合物，它在工業、商貿及家庭中被

廣泛地應用。氯化烴是與環境問題有著密切關係的一類重要化合物。一
般它可溶於脂肪中，能在*食物鏈*中累積，並且對生態系統和人具有毒
性。

Hydrocarbons with one or several atoms of chlorine with a very wide range of
applications in industry, trade and households. As a group, chlorinated hydrocarbons are particularly important in connection with environmental questions.
Chlorinated hydrocarbons are generally fat soluble lipophilic, they may accumulate
in the *food chain*, and have ecotoxic or human toxic properties.

chlorination 氯化 *lühua*

見：effluent disinfection。

See *effluent disinfection.*

chloro fluoro carbons (CFC) 氯氟烴 *lüfuting*

家用或工業致冷機中作為致冷劑的一類物質。致冷劑的化學結構
為：R11三氯氟甲烷、R12二氯二氟甲烷、R113三氯三氟乙烷、R114二
氯四氟乙烷、R115一氯五氟乙烷。

氯氟烴釋放進入大氣，從環境觀點看是有問題的，因為它們會引起
*溫室效應*和*臭氧耗竭*，但不同的氯氟烴產生的環境影響是不同的。1987
年通過並於1990年修正的*蒙特利爾議定書*，對某些化合物的未來使用作
出了限制。

A group of substances used as refrigerants in domestic and industrial cooling
machines. The chemical structures of the refrigerants are: R11 trichloro fluoro
methane; R12 dichloro difluoro methane; R113 trichloro trifluoro ethane; R114
dichloro tetrafluoro ethane; R115 chloro pentafluoro ethane.

The release of CFCs into the atmosphere is problematic from the environmental point of view because they contribute to the *greenhouse effect* and to *ozone
depletion.* The environmental impact varies for different CFCs. The future use of
certain compounds is limited by the *Montreal Protocol* of 1987, amended in 1990.

chromatography 色譜法 *sepufa*

一類用來分離不同化合物的方法。它是基於物質和吸附劑之間吸附

度（吸附力、結合力）的不同而進行分離的。操作時，用惰性溶劑或惰性
氣體來載送物質，使其通過裝填在柱子中或塗在平板上的吸附劑，由於
物質在柱子中（或平板上）具有不同的保留時間，因而可得到分離。

A group of methods used to separate different chemical compounds. The
separation is effected as a result of different degrees of adsorption (adsorption
forces, binding forces) between the substance and the adsorbent. The substance is
passed through the adsorbent in a column or on a plate. An inert solvent or an inert
gas serves to transport the substances which are characterized by lower or higher
retention times in the column (or on the plate).

chromatography, high pressure, liquid 高壓液相色譜法 *gaoyayexiang sepufa*

見：high pressure liquid chromatography。

See *high pressure liquid chromatography.*

chromatography, thin layer 薄層色譜法 *baoceng sepufa*

見：thin layer chromatography。

See *thin layer chromatography.*

chronic effect 慢性效應 *manxing xiaoying*

開始時不明顯又發展緩慢的效應，它是*靶生物*在進行較長時間接觸
後才顯現出來的。有時，這種效應的發作可能是滯後和很難辨認的。

A slow effect with a beginning which is not clearly defined. Chronic effects
manifest themselves after a relatively long exposure of the *target organisms*. The
onset of the effect may sometimes be delayed and difficult to recognize.

ciliata 纖毛蟲 *qianmaochong*

顯微鏡鏡檢可看到的微小的好氧生物，它們具有特殊的結構來進行
移動和攝取食物。纖毛蟲屬於*原生動物族或藻類*。可通過對*活性污泥*進
行顯微鏡鏡檢由它們的似頭髮狀的結構，即所謂"纖毛"來加以鑒別。這
種纖毛結構可用來攝取食物和進行移動。與其他的原生動物一起，纖毛
蟲能攝取和消化各種不同的有機廢物，包括細菌在內。因此，它們在活
性污泥法廢水處理場中部分地起到*生物降解*過程的作用。

某些特殊的纖毛蟲（和某些鞭毛蟲）還可在河流、湖泊和池塘水質的分類中作為*指示生物*。

Microscopically small aerobic organisms which possess specific structures for their displacement and ingestion of food. Ciliata belong to the group of *protozoa* or *algae*. Ciliata can be identified by careful microscopy of *activated sludge* by their hair-like structures, called 'cilia'. These structures are used for the intake of food and/or for their own displacement. Together with some other protozoa, ciliata are able to ingest and digest a variety of organic waste compounds, including bacteria. They are thus responsible in part for the process of *biodegradation* in an activated sludge effluent plant.

Specific species of ciliata (and some flagellata) are also used as *bioindicator* organisms in the classification of the water quality of rivers, lakes and ponds.

clarifier 澄清池 *chengqingchi*

廢水處理中用於去除污泥和懸浮物的設備。它一般通過*沉降*進行分離，但有時也採用*浮選法*。根據澄清池在整個處理流程中的位置，分別稱為*一次澄清池*，*最終澄清池*或*二次澄清池*。

A processing unit in effluent treatment, serving to remove sludge and suspended solids. Separation generally takes place through *sedimentation*. In some cases, *flotation* is also used. Depending on the position of the clarifier unit in the overall process flow, the terms *primary clarifier*, *final clarifier* or *secondary clarifier* are used.

classes of off-gases 廢氣分類 *feiqi fenlei*

見：off-gas types。

See *off-gas types*.

clean chemistry 清潔化學 *qingjie huaxue*

見：clean technologies。

See *clean technologies*.

clean production processes 清潔生產過程 *qingjie shengchan guocheng*

見：clean technologies。

See *clean technologies*.

clean technologies 清潔工藝 *qingjie gongyi*

從環境保護、能源利用和持續發展的觀點選用的最佳生產工藝。清潔工藝是通過*工藝措施*，而不是通過*末端處理措施*，來盡可能實現在生產過程中解決污染控制問題。

A selection of production technologies which have been optimized from the viewpoint of environmental protection, energy consumption and sustainable development. In clean technologies, pollution control problems are solved as much as possible in the manufacturing process itself by *in-process measures*, instead of *end-of-pipe measures*.

clean up costs 清除淨化費用 *qingchu jinghua feiyong*

對污染的地下水或污染的土地和土壤進行*恢復補救*所需的費用。這些補救活動可能需要基建投資來建設*現場恢復*的技術設施，此外也還將需要長期的*運行費用*來運行這些設施。許多案例的經驗表明，清除淨化活動的費用要比開始時就對廢物進行正確處置所需的費用高得多。

The costs for *remediation* activities for polluted groundwater or for contaminated ground and soil. These activities may require a capital investment into technical facilities for *on-site treatment in remediation*. In addition, there will be long term *operating costs* for the operation of these facilities. Experience in many cases has shown that the costs of cleanup activities are much higher than the costs of proper disposal at the beginning.

climatic changes 氣候變化 *qihou bianhua*

全球氣候的變化和變動在許多方面與環境問題有關。氣候對植物、動物的生命和人類有很大影響，而*人為的排放*又會對全球氣候的變化產生影響。

全球氣候是投射到地球上的太陽能和由地球逸散的能量之間平衡的結果。決定氣候的重要因子是太陽輻射強度和通過*溫室效應*大氣所截留的那部分能量。其他起重要作用的因子還有海洋溫度和洋流、風的流型、火山活動、極地冰冠的大小、植物、尤其是熱帶雨林的作用等等。

作為自然現象,全球氣候在不斷緩慢地變化著。對於這種變化的確切原因,仍然是科學研究和推測的一個問題。有跡象表明,太陽活動(太陽耀斑週期)的變化具有重要影響。此外,還有研究認為大氣二氧化碳濃度的增加會強化溫室效應,並且正導致全球變暖。事實上,在過去一百年內二氧化碳的人為排放,已使得二氧化碳的濃度有了明顯的增加。其他所謂的 *溫室氣體* 也起著附加作用。

需要考慮的眾多因子及其相互關係,使準確預測氣候的變化變得複雜化和具有不確定性。一些數學模擬模式預測,在下一世紀內全球溫度將增高 1-2°C。然而,對此預測的看法並不一致。

Environmental aspects are related to changes and fluctuations of the global climate in many ways. The climate is strongly influencing plant- and animal life and mankind. In turn, *anthropogenic emissions* are believed to affect the change of the global climate.

The global climate is the result of a balance between the input of solar energy onto the earth and the dissipation of energy from the earth. Important factors determining the climate are the intensity of solar radiation and the fraction of energy which is held back by the atmosphere through the *greenhouse effect*. Additional factors playing an important role are oceanic temperature and oceanic currents, wind patterns, volcanic activity, magnitude of the polar ice caps, vegetation, in particular the function of tropical rain forests and others.

The global climate is constantly changing very slowly as a result of natural phenomena. It is still a question of scientific investigation and speculation what exactly caused these changes. Indications are that the fluctuation of solar activity (solar flare cycles) has a significant influence. In addition, it is believed that the increase of the atmospheric CO_2 concentration has intensified the greenhouse effect and is leading to a slow global warming. Anthropogenic emissions of carbon dioxide have in the past century led to a clearly measurable increase of the atmospheric CO_2 concentration. Other so-called *greenhouse gases* may exert an additional effect.

The number of factors which need to be considered and their interrelation make an exact prediction of climatic changes complicated and uncertain. Some mathematical simulation models predict an increase of the global temperature of between 1–2°C in the next century. There is, however, no consensus on these predictions.

closed cycle activity 閉環活動 *bihuan huodong*

控制論中應用的術語。在閉環活動中，要考慮對其周圍及環境的不良影響，並用它作為控制信息，通過反饋來修正該活動，以逐步減少其不良影響。（參見：線性活動）

A term used in *cybernetics*. In a closed cycle activity the negative impact on its own surroundings and environment is taken into consideration and used as a control information and feedback for shaping the activity. The integration of this feedback gradually leads to a reduction of the negative effects. (See also *linear activity*)

closed mass flow system 閉合物流系統 *bihe wuliu xitong*

無物料損失到*系統邊界外*的物流循環。例如，*碳循環*、*最終去所*。

A mass flow cycle in which no mass is lost to the outside of the *system boundaries*. Example: the *carbon cycle*; *final sink*.

Club of Rome 羅馬俱樂部 *luoma julebu*

於1968年在全球開始對環境問題進行爭論時，由意大利實業家佩塞先生（Aurelio Peccei）倡導，在意大利羅馬成立的一個組織，它大約有30個成員，包括科學家、商界領導人和政治家。與此同時，在美國的麻省理工學院建立了一個有關地球上相互作用的簡化數學模式，其結果發表在《增長的極限》一書中。這本書使人們提高了認識，了解自然資源是有限的。自此以後，對這第一個簡化的數學模式已進行了許多改進。

An organization founded at the beginning of the world-wide environmental debate in 1968 at the initiative of Aurelio Peccei, an Italian industrialist, in Rome, Italy. The membership of about 30 included scientists, business leaders and politicians. In conjunction with this initiative, at the American Massachusetts Institute of Technology (MIT), a simplified mathematical model for the interactions on the planet earth was established. The results were presented in the book *Limits to*

Growth, which raised the awareness that natural resources are limited. The first simplified mathematical model has since been improved.

co-current 並流 *bingliu*

兩種介質（如液體和氣體，或兩種不混溶的液體）以同一方向流經加工過程的塔設備。（參見：逆流）

Two media (e.g. a liquid and a gas, or two immiscible liquids) flowing in the same direction through a processing column. (See also *counter-current*)

co-generation of energy 能量共生 *nengliang gongsheng*

由一次能源如礦石燃料同時協調地產生熱能（如工藝用熱量）和電能。按照熱力學觀點，共生是提高能量轉換總效率的一種方法。

The simultaneous and coordinated production of thermal energy (e.g. process heat) and electrical energy from a primary energy source, e.g. a fossil fuel. For thermodynamic reasons, co-generation is a means to increase the overall efficiency of energy transformation.

coagulating agents 混凝劑 *hunningji*

投加到廢水中能使懸浮超細顆粒或膠體混凝的助劑。根據實際應用情況的不同，可以使用不同類型的混凝劑，常用的有三價鐵鹽、二價鐵鹽、鋁鹽以及它們的混合物。

Auxiliary agents which, when added to a suspension of submicron particles or colloids in an aqueous effluent, will lead to their coagulation. Depending on the application, different types of coagulants are used, often iron-3 or iron-2 or aluminum salts, or mixtures of them.

coagulation 混凝 *hunning*

使超細顆粒（膠體）變成較大顆粒的過程。混凝這一術語通常用於液體系統，但也可用於氣體系統。在氣體中，通過混凝可使煙霧或氣溶膠中的小懸浮顆粒或小液滴凝結成較大的顆粒或液滴。通常，混凝後緊隨著產生絮凝作用而生成較大的顆粒，然後通過沉降、離心或過濾將絮凝

物去除。混凝是從廢水中分離去除各種懸浮物的重要過程，投加一些特定的混凝劑會促進混凝作用。(參見：絮凝)

The process of forming larger particles from submicron particles (colloids). The term, generally used for liquid systems, can also be applied to gases. In gases, small suspended particles or droplets in mists or aerosols may be coagulated to larger ones. Coagulation is usually followed by *flocculation* to form larger particles. The flocculated matter can then be removed by sedimentation, centrifugation or filtration.

Coagulation is an important process to separate suspended solids of various kinds from the effluent. Coagulation may be promoted by the addition of special coagulating agents. Compare also with *flocculation*.

coalescence 聚結 *jujie*

使乳化的小液滴合併或凝聚成較大液滴而被分離去除的過程。現在市場上有專門設計的*聚結器*供應。

The process of uniting or agglomerating small droplets of an emulsion to larger drops which can then be separated. Specially designed *coalescer units* are available for this process.

coalescer unit 聚結器 *jujieqi*

化工過程中用於改進連續運行系統內兩不混溶相進行*相分離*的設備，它將促使小液滴混凝和聚集為較大的液滴，然後更有效地進行分離。當液體/液滴系統流過安設有*折板*的聚結器時，由於液體流向發生轉折而使液滴趨於聚集，從而達到分離的目的。

A unit used in chemical processing to improve the *phase separation* of two immiscible phases in a continuous system. The coalescer unit will promote the coagulation and aggregation of small droplets to larger ones which will then separate more efficiently. This is achieved by a flow of the liquid/droplet system through the coalescer unit which is equipped with a system of *deflection plates* against which the liquid is flowing and which tend to aggregate the droplets.

Coalition for Environmentally Responsible Economies (CERES) 環境負責組織
聯盟 *huanjing fuze zuzhi liangmeng*

以重要的社會投資者、環境組織、工會和教會團體為成員的一個非
盈利組織。它成立於1988年，以美國麻薩諸塞州波士頓為總部。其目的
是遵循*可持續發展*原則來推動企業實施環境責任管理和運作。為此，它
提出了*瓦爾德之原則*，其中概述了環境責任管理的要點。

A non-profit organization with members from leading social investors, environmental organizations, labor unions and church groups. CERES was founded in 1988
and is based in Boston, Mass. USA. The objectives are to promote responsible
environmental management and operation of enterprises following the principles of
sustainable development. To this end, CERES has developed the *Valdez Principles*,
outlining key points of environmentally responsible management.

COD 化學需氧量 *huaxue xuyangliang*

Chemical oxygen demand的縮寫。它是用以表徵廢水特性的一項綜
合指標。COD值是用標準方法，即在酸性介質中用重鉻酸鹽為氧化劑將
所有污染物氧化所消耗的總氧量。在測定時，添加銀鹽作為催化劑，同
時添加汞鹽以防止氯離子的氧化。

Chemical oxygen demand. Collective parameter for the characterization of an
aqueous effluent. The COD value relates to the total amount of oxygen consumed to
oxidize all pollutants by a standardized method using bichromate as the oxidant in
acidic medium. Mercury salts are added to prevent the oxidation of chloride ion.
Silver salts are used as a catalyst.

COD analysis 化學需氧量分析 *huaxue xuyangliang fenxi*

用標準方法，即在酸性溶液中用重鉻酸鉀氧化污染物來測定
COD值。

The measurement of the COD value by a standardized method, using potassium
bichromate in acidic solution.

COD load 化學需氧總量 *huaxue xuyang zongliang*

用重鉻酸鹽對廢水中污染物進行化學氧化所需的總氧量，以氧當量

表示。作為一項綜合指標,它包括大多數有機化合物和其他可被氧化的化合物,如氨、亞硫酸鹽等。(參見:綜合指標、COD)

The total quantity of oxygen equivalents needed for chemical oxidation of pollutants in an effluent with bichromate. The COD value is expressed as oxygen equivalents. As a group parameter, it includes most organic compounds and other oxidizable compounds, e.g. ammonia, sulfite and others. (See also *group parameters*; *COD*)

codes for plastic wastes 塑料廢物標記 *suliao feiwu biaoji*

在消費品的塑料部分上貼有的標記系統,以方便辨識、分離和回收利用使用後的塑料。例如,某些塑料的標記和標記號碼為:聚對苯二甲酸乙酯,PET,01;高密度聚乙烯,PE-HD,02;聚氯乙烯,PVC,03; 低密度聚乙烯,PE-LD,04;聚丙烯,PP,05;聚苯乙烯,PS,06。此標記系統基於德國標準,DIN6120。目前,正試圖制定和頒佈國際標準組織的國際標準。

A system of codes imprinted on the plastic parts of consumer goods. The codes allow for proper identification, separation and recycling of plastic wastes after use. Some codes and code-numbers are: Polyethylene terephthalate, PET, 01; High density polyethylene, PE-HD, 02; Polyvinyl chloride, PVC, 03; Low density polyethylene, PE-LD, 04 ; Polypropylene, PP, 05; Polystyrene, PS, 06.

The codes are based on a German Norm, DIN 6120. Current activities aim to develop and issue an international ISO norm.

cold plasma electrode 冷等離子體電極 *lengdengliziti dianji*

見:negative ionizing electrode。

See *negative ionizing electrode.*

cold wall incinerator 冷壁式焚燒爐 *lengbishi fenshaolu*

見:incinerator, cold wall。

See *incinerator, cold wall.*

coliform count 大腸菌數 *dachangjunshu*

水中大腸菌(主要是大腸桿菌)數的計數。它是通過細菌培養試驗進行測定的。在100毫升的飲用水水樣中,應該不存在任何大腸菌。作為游泳和洗澡用的地面水(湖泊、河流),可以根據大腸桿菌這一指標進行分類。對於處理後的廢水,一般沒有規定大腸菌數的限值,但實際總數在每毫升100至1000個細菌範圍內。(參見:廢水消毒)

An indicator for the number of coliform bacteria, mainly escherichia coli, in water. The determination is carried out by a bacteriological culture test. In a drinking water sample of 100 ml, coliform bacteria should be absent. Surface water (lakes, rivers) used for swimming and bathing may be classified according to this parameter.

The coliform count for treated effluents is generally not limited. Actual total counts are in the range of 100 to 1,000 bacteria per milliliter. (See also *effluent disinfection*)

collective parameter 綜合指標(參數) *zonghe zhibiao (canshu)*

見:group parameter。

See *group parameter*.

colorimetric analysis 比色分析 *bise fenxi*

對廢水中化合物的濃度進行目視或儀器測定。在比色分析中,首先,必須將無色物質經化學改性變為有色物質。

在目視比色法中,再將溶液的顏色與參比樣品或色標相比較。在儀器分析中,則測定由濾光器產生的有色光的吸收作用,並用分析程序的校准來進行估算。如要進行更準確的測定,最好採用單色光的分光光度測定法。

Visual or instrumental measurement of the concentration of compounds, for instance in an effluent. Colorless substances must first be modified chemically to change them into colored species.

In visual colorimetry, the color of the solution is then compared with a reference sample or color scale. In instrumental analysis, the absorption of colored light, produced by a light filter, is measured and evaluated using *calibration of analytical procedures* . For more accurate measurements *spectrophotometry*, using mono-chromatic light, is a preferable method.

column packing 塔填料 *tatianliao*

塔式設備接觸層中的填料。按照不同過程的類型，可以選用不同類型的填料。(參見：填料塔、填充段、無規填料、有規填料)

The material in the contact zone of a processing column. Depending on the type of process, several types of column packing may be used. (See also *packed column*; *packing zone*; *random packing*; *structured packing*)

combined APC-systems 組合空氣污染控制系統 *zuhe kongqi wuran kongzhi xitong*

見：APC-systems, combined。

See *APC-systems, combined.*

combustion 燃燒 *ranshao*

用熱氧化法使有機物礦質化和降解的過程。在高溫、足夠的停留時間和過量氧存在的條件下，有機物(如廢物)可被完全氧化和礦質化。現代焚燒爐的燃燒溫度可高達1200°C。

The process of mineralization and degradation of organic material by thermal oxidation. Organic matter (e.g. wastes) is subjected to a high enough temperature for a long enough time in the presence of excess oxygen to ensure complete oxidation and mineralization. In modern incinerators, the combustion temperature goes up to 1200°C.

combustion chamber 燃燒室 *ranshaoshi*

使有機物在其中進行燃燒的燃燒爐的主要部件。(參見：回轉窯、冷壁式焚燒爐、靜態燃燒室)

The main part of an incinerator in which combustion of organic matter takes place. (See also *kiln, rotary*; *cold wall incinerator*; *static combustion chamber*)

combustion of off-gases in a boiler house 鍋爐房中廢氣的燃燒 *guolufangzhong feiqide ranshao*

它是廢氣燃燒處理的一種特殊應用方式，即可將廢氣引入工廠現有鍋爐房的燃燒室來分解破壞廢氣中的污染物。在這種情況下，可用輕度污染的空氣來代替等量的正常燃燒空氣。此技術主要應用來處理含低濃度但具有惡臭的污染物的廢氣。必須仔細監測煙氣中被氧化的污染物如氮氧化物、二氧化硫和氯化氫的濃度。

A special application of off-gas combustion. A possibility to destroy pollutants by feeding off-gases into the firing chamber of an existing boiler house in a factory. In such a case, the slightly polluted air replaces an equivalent volume of normal combustion air. The technology can be applied mainly to treat off-gases with low concentrations of highly odorous pollutants. Care must be taken to monitor the concentrations of oxidized pollutants like NOx, SO_2 and HCl in the flue gas.

community, ecological 生態群落 *shengtai qunluo*

在給定的生態系統中所有不同的個別物種的生態*群體*之總和。這些群體彼此相互作用，而且往往彼此相互依存。*共生現象或捕食者與被捕食者之間的關係*，就是這種相互作用的例子。（參見：生態學、生態系統）

The total of all different ecological *populations* of individual species in a given ecosystem. These populations interact with each other and often depend on each other. Examples of such interactions are *symbiosis* or a *predator–prey relationship*. (See also *ecology*; *ecosystems*)

company environmental policy 公司環境政策 *gongsi huanjing zhengce*

簡要說明公司在運作管理上所制定的環境保護目標的文件。綜合的環境政策必須包括與產品和生產過程有關的環境問題，並應符合*可持續發展*的目標。環境政策是管理體系和管理程序的基礎，它必須由公司領導批准生效，並對其實施加以監督。為了積極進行溝通與交流，應將環境政策公之於眾。

A set of concise statements outlining the objectives and goals on environmental protection which the management of a company has adopted for its operations. A comprehensive environmental policy must include relevant environmental aspects of the products and of the production processes. It should be in line with the objectives of *sustainable development*. The policy forms a base for management systems and procedures. It must be put into force by top management and its implementation should be supervised. In line with good communication practices, the policy should be made public.

company internal impact assessment 公司內部的影響評價 *gongsi neibude yingxiang pingjia*

以*綜合環境管理體系*為基礎的公司內部管理的一項規程，它類似正式的*環境影響評價*，適用於對環境具有潛在影響的一切內部工程項目。它包括兩方面的內容：在項目文件中對該項目的環境問題加以說明；由環保人員對該項目進行獨立的評價。通過評價提出的建議，必須在該項目進行的各階段工作中加以落實。

A company internal management procedure which is based on an *integral environmental management* system. The procedure is similar to the official *environmental impact assessment*. It is applied to all internal projects with a potential impact on the environment. The procedure includes two phases: the documentation of environmental implications of a project in the project documen-tation, and an independent assessment of the project by the environmental officer. Recommendations resulting from this process must then be included in the subsequent steps of the project.

company internal levy 公司內部收費 *gongsi neibu shoufei*

將公司內部收費的辦法作為一種可能的管理手段，以指導企業的運營朝一定方向發展。例如，對生產過程排放的不可生物降解的總有機碳，採取特殊收費的方式進行人為的經濟懲罰。在*分攤污染控制費用*上，可以提高對*不可生物降解*的*總有機碳*的收費。從長遠經濟上考慮，這將會促進生產水平的提高。（參見：收費、排污收費）

The application of levies within a company as a possible management instrument with the objective of directing the operations of an enterprise in a certain direction. The artificial economic penalization of the discharge of non-biodegradable TOC from production through a special levy may be taken as an example. In the context of the *allocation of pollution control costs* it is possible to increase the charge on *non-biodegradable TOC*. In the long term, through economic motivation, production will change to better alternatives. (See also *levy*; *levy on discharges*)

compartments, environmental 環境的組成部分 *huanjingde zucheng bufen*

見：environmental compartments。

See *environmental compartments*.

compartments in biological basins 生物處理池的隔間 *shengwu chulichide gejian*

把廢水處理場*活性污泥池*分隔成若干小池，這樣可使操作更具靈活性，更便於維修，適應於流量的變化，以及可能將活性污泥池改變成為兩段處理過程。

The construction of the *activated sludge basin* of an effluent plant in the form of a number of separate basins. This allows for more flexibility in the operation, better servicing and maintenance, adaptation to variable flow volumes and possibly the switching of the basins to divide the treatment into two sequential process steps.

compliance audit 法規符合審計 *fagui fuhe shenji*

僅在工廠進行涉及符合法規要求狀況的(環境)審計。

The (environmental) audit dealing exclusively with the legal compliance situation at a factory.

compliance with company internal norms 符合公司內部標準 *fuhe gongsi neibu biaozhun*

重視並遵循公司內部的一切政策和標準。

To be in a situation where all valid company internal policies and norms are respected and adhered to.

compliance with legal requirements 符合法律要求 *fuhe falü yaoqiu*

符合一切法律的要求，包括適用的各級法規中規定的要求和生產運營許可證中規定的特殊要求。

To be in a situation where all legal requirements are fulfilled. Legal requirements include the applicable legislation of different levels, as well as specific requirements which are laid down in the permits for the operation of a site.

compost 堆肥 *duifei*

見：composting of wastes。

See *composting of wastes*.

composting of wastes 廢物堆肥 *feiwu duifei*

在高濃度和高水分系統中廢物的生物處理和改性。它是基於有機物在各種微小生物體和微生物的作用下進行的*好氧生物降解*和新陳代謝。氧和空氣必須能自由地進入到系統中以防止發生厭氧過程。在傳統上，堆肥被應用於小規模處理自然有機廢物，但也可應用於較大規模的處理裝置。處理時必須採取措施控制臭味。

The biological treatment and modification of waste material in a highly concentrated and moist system. Composting is based on *aerobic biological degradation* and metabolization of organic material by various small organisms and microorganisms. Oxygen and air must have free access to the system to prevent anaerobic processes. Traditionally, composting is carried out on a small scale to eliminate natural organic material, but the process can also be applied to larger scale units. Care must be taken to keep odors under control.

concentrate 濃縮液 *nongsuye*

膜分離過程中得到的濃液相，它含有不能透過分離膜而被截留下來的物質。在同一過程中，所得到的*滲透液*是較稀的水相。

The concentrated aqueous phase obtained in membrane separation processes. The concentrate contains the substances which do not pass the membrane. In the same process, the *permeate* is obtained as the more dilute fraction.

concentration limits 濃度限值 *nongdu xianzhi*

對排放到環境中某污染物的濃度所規定的限值。濃度限值還應包括統計數據來指明是否所有測定值都必須達到要求或者百分之多少達到要求，它既可適用於*單一要素指標*，也可適用於*綜合指標*。單一要素指標常常應用於單分子物質。如果同時不限制排放體積的話，那麼濃度限值並不能限制所排放污染物的量。另一種控制辦法是實行*總量限值*。

The specification of a concentration which limits the discharge of a pollutant into the environment. The concentration limits should also include statistical information to indicate whether it must be met by all measured values or by what fraction. Either *single element parameters* or *group parameters* may be specified. Single element parameters often apply to a single molecular species. If the discharge volume is not limited at the same time, then concentration limits do not limit the quantity of pollutant discharged. As an alternative, *mass flow limits* are used.

concentration limits versus mass flow limits 濃度限值與總量限值 *nongdu xianzhi yu zongliang xianzhi*

用濃度單位(如廢水中X污染物的mg/l或廢氣中Y污染物的mg/m³)規定的排放限值。它不能直接表示環境中所排放的污染物的影響，因為其影響主要取決於污染物的量。而*總量限值*由於限制了污染物的量，也就直接限制了排放量。所以，限制總量直接與環境影響有關。總量限值的控制可以根據同時進行的濃度測定與體積測定的結果來進行。(參見：總量限值)

The definition of limits for discharges in terms of concentration units (e.g. mg/l of X in an effluent, or mg/m³ Y in off-gas). It does not give a direct indication of the impact of the discharged pollutant in the environment. This impact depends primarily on its quantity. Limiting the quantity of a pollutant in *mass flow limits* directly relates to the emitted quantity. The rate of mass flow is therefore directly linked to the environmental impact. The control of compliance with mass flow limits may be based on concentration measurements combined with simultaneous volume measurements. (See also *mass flow limits*)

conductivity 電導率 *diandaolü*

與水樣或廢水中*含鹽量*有關的一項*綜合指標*。可使用儀器來測定標準電池兩極之間水樣或廢水的電阻,其測定結果給出的是溶液中存在的總鹽含量,即離子化物質的量。這種測定方法不是專門針對各個不同的鹽類的。對於特定離子的分析,可參見*離子分析*。

A *group parameter* correlating with the degree of *salinity* in an aqueous sample or an effluent. The instrument measures the electrical resistance of a water sample or an effluent in a standardized cell between two electrodes. The result gives an indication of the total salinity, i.e. the amount of ionic substances present in a solution. The method is not specific for individual salts. For analysis of specific ions, see *ion analysis*.

confidence limits 置信界限 *zhixin jiexian*

統計中用來衡量一組數據的質量的術語。置信界限是具有某一概率(如95%)的真實平均值的範圍。它可以由數據的平均值、*標準偏差*和測量次數,並按所要求的概率水平計算得到。

A term used in statistics as a measure of the quality of a set of data. The confidence limits describe a range within which the true mean value will be found with a certain probability, e.g. 95%. Confidence limits are calculated from the mean value, the *standard deviation* of the data and the number of measurements. Confidence limits are calculated for a desired percentage level of probability.

construction waste 建築廢物 *jianzhu feiwu*

建造和拆建產生的廢物,雖然其主要成分(礫石、磚頭、碎塊、木材和鋼材)是無害的,也不會造成填埋處置的問題,但必須小心地將受污染而可能會產生問題的那一部分物料如化學防腐木材、油漆、汽油、塑料建材、化學添加劑等加以分開。建築廢物的現代管理方法包括將可重複利用的廢物進行分離和循環使用。

Waste material resulting from construction and demolition activities. While the main components are inoffensive (gravel, brick, debris, wood, steel) and cause no problem in a landfill, care must be taken to separate contaminated fractions which are potential sources of problems. Such fractions include chemically conserved wood, paint, motor oils, plastic construction material, chemical additives etc. Modern management of construction waste includes careful separation and recycling of reusable fractions.

contact zone 接觸層（帶）*jiechucheng (dai)*

見：packing zone。

See *packing zone.*

containment measures 事故遏制措施 *shigu ezhi cuoshi*

限制事故性潛在釋放所造成嚴重影響的措施。這種影響可能是由火災、爆炸、有害物質事故性釋放到環境和周圍居民區造成的。

事故遏制措施包括設立隔離屏障、貯存池、雙層包封反應器等技術性措施，以及在發生事故時減小事故釋放影響的措施。它是現代緊急響應管理的一個組成部分。（參見：預防性安全措施）

Measures to limit the negative impact of a potential release resulting from an accident. The impact can be the result of a fire, of an explosion, or of an accident leading to harmful emissions to the environment, including a populated neighborhood.

Containment measures include technical measures such as various barriers, holding basins, double containment reactors etc., as well as measures to combat and reduce the effect of accidental releases in case of an emergency. Containment measures are part of modern *emergency response management.* (See also *preventive safety measures*)

contaminant 污染物 *wuranwu*

所研究的環境介質中外來的而且會對環境介質產生不良影響的化合物。它是pollutant的同義詞。

A compound which is foreign in the medium under investigation and which is influencing the medium negatively. A synonym for pollutant.

continuous analysis 連續分析 *lianxu fenxi*

見：analysis, continuous。

See *analysis, continuous.*

continuous measurement of biodegradability (Husmann Test) 可生物降解性的
連續測定(哈斯曼試驗) *keshengwu jiangjiexingde lianxu ceding (hasiman shiyan)*

模擬生活污水處理場的實驗室試驗或中間試驗。*好氧生物降解試驗*
是在特殊設計的玻璃設備中進行的，它包括一根曝氣管、一個沉降柱和
這兩部分的連接部件。試驗的目的是研究廢水在所控制條件下的生物降
解性。它通過比較進水和出水的*總有機碳*來測定生物降解的程度。

A laboratory or pilot plant test to simulate a sewage water treatment plant. The
test (*aerobic biodegradation*) is carried out in a specially designed glass apparatus
including an aeration tube, a sedimentation part and interconnections. The test is to
study the biodegradation behavior of an effluent under controlled conditions. The
degree of biodegradation is measured by comparing the *TOC* values of the inflow
with the outflow.

contracts 合同 *hetong*

就某一事項，雙方或多方所達成的共同協定。以合同方式規定某特
定行業的空氣污染控制的排放限值，見*行業合同*。

An agreement made between two or more parties on a certain subject matter.
For branch specific emission limitations in air pollution control based on contracts,
see *branch contracts*.

controlled disposal site 受控處置場 *shoukong chuzhichang*

見：controlled landfill。
See *controlled landfill*.

controlled landfill 受控填埋場 *shoukong tianmaichang*

以保護*地下水*、衛生操作以及*保護景觀*為目的而設計的廢物*處置
場*。它周圍用圍牆隔離，而且以專門業務方式運作與管理，包括入場廢
物的鑒別和控制、*浸出液*和地下水的監測以及*填埋氣體*排放的監測。現
代的廢物管理和填埋概念，將可接納廢物的範圍限定為非反應性廢物、
非流動性廢物和/或無機廢物。

A *disposal site* for wastes which has been designed with the objectives of
groundwater protection and hygienic operation and which also integrates the

objectives of *landscape protection*. Controlled landfills are fenced off. They must be operated and managed in a professional manner. This includes specification and controls of incoming wastes and monitoring of liquid *leachate*, of groundwater and of the emission of *landfill gas*. Advanced waste management and landfill concepts limit the range of acceptable wastes to non-reactive, immobilized and/or inorganic wastes.

controlling, environmental 環境控制 *huanjing kongzhi*

這一術語通常當作是環境管理的同義詞。它包括一般管理控制，但重點放在環境控制方面。（參見：環境控制）

This term is generally used as a synonym for environmental management. It includes the activities of general controlling with focus on environmental aspects. (See also *environmental controlling*)

controlling and reporting systems 控制和報告體系 *kongzhi he baogao tixi*

應用於*一般管理中的控制*尤其是環境控制方面的*管理體系*。（參見：環境行為指標）

See *management systems* which are applied in the context of *controlling in general*, and specifically environmental controlling. (See also *environmental performance indicators*)

controlling in general management 一般管理中的控制 *yiban guanlizhongde kongzhi*

用於概述某些管理要素的術語，其定義是各不相同的，偶爾也用作為管理的同義詞。控制具有十分重要的作用，最初係著重於成本控制。控制者的傳統任務包括對工廠中的信息管理和數據獲取工作進行協調，為管理決策做好準備和為決策提供依據。其重點最初放在行為控制數據上。（參見：環境控制）

A term used to summarize certain elements of management. The definition of the term varies and is occasionally used as a synonym for management.

Controlling is a staff function, focusing initially on cost controlling. The traditional tasks of the controller include the coordination of information management and data acquisition in a factory to prepare and provide a decision base for management decisions. The focus was initially on performance control data. (See also *environmental controlling*)

cooling and liquefaction 冷卻和液化 *lengque he yehua*

應用於空氣污染控制的一般化工過程。冷卻和液化可從氣流中去除和回收揮發性有機化合物，但過程單元和過程條件必須適合於空氣污染控制的具體情況與要求。（參見：再生技術）

Processes adapted from general chemical technology to air pollution control. Cooling and liquefaction may serve to eliminate and recover volatile (organic) compounds from a gas stream. Processing units and processing conditions must be adapted to the specific situation and requirements of air pollution control. (See also *regenerative technologies*)

cooling water 冷卻水 *lengqueshui*

在單獨的閉合冷卻系統中用於冷卻的工藝用水。冷卻水不應與可能污染的物質（如化學品）直接接觸，不應被污染。儘管如此，對冷卻水排放到受納水體，仍必須加以控制。

Process water of a factory which is used for cooling purposes in a closed and separate cooling circuit. It should not be in direct contact with potentially polluting material (e.g. chemicals) and should therefore not be contaminated. Nevertheless, its discharge to a receiving waterbody must be kept under control.

corrective measures 修正措施 *xiuzheng cuoshi*

對某項活動產生的不良影響進行修正而採取的措施，它與為防止不良影響而採取的*預防措施*相反。（參見：閉環活動、線性活動）

Measures to correct the negative aspects of an activity. As opposed to *preventive measures*, which are taken first place to prevent such a negative impact. (See also *closed cycle activity*; *linear activity*)

correlation coefficient 相關系數 *xiangguan xishu*

表示呈正態分佈的兩組數據（如A和B）之間關係的一項統計參數。相關系數大，表示兩組數據之間存在一定關係。但當這兩組數據可能與第三組數據（C）有關時，其相關系數不能表明存在直接的因果關係。（參見：因果關係）

A statistical parameter measuring the relationship between two sets of data (e.g. A and B), assuming a normal distribution of both data sets. A high correlation co-efficient indicates that there is a relationship between the two sets of data. However, as the two sets of data may depend on a third set (C), the correlation coefficient does not provide any direct evidence of cause and effect. (See also *cause-effect relationship*)

cost allocation of pollution control costs 污染控制費用的分攤 *wuran kongzhi feiyongde fentan*

將環境保護費用（*損害預防費用*）納入與包括在某產品或某服務協議的總生產成本中的過程。其目的是計算得到*真實的總產品成本*，包括*外部費用*的分攤在內。作為費用分攤的依據，需要有準確的排放量和實際的*比處置費用*的數據和資料。

The process of integrating and including environmental protection costs (*damage prevention costs*) in the total manufacturing costs of a product or of a service agreement. The objective is to calculate *true total product costs* that include an allocation of *external costs*. As a base of cost allocation, true quantities of emissions and true *specific disposal costs* should be used.

cost calculation procedure 成本計算方法 *chengben jisuan fangfa*

計算某產品或所提供的某種服務的真實總成本的標準化方法。它的數據基礎是所使用的物料量和所付的實際價格。此外，還必須包括能耗、勞動力、設備與建築物的使用與維護、基礎設施和*環境保護費用*。後者應像其他費用因素一樣給與同等的注意。

A standardized procedure to calculate the true total costs of manufacturing of a product or of a service rendered. The data elements of the procedure are based on quantities of materials used and actual prices paid. The procedure must also include elements of energy consumption, labor, use and maintenance of equipment and buildings, infrastructure and *environmental protection costs*. The latter should receive the same attention as the other cost elements.

cost center 費用中心 *feiyong zhongxin*

與公司內部會計系統有關的會計學技術術語。在費用中心,各單位一切活動所產生的所有費用都要加以確認和進行累計。它採用計算機費用計算方法。這樣,不同活動的費用可以分攤到某一特定產品的生產成本中去。企業通常都制定有它的費用中心的計劃。

A technical term for an account in connection with the company internal accounting system. In a cost center, all costs resulting from all different activities of the organizational unit are identified and accumulated. Cost centers use computerized cost calculation procedure. The costs for different activities can thus be allocated to the manufacturing costs of a particular product. An enterprise generally has defined a plan of its cost centers.

cost center structure (plan) of a factory 工廠的費用中心結構(計劃)
gongchangde feiyong zhongxin jiegou (jihua)

工廠會計和成本計算系統的一個組成部分,目的在於使工廠的成本結構透明化。費用中心的結構與工廠的組織機構有密切關係。

An element of the accounting and cost calculation system of a factory, with the objective of making the cost structure of a factory transparent. The structure of the cost centers is closely related to the organization of the factory.

cost covering prices 成本價格 *chengben jiage*

包括一切正常生產成本以及污染控制所有費用在內的真正的總產品價格。它們是由成本計算方法計算得到的。

True total product prices which include all normal manufacturing costs and also all aspects of pollution control. They are calculated by a *cost calculation procedure*.

cost-benefit analysis 費用-效益分析 *feiyong-xiaoyi fenxi*

環境經濟學的一方面內容。它將污染控制或污染預防的費用與由污染控制對社會產生的效益進行比較分析。效益即等於所*避免損害的費用*。就工廠的極端情況而言，其損害可能是由於重大災難造成工廠的關閉。總費用由*損害預防費用和損害恢復費用*構成。

費用和效益是多方面的，因此對它們的分析需要對一切因素作認真考慮。由於效益與個人或社會的喜好與價值有關，所以也帶有很強的主觀性。

An aspect of environmental economics. The comparative analysis of the costs of pollution control or pollution prevention versus the benefits to society resulting from pollution control. Benefits are identical with the *costs of avoided damage*. In the extreme case of a factory, the damage may be its closure as a result of a major disaster. Overall costs are made up of *damage prevention costs* and *damage repair costs*.

Costs as well as benefits are multi-faceted and their analysis therefore demands careful consideration of all factors. Benefits are also very much subjective since they are dependent to personal or society driven preferences and values.

costs 成本 *chengben*

企業會計系統中所包括的會降低企業利潤的要素。它包括*借貸的基建費用*和工廠的運作費用。運作費用又包括*直接和間接費用* —— 而後者意指非特殊的*一般管理費用*（管理費用又可以分為一般管理費用和生產管理費用）。各個費用要素可以按照會計規程分類成為上述費用類別之一。然而，有相當多的要素並不明顯屬於直接或間接費用。在這種情況下，必須採用一個合理的定義。企業必須確定最適於它自身的會計系統。從環境保護觀點看，也應盡可能將所有廢物處置產生的費用包括在內。其基礎是要有*比廢物量*和廢物的*比處置費用*的數據與資料。

Elements included in the accounting system of an enterprise which reduce the profit of the enterprise. Costs include *costs of borrowed capital* and costs of the operation of the factory. Operating costs in turn include *direct costs* and *indirect costs* which are unspecific *overhead costs*. (Overhead, in turn, can be split up into general overhead costs and manufacturing overhead costs). The classification of individual cost elements into one of the above categories follows certain rules of accounting. However, quite a few elements are not clearly direct or indirect. In these cases, a reasonable definition must be adopted. The enterprise must define the system which is best applicable for its own accounting. From the environmental protection point of view, the costs resulting from the disposal of all wastes should also be included as much as possible. The bases are the *specific waste quantities*, and their *specific disposal costs*.

costs, environmental protection (EP) 環境保護費用 *huanjing baohu feiyong*

由於環境保護活動或未採取環境保護措施造成企業或社會所負擔的各種不同費用的要素。環境保護費用可按照不同的準則來確定：(1) *損害費用*，即由環境破壞所產生的費用；(2) *損害預防費用*或污染控制費用，即由採取減少排放和預防損害的行動所產生的費用；(3) *損害恢復費用*或恢復補救費用。

對於(2)和(3)兩類費用，必須區別污染控制設施的*基建投資*和這些設施的運行費用。環境友好產品和工藝(*完整的工藝過程開發*)的研究開發費用也屬於此類費用，它還包括如分析和行政管理之類的支持活動的費用。損害費用(1)又可分為即刻費用和*滯後費用*。

此外，在討論環境經濟學和環境費用時，還需區別*內部費用*和*外部費用*。然而，需要指出的是，明確地鑒別環境保護費用常常是很困難的，因為符合總的技術發展要求的新工藝，本身就是清潔工藝。

All different elements of costs accruing to an enterprise or to society as a result of environmental protection activities or the lack thereof. EP costs can be structured according to different criteria: (a) *damage costs*, resulting from environmental damages; (b) *damage prevention* or pollution control costs, resulting from activities to reduce or moderate emissions and to prevent damages; and (c) *damage repair costs* or remediation costs.

For the two categories (b) and (c) one has to distinguish between *investments costs* in pollution control facilities and the operating costs for these facilities. Research and development costs for environmentally friendly products and processes (*integral process development*) also belong to this group, as well as the costs of support activities like analysis and administration. Damage costs (a) in turn can be classified into immediate costs and *delayed costs*.

In the discussion of environmental economics and environmental costs one also distinguishes between *internal costs* and *external costs*. A clear-cut identification of environmental protection costs is often difficult since new processes which are in line with the general development of technology are also cleaner processes.

costs, external 外部費用 *waibu feiyong*

*環境保護費用*的一部分。外部費用並不包括在污染產生者如排放污染的企業的會計系統中。可用兩個例子來說明這一點：在離森林有一定距離的某些工廠所排放的污染物造成的森林破壞，工廠是不負責賠償的，它們是外部費用；由於污染使湖泊的娛樂價值受到損失，這是另一種外部費用要素，它是由整個社會而不是由污染產生源來承擔的。

A portion of the total *costs of environmental protection.* External costs are not included in the accounting system of the generator of a pollution, e.g. the enterprise emitting a pollutant. Two examples serve to illustrate the point: the damages to a forest resulting from emissions of several factories at a distance are not paid for by the factories, they are external; the loss, as a result of pollution, of the recreational value of a lake is another external cost element and is born by society as a whole and not by the sources of pollution.

costs, internal 內部費用 *neibu feiyong*

被納入企業會計系統並包括在產品或所提供服務的價格計算中的環境費用的要素。內部費用最初是由生產者承擔的，然而其中至少有一部分反映在該產品的價格中。(參見：費用分攤、外部費用)

Elements of environmental costs which are integrated into the accounting system of an enterprise and which are included in the calculation of the price of products or services rendered. Internal costs are initially born by the producer, however, at least a part of them reflects itself in the product price. (See also *cost allocation*; *costs, external*)

costs of avoided damage 避免損害的費用 *bimian sunhaide feiyong*

污染控制的效益。它是通過採取適當的污染控制措施所避免的可能因環境破壞而產生的潛在費用。(參見：損害預防費用)

The benefit of pollution control. The potential costs which would have resulted from environmental damages, but which have been avoided through proper pollution control measures. (See also *damage prevention costs*)

costs of borrowed capital 借貸的基建費用 *jiedaide jijian feiyong*

與借貸的基建投資有關的費用要素。這種基建投資可能用於污染控制的廢水處理場或大型焚燒爐。基建費用包括利息和與貸款有關的其他費用要素。借貸的基建費用是應以某種方式包括在企業的利潤或虧損中的那些要素。它們構成設施運行費用和產品總生產成本的一部分。

These cost elements are associated with borrowed capital, used for an investment. The capital investment may be an effluent plant or a large incinerator, used for pollution control. The capital costs include interest and other elements associated with the loan. Costs of borrowed capital are elements which should be included in the profit and loss of the enterprise in one way or another. They contribute to the operation costs of a facility and the total manufacturing costs of a product.

costs of remediation 恢復補救費用 *huifu bujiu feiyong*

見：damage repair costs。

See *damage repair costs.*

Cottrell separator, electrostatic 科特雷爾靜電分離器 *ketelei' er jingdian fenliqi*

一種乾式靜電分離器。（參見：靜電分離器）

A type of dry electrostatic separator. (See also *electrostatic separator*)

counter-current 逆流 *nilu*

兩種介質（如液體和氣體，或兩種不混溶的液體）以相反方向流經加工過程的塔設備。（參見：並流）

Two media (e.g. a liquid and a gas or two immiscible liquids) flowing in the opposite direction through a column. (See also *co-current*)

cPAH 致癌多環芳烴 *zhi'ai duohuanfangting*

Carcinogenic polyaromatic hydrocarbons的縮寫。（參見：多環芳烴）

Carcinogenic polyaromatic hydrocarbons. (See also *PAH*)

cradle-to-grave control "從搖籃到墳墓"控制 *cong yaolan dao fenmu kongzhi*

應用於有害廢物管理的控制原則，即從生產過程產生廢物開始，直到廢物的最終處置，對廢物進行的全過程管理控制。該原則是現代廢物管理立法的基礎。(參見：廢物多聯單)

A control principle applied in the management of hazardous wastes. The principle implies total management control over the wastes from the moment wastes are generated in the production process to the moment of their final disposal. The principle of 'cradle-to-grave' control is the basis of modern legislation on waste management. (See also *waste manifest*)

cybernetic management approach 控制論管理方法 *kongzhilun guanli fangfa*

將企業活動對環境的影響作為反饋信息來控制企業管理體系的管理方法。與線性體系相反，這種管理方法是一種*閉環活動*。(參見：線性活動)

A management approach which integrates the impact of the activities of the enterprise on its environment as a feedback for steering its management system. Management is taken as a *closed cycle activity*, in contrast to an open, linear system. (See also *linear activity*)

cybernetics 控制論 *kongzhilun*

研究控制機理和系統的學科。此詞匯來源於希臘詞匯"控制"之意。控制論已應用於生物系統和電力系統，其基本概念也可應用於管理上。

The science and study of control mechanisms and systems. The word is derived from the Greek word for 'steering'. Cybernetics is applied to biological systems as well as to electrical systems. The basic concept can also be applied in management.

cyclone separators 旋風分離器 *xuanfeng fenliqi*

利用離心力從廢氣中分離粒子的技術。它使污染氣流通過一錐形工藝設備產生迅速的圓周運動，由於離心力的作用使粒子從氣流中分離出來。所收集的粉塵掉落到圓錐底部而被連續取出。旋風分離器常常用於回收乾產品粉塵，也可用作為簡單的空氣污染控制設備。

A technology to separate particles from an off-gas stream by centrifugal forces.

The polluted air stream is passed through a conical process unit designed in such a way as to impart a rapid circular movement to the air. Particles are separated from the air stream due to centrifugal forces. The collected dust falls into the lower end of the conical section and can be taken out continuously. Cyclones are often used in the recovery of dry product dusts and also as simple air pollution control units.

D

damage costs, environmental 環境損害費用 *huanjing sunhai feiyong*

由於環境污染所產生的企業或社會的費用。只有在特殊情況下，污染源、損害程度和*因果關係*才是顯而易見的。由於種種原因，這種關係常常是很複雜的。損害往往會滯後相當時間才顯現，何時開始並不明顯。污染源、污染和損害類型之間的關係一般是很複雜的，而且損害常常出現在遠離污染源的地方。

在某些情況下，有可能將損害費用定量化。從污染的河流生產飲用水需要較高的成本，就是一個例子。然而在其他許多情況下，進行鑒別卻不是那麼簡單明白的；評估經濟上的損失會受個人或社會因素很大的影響。有些損害，如文化古迹的損害是難於用金錢來衡量的。（參見：費用－效益分析、損害預防費用、損害恢復費用）

A term summarizing costs to an enterprise or to society resulting from a polluted environment. Only in exceptional cases are the source of a pollution, the extent of damage and the *cause-effect relationship* clear. Very often, the relationship is complex for several reasons. Often damages are delayed for a considerable time and the beginning of damages is not clear. The relationship between source of pollution, type of pollution and damage is generally complex. Damages frequently emerge far away from the source of pollution.

In certain cases, it is possible to quantify the cost of damage. Higher costs for the production of drinking water from a polluted river serve as example. In many other cases the identification is not straightforward; the monetary damage is subject to personal or societal preferences. Some damages, e.g. the damage to cultural monuments can hardly be defined in terms of money. (See also *cost-benefit analysis*; *damage prevention costs*; *damage repair costs*)

damage prevention costs 損害預防費用 *sunhai yufang feiyong*

因採取行動防止對環境造成破壞而產生的由潛在的污染者負擔的費用。其典型的例子是*廢水處理、空氣污染控制*和*廢物管理*所產生的費用。*清潔技術開發*的費用可能只有部分包括在內。這種*完整的工藝過程開發*與技術更新的目標是相一致的。

Costs accruing to a potential polluter (an enterprise or a municipality) as a result of activities carried out to prevent damages to the environment. Typical cases of damage prevention costs are costs from *effluent treatment*, from *air pollution control* and from proper *waste management*. Costs arising from the development of *clean technologies* can be included only partially. Such an *integral process development* is also in line with objectives of a general renewal of technology.

damage repair costs 損害恢復費用 *sunhai huifu feiyong*

因採取行動對生態系統或其一部分的破壞進行恢復補救所產生的費用。典型的例子如採取行動對老的無控廢物*堆放場*和地下水體消除污染。另外，對污染河水進行淨化來生產飲用水而產生的額外費用，也可以包括在這類費用中。

根據*污染者付費原則*，損害恢復費用應由各自的污染者負擔。由於污染常常是滯後顯現的，因此確定承擔污染責任的當事者可能會是一個問題。在許多情況下，污染是由若干污染者造成的，因此需要認真考慮損害恢復費用的支付問題。（參見：超級基金計劃）

Costs which arise from activities to repair damages to an ecosystem or a part thereof. The activities to decontaminate old uncontrolled waste *dumps* and ground-water bodies underneath can be taken as typical examples. Also, additional costs for the purification of polluted river water into drinking water can be included in this group.

According to the *polluter-pays-principle*, damage repair costs should be born by respective polluters. Because pollution is often delayed, there may be a problem of finding the party which is or was responsible for the pollution. In many cases, several polluters were contributing. The question of payment of damage repair costs therefore needs careful consideration. (See also *SUPERFUND program*)

data, environmental 環境數據 *huanjing shuju*

與環境管理各方面有關的定量與定性數據。全面完整的數據包括涉及*環境的組成部分如空氣、水和土壤等要素*，以及*產品整個生命周期中各個階段的有關資料*。

Numerical, quantitative and qualitative data on different aspects of environmental management. Comprehensive data must include elements dealing with the *environmental segments* air, water, soil, and covering for each segment all steps of the full *life cycle of a product*.

data logger 數據記錄儀 *shuju jiluyi*

　　對於獨立應用而不能直接與計算機連接的分析儀器所採用的數據收集設備。其數據的處理可在以後進行。

A data collection device for stand-alone analytical instruments which cannot be connected directly to a computer. The data can be transferred for processing later.

data management systems 數據管理系統 *shuju guanli xitong*

　　定量與定性數據的系統管理方法。它包括數據的收集、數據的處理、數據統計意義的評價、數據的壓縮與集合、數據儲存及檢索。現在，數據管理一般均以計算機數據庫為基礎。對於數據管理的每一階段及其任務，必須規定明確的責任和信息流渠道。

A systematic approach to management of numeric and qualitative data. Today, data management is generally based on computer databases. Data management includes clear provisions for collecting data, processing of data for subsequent use, evaluation of the statistical significance of data, condensation and concentration of data for later evaluation and storing data for later retrieval. For all steps and tasks, clear responsibilities and information flow pathways are necessary.

data types, air quality 數據類別，空氣質量 *shuju leibie, kongqi zhiliang*

　　為報告環境空氣質量所提供的不同類別的數據和平均值。它們往往是在不同期間測定的結果；測定時間可從半小時到24小時。因此，所報告的數據是在不同期間的平均值。"年平均"是按年的天數由每日"24小時平均"計算得到的平均值。

　　當報告、比較或確定這些數值的限值時，必須認真地詳盡說明計算平均值所採用的原則。依一次測定時間的不同，數值會具有不同的意義。短期平均表示一天期間的變化,而長期平均可用來分析多年間環境空氣質量變化的趨勢。

　　以上這些考慮適用於一些重要參數(如二氧化硫、氮氧化物、臭氧和顆粒物)的數據,也適用於為進行專門調查研究而監測的其他許多補充參數。

Different types of data and average values are available to report the quality of ambient air. The values are usually the result of measurements over different periods of time, with measurement periods from half an hour to 24 hours. The reported data are therefore averages over different periods of time. The 'one year average' is calculated as the average value of the daily '24h averages' for the days of one year.

When reporting, comparing or defining limits for such values, care must be taken to specify exactly the principles used for averaging. Depending on the time of one single measurement, the values have a different meaning. Short term averages give an indication of variations during a day. Long term averages serve to analyze trends of the ambient air quality over periods of years.

These considerations apply to data for key parameters, (such as sulfur dioxide (SO_2), oxides of nitrogen (NOx), ozone, and particulate matter) as well as to a wide range of additional parameters monitored for special investigations.

dB(A) 分貝(A聲級) *fenbei (A shengji)*
　　見:decibel (A)。
　　See *decibel (A)*.

decibel(A) (dB(A)) 分貝(A聲級) *fenbei (A shengji)*
　　測量聲音的單位。(參見:聲壓級)
　　A unit for measuring sound. (See also *sound pressure level*)

decision factors 決策因子 *juece yinzi*
　　決策方法中會影響決策結果的因子。有些決策因子是強制性的,即它們是必須滿足的,如上限成本。另一些決策因子如說明顏色之類是最

好能具有的性質，這些因子是非強制性的並且是可以優化的。（參見：排序判據）

Factors used in *decision techniques*, influencing the outcome of a decision. Some decision factors are compulsory, i.e. they must be met at all rates. Example: the upper limit of costs. Others describe aspects which would be nice to have (e.g. the color), but which are not compulsory and which can be optimized. (See also *ranking criteria*)

decision techniques 決策方法（技術）*juece fangfa (jishu)*

在有若干方案存在的情況下，能有助於選擇最適合方案的管理方法（技術），它們為決策過程提供了系統、合理的依據，使得決策更透明，但選擇仍然會受管理者主觀因素所影響。（參見：排序方法、排序判據）

Management techniques to assist in the selection of the most suitable option in cases where several options are available. Decision techniques provide a systematic and rational base for the decision making process. They make decisions more transparent, but the choice is still influenced by subjective factors of the manager. (See also *ranking methods*; *ranking criteria*)

decolorization of effluent 廢水脫色 *feishui tuose*

去除廢水的顏色。可以用於廢水脫色的污染控制單元過程有：（1）用特定的化學藥劑還原或氧化分解有顏色的污染物；（2）用高壓*濕式氧化法*或高溫焚燒法分解帶色濃母液。為此，對於稀廢水可用*逆滲透技術*進行*預濃縮*，以降低處理費用；（3）用鐵或鋁鹽沉降處理低濃度的染料廢水；（4）用專門的高分子絮凝藥劑沉降處理某些類型陰離子染料廢水；（5）用活性碳吸附。

處理過程的選擇取決於有顏色污染物的化學性質及其溶液的狀態。這些過程可以應用作為*末端工藝過程處理*或*末端處理措施*。

Removal of color from an effluent. Several pollution control unit processes can be used to remove color: reductive or oxidative destruction of colored contaminants with specific chemical reagents; oxidative destruction of concentrated colored mother liquors by high pressure *wet air oxidation* or high temperature incineration. In this case, a *preconcentration* of dilute streams by *reversed osmosis* serves to optimize costs; precipitation with iron or aluminium for low concentrations of dyes; precipitation of certain types of anionic dyestuff molecules with specific polymeric reagents; and adsorption on activated carbon.

The processes depend on the chemical nature of the colored contaminants as well as on their state of solution. The processes can be applied as *end-of-process treatment* or as *end-of-pipe measures.*

decompression stage 減壓段 *jianyaduan*

濕式空氣氧化裝置的一個組成部分，它是高壓設備與其他設備之間的過渡段，以將反應塔的高壓狀態降到常壓狀態。

An element of a wet air oxidation unit. It is the process stage between the high pressure elements and the other units. In this process stage, the high pressure of the actual reaction column is lowered to atmospheric pressure.

deep shaft technology 深井曝氣技術 *shenjing baoqi jishu*

一種廢水生物處理技術，其活性污泥曝氣池建造在地下，為一圓形深筒。深井曝氣筒的直徑可達2米，深度可達地下100米或更深。壓縮空氣在離筒頂不遠處引入，並使*混合液*連同氣泡一起在筒內下流循環，因此空氣利用效率高；而且深井曝氣氣泡停留時間長，氧幾乎可被全部吸收。該技術的優點是佔地空間小，氧利用效率高，在極寒冷的天氣下系統仍可維持運行；缺點是難於檢查深井的泄漏情況。（參見：廢水生物處理）

A type of biological effluent treatment plant in which the activated sludge basin is constructed as a deep circular basin below ground. The dimensions of the 'deep shaft' basin may range in diameter up to 2 meters and below ground down to 100 and more meters. Compressed air is introduced not far from the top of the basin. The introduction of air is efficient and economical due to the downward flow of the circulating *mixed liquor* which transports air bubbles along. The retention time for air bubbles is high and oxygen is almost completely absorbed. Advantages are the reduced need for space, the efficient oxygen introduction and the insensitivity to extremely cold weather. Problems may result from the difficulty to check the deep shaft for leaks. (See also *biological effluent treatment*)

deflector elements 折板 *zheban*

主要應用在空氣污染控制中，用以改變（折轉）污染空氣流方向的部件。在此過程中，氣流中的微細液滴被聚集成較大的液滴，而可作為液體除去。

Elements used mainly in air pollution control which change (deflect) the direction of a polluted air stream. In this process fine droplets in the gas stream are agglomerated to larger droplets and can be removed as a liquid.

degassing zone 脫氣區 *tuoqiqu*

廢氣生物處理的一個步驟。脫氣區體積較小，它設置在曝氣池後和二沉池前，目的是在混合液進入沉降池前使氣泡從混合液中溢出，以分離活性污泥和進行最終澄清。

A process step in biological effluent treatment. The relatively small degassing zone is placed after the aeration basin and before the secondary clarifier. Its purpose is to allow air bubbles to escape before the mixed liquor is passed to the sedimentation basin for separation of activated sludge and final clarification.

delayed damage costs 滯後損害費用 *zhihou sunhai feiyong*

只有在實際污染發生後多年才能顯現的環境損害所產生的費用。例如，不正確地填埋處置廢物，多年後會造成地下水污染，而為治理地下水污染採取補救恢復措施所需的費用，將比在開始時正確地處置廢物所需的初始費用要高得多。對其他許多環境問題，情況也是如此。（參見：損害費用）

Costs resulting from environmental damages which manifest themselves only years after the actual pollution took place. Example: improper disposal of wastes in a landfill may cause groundwater pollution with a delay of many years. The costs for the remediation of this groundwater pollution will be much higher than the initial costs for proper waste disposal. Similar examples can be found for many other environmental sectors. (See also *damage costs*)

delayed effects 滯後效應 *zhihou xiaoying*

時間上滯後顯現的效應，環境損害常常是由滯後效應造成的。

An effect which manifests itself only after a delay time. Environmental damages are often resulting from delayed effects.

demister 除霧器 *chuwuqi*

多段空氣污染控制裝置中的分離段，用於從尾氣中去除微細霧滴。在多段空氣洗滌裝置中，它可防止洗滌液從前一段被夾帶到下一段。

A separation stage in a multi-stage air pollution control device. A demister serves to remove mist (fine droplets) from an off-gas stream. The demister stage reduces carry-over of scrubbing liquid from one stage to the next stage of the multi-stage air scrubbing device.

demolition waste 拆建廢物 *chaijian feiwu*

見：construction waste。

See *construction waste*.

denitrification 反硝化 (作用) *fanxiaohua (zuoyong)*

為減少廢水中氮污染物所採用的硝化 / 反硝化過程的第二階段。在第一階段，氨氮在*亞硝化桿菌*和*硝化桿菌*的作用下被氧化為硝酸鹽氮。在第二階段，在缺氧條件下硝酸鹽氮被還原為元素氮。為實現這一轉化，可將傳統廢水處理場中的生物處理裝置置於有限供給外部氧的情況下運行。另一種辦法是，在處理裝置中設缺氧區。（ 參見：硝化、缺氧生物降解)

A second step in the nitrification/denitrification process used to reduce the quantity of nitrogen pollutants in an effluent. In the first step, ammonia nitrogen is oxidized to nitrate nitrogen, using *nitrosomonas* and *nitrobacter*. In the second process step, anoxic biodegradation serves to reduce nitrate nitrogen to elemental nitrogen. To achieve this transformation, the biological stage of a conventional effluent treatment station is run under a limited supply of external oxygen. Alternatively, oxygen deficient zones may be incorporated in the treatment plant. (See also *nitrification; anoxic biodegradation*)

denitrification-nitrification process 反硝化–硝化過程 *fanxiaohua-xiaohua guocheng*

從廢水中去除氨氮和硝酸鹽氮的過程。（參見：反硝化、硝化）

The process of removing ammonium nitrogen and nitrate nitrogen from an effluent. (See also *denitrification*; *nitrification*)

DENOX converter 除氮氧化物轉化器 *chudanyanghuawu zhuanhuaqi*

對降低氮氧化物濃度方法所廣泛使用的一般術語。（參見：催化還原去除氮氧化物、選擇性非催化還原（SNCR）法）

A widely used general term for processes to reduce the concentration of oxides of nitrogen. (See also *catalytic reductive removal of nitrogen oxides*; *selective non-catalytic reduction (SNCR) process*)

deodorizing processes 脫臭方法 *tuochuo fangfa*

從空氣流中去除有氣味污染物的方法，如氧化洗滌法、生物去除法和焚燒熱降解法，其選擇取決於廢氣量和惡臭污染物的類別與性質。

Processes to remove odorous pollutants from an air stream. Basically, deodorization can be carried out by different processes: oxidative scrubbing; biological removal processes; thermal degradation by incineration. The selection of the process depends on the volume of the off-gas stream and the type and properties of the odorous pollutants.

desorption from activated carbon 活性碳解吸 *huoxingtan jiexi*

吸附的相反過程。如果利用活性碳作為緩衝系統並在飽和點下運行，則當氣流中污染物的濃度低於其平衡濃度時就會發生解吸。吸附於活性碳上的物質幾乎可以全部被解吸分離出來而加以回收。如果將吸附飽和的活性碳用蒸汽或（熱）氮氣處理，也會發生解吸。在液相中也同樣會發生吸附與解吸過程。

The opposite process of adsorption. If activated carbon is used as a buffer system and operated at the saturation point, then desorption takes place when the pollutant concentration in the gas stream is lower than the equilibrium concentration. Substances adsorbed on activated carbon can also be almost fully desorbed, i.e. separated, from the loaded carbon and be recovered. Desorption also takes place if the loaded carbon is treated with steam or with (hot) nitrogen gas. Adsorption and desorption can also take place in a liquid phase.

detection limit of a method 方法的檢測極限 *fangfade jiance jixian*

表示某分析方法的質量和靈敏度的參數。檢測極限是可分析測定出的某未知物的最低濃度（其*分析的信噪比為*2）。不同的分析方法具有不同的檢測極限。

A parameter to describe the quality and the sensitivity of an analytical procedure. The detection limit is given as the lowest concentration of an unknown substance which can be determined analytically with a *signal-to-noise ratio* of two. The detection limit depends on the analytical procedure.

detection methods in analysis 分析中的檢測方法 *fenxizhongde jiance fangfa*

物質分析的最後步驟。在用氣相色譜、高壓液相色譜或薄層色譜技術將混合物分離成各個化合物後，應用檢測設備產生出與被測化合物的量成正比的信號。此檢測器的初始電信號可以被記錄下來並進一步進行加工處理。檢測方法應針對被分析的化合物來具體選擇，並且應不受其他化合物的干擾。它可以是儀器模件、化學反應或者二者之組合。幾乎每一種分析方法都有它自己的檢測技術。在氣相色譜法和高壓液相色譜法中，有火焰離子化檢測器、電子捕獲檢測器或氮的特定檢測器等不同檢測技術。對於吸收紫外線的化合物，在高壓液相色譜法中採用紫外吸收進行檢測。

The last step in the analysis of a substance. After the separation of mixtures into the individual compounds, (with techniques such as GC, HPLC or TLC) a detection device serves to generate a signal which is proportional to the quantity of the compound being measured. The primary electric signals of the detector can be recorded and processed further. The detection method should be specific for the compound in question and interference from other compounds should be negligible. Detection methods are instrumental modules, chemical reactions, or a combination of both. Nearly each analytical method uses its own technique. In GC and HPLC, there are different techniques of detection, e.g. FID (flame ionization), EC (electron capture), or chlorine specific detection. UV absorption is used in HPLC for UV-absorbing compounds.

detector signal 檢測器信號 *jianceqi xinhao*

檢測設備發出的初始信號。

The primary signal from a detecting device.

detoxification of waste 廢物去毒 *feifu qudu*

用物理、化學和生物方法改變廢物中組分的毒性或去除廢物中有毒的組分。例如，用氧化法分解*電鍍廢物*污泥中的氰離子。

The modification of toxic properties of waste components or the removal of toxic components by physical, chemical and biological processes. An example of a detoxification is the destruction of cyanide ion in *galvanic waste* sludge through oxidation.

diesel engines 柴油發動機 *chaiyou fadongji*

見：catalytic converters for diesel engines。

See *catalytic converters for diesel engines.*

diffuse emissions 擴散排放 *kuosan paifang*

與*點排放*相反，它是由難以鑒別的不確定擴散源（例如敞口儲槽、敞口容器、設備泄漏等）向大氣的無組織排放。

Emissions into the atmosphere which originate from badly defined, diffuse sources (e.g. open vessels, open containers, spills, leaks in equipment, etc.) and which are difficult to identify. In contrast to *point emission.*

diffuse emission source 擴散排放源 *kuosan paifang yuan*

空氣污染的*擴散排放源*。

A source of *diffuse emissions* into the air.

diffusors 擴散器 *kuosanqi*

廢水生物處理場使用的一種曝氣設備，用以將空氣或氧引入曝氣池液面下的*混合液*中進行曝氣。擴散器中裝有一平板或半球形多孔陶瓷部件，能生成非常小的空氣或氧氣氣泡進入液體中。此外，也還有用其他材料製成的擴散器。擴散器的能量利用效率很高，但存在孔易堵塞的問題。

A type of aeration equipment in a biological effluent treatment plant. The introduction of air or oxygen into *mixed liquor* takes place below the surface. Diffusors incorporate a ceramic element, shaped as a plate or a half-dome. The diffusor is porous or contains very small holes through which very small bubbles of air or oxygen enter the liquid. Similar types made of other materials are also available. The energy efficiency of diffusors is high, but plugging of the pores may pose a problem.

digital recorder 數字記錄儀 *shuzi jiluyi*

以數字形式將信息收集與貯存在磁帶或磁盤上的數據收集設備。其數據可在以後用計算機進行加工處理。（參見：數據記錄儀）

A data collecting device to collect and store information in a digital form on a tape or disc. Data can later be processed by computer. (See also *data logger*)

digital signal 數字信號 *shuzi xinhao*

由某種作用產生的信號，它以數字形式獲得以便於在計算機上直接進行處理。例如，開關處於開（信號為1）或關（信號為0）的位置，就是一種二元數字信號。通常可以通過不同的電子信號處理方法把模擬信號轉換為數字信號。

A signal, derived from a certain action, which is available in a digitized form for direct processing in a computer. The position of a switch which is either open (Signal = 1) or closed (Signal = 0) is an example of a binary digital signal. Digital signals usually are derived from analog signals by various electronic signal processing methods.

dilemma, time horizon 時間範疇上的難題 *shijian fanchoushangde nanti*

見：time horizon dilemma。

See *time horizon dilemma*.

direct costs 直接費用 *zhijie feiyong*

　　某產品生產中那些與生產量成正比關係的費用要素。典型的要素為原材料、輔助材料、能源、與產品有關的特殊基礎設施要求。在現代的產品成本會計系統中，應根據真實的處置費用和實際數量，將*污染控制費用*包括在直接費用之中。

Those cost elements in the manufacturing of a product which are directly proportional to the quantity of production. Typical elements are: raw materials, auxiliary materials, energies, product specific infrastructure requirements. In a modern product costing system, *pollution control costs* are included in the direct costs on the basis of true disposal costs and real quantities.

direct pollution control costs 直接污染控制費用 *zhijie wuran kongzhi feiyong*

　　工廠中污染控制活動所產生的直接費用要素。這類費用包括工廠中正確處置各種廢物的費用：特殊廢物處置費用及廢水和空氣污染控制費用。對於分析和控制的費用，可按一定的比例進行分攤。

Direct cost elements resulting from pollution control activities in the plant. This type of costs includes the costs of correct disposal of all waste streams in a factory: disposal of special wastes, of wastewater and air pollution control. It is possible to allocate costs of associated analyses and controls on a proportional basis.

disc reactor, rotating 轉盤式反應器 *zhuanpanshi fanyingqi*

　　見：rotating biological contactor。
　　See *rotating biological contactor.*

discharge fees 排污費 *paiwufei*

　　見：fee; fees for discharges。
　　See *fee; fees for discharges.*

discharge levy for CO₂ 二氧化碳排放收費 *eryanghuatan paifang shoufei*

　　採用收費辦法作為一種*經濟刺激*手段來減少向大氣排放二氧化碳。對不同的收費模式正在進行討論；有些人也稱它為二氧化碳稅。減少二氧化碳排放應與減少*溫室效應*以防止*人類活動*造成未來氣候變化的目標相一致。

The application of levies as an *economic incentive* to reduce emission quantities of carbon dioxide into the atmosphere. Different models are under discussion; the word CO_2 tax is also used in certain groups. The CO_2 reduction would be in line with the objectives of reducing the *greenhouse effect* to prevent *anthropogenic* climatic changes in the future.

discharge levy in legislation 立法中的排污收費 *lifazhongde paiwu shoufei*

環境立法中採用經濟刺激手段來指導污染控制的有關法規。它不明確限制排放某種污染物到某一水平，但是污染者必須根據所排放的污染量向政府主管當局付費。有不同的模式用來說明收費與排放量之間的關係，收費可以隨排放量呈線性遞增，或者也可以呈累進遞增。此外，收費水平還可以隨時間遞增。（參見：聯邦德國廢水收費法、二氧化碳排放收費、公司內部收費）

An element in environmental legislation making use of economic incentives to direct pollution control. In such legislation, the emission of a pollutant up to a certain level is not explicitly forbidden but the polluter has to pay a levy to the authorities depending on the quantity of a pollution. Different models are used to specify the relationship between the level of the levy and the level of the discharged quantity. The levy may increase linearly with the quantity, or it may increase progressively. The level may also increase with time. (See also *law on effluent levies* (Federal Republic of Germany); *discharge levy for CO₂*; *company internal levy*)

discharge limit 排放限值 *paifang xianzhi*

通常指廢液或廢水*排放的限值*。

Generally, an *emission limit* which applies to liquid, aqueous discharges.

disinfection, general 消毒（一般的）*xiaodu (yibande)*

去除接觸傳染、感染、致病和有害的生物或使它們失活的方法。要

使某一系統無菌或衛生化，需要進行不同程度的消毒。(參見 : 滅菌、衛生化)

The process of removing or inactivating contagious, infectious, pathogenic or otherwise unwanted organisms. Different levels of disinfection are needed to render a system sterile or hygienic. (See also *sterilization*; *hygienization*)

dispersion models 擴散模式 *kuosan moshi*

用於計算某污染物在受納介質 (大氣或水體) 中擴散的數學模擬模式。它把污染物濃度作為時間和距離的函數進行計算。不同模式在不同程度上考慮了受納介質中流動和擾動產生的干擾影響。可利用擴散模式來處理*點源*的排放，計算*受納河流*中污染物的擴散以及*地下水體*中污染物的擴散。其特殊模式還可以用來估算和預測事故引起的有毒物質釋放 (如氯氣釋放) 的有害影響。

Mathematical simulation models which serve to calculate the dispersion of a pollutant in a receiving medium (the atmosphere or a waterbody). The models calculate pollutant concentrations as a function of time and distance. Different models integrate to a variable degree disturbing effects due to flow and turbulence in the receiving medium. Dispersion models are available for the treatment of emissions from a *point source*, for the dispersion of a pollutant in a *receiving river*, as well as for the dispersion of contamination in a *groundwater body*. Special versions are also used to estimate and predict harmful effects of toxic releases (e.g. chlorine) resulting from accidents.

dispersion of a pollutant 污染物擴散 *wuranwu kuosan*

在空氣污染控制中，指污染物從某一點源排出後被氣流攜帶，在大氣中稀釋與擴散的過程。同樣，污染物在排放到受納水體後也會被稀釋與擴散。雖然通過擴散污染物的濃度會降低，但其總量保持不變。

As used in air pollution control, the dilution and spreading of a pollutant in the atmosphere after it is emitted from a point source and is carried along with the movement of the air. Similarly, a pollutant is also dispersed after discharge to a receiving waterbody. Although the concentration of the pollutant is reduced through dispersion, its total quantity remains constant.

dispersion of analytical results 分析結果的離散性 *fenxi jieguode lisanxing*

當重複進行多次測定(如對某廢水樣BOD值進行五次測定)時,所得到的結果是有差別的。在高質量分析中,這種差別應很小,或者說結果的離散性應很小。因此,可用結果的離散性來表明一組數據的質量。它可以若干方法進行計算,通常以標準偏差表示。在統計分析上,也可以一些其他的方式涉及結果的離散性問題。

Whenever a measurement is repeated a number of times, (e.g. five measurements of the BOD-value of an effluent sample) the observed results differ. In high quality analysis, this difference will be small, or, the dispersion of results will be small. The dispersion of results is therefore an indication of the quality of a set of data. Such a quality indicator can be calculated in several ways. Frequently, the standard deviation of the results is calculated. Statistical analysis deals in several additional ways with this aspect.

disposal permit for wastes 廢物處置許可證 *feiwu chuzhi xukezheng*

允許廢物產生者或擁有者在特定的設施中對其廢物進行處置的許可證。這種許可是以廢物的組成和處置設施類型為依據的。處置設施的能力必須與廢物的性質相匹配。許可證是由政府主管部門或處置設施的運營者頒發的。它只對許可證中規定的一種特殊廢物適用。廢物處置許可證是*特殊廢物的管理*的一個組成部分。

A permit issued to a generator or owner of a waste to dispose of the waste in a specified facility. Such a permit is based on the waste composition and the type of the disposal facility. The capabilities of the facility must match the properties of the waste. The permit is issued by the competent authority or the operator of the disposal facility. It is valid for a particular waste which is specified in the permit. Disposal permits are one part of the *management of special wastes*.

disposal site 處置場 *chuzhichang*

廢物處置的場地。這是一般性術語,泛指各種處置場,包括從無控制的廢物堆放場到受控填埋場,以及在鹽礦地下層中堆存廢物的特殊處置場。

A site for disposal of wastes. This is a general term including a wide range of sites, from uncontrolled waste *dumps* to *controlled landfill* sites and special sites like e.g. *underground waste deposits*, e.g. in salt mines.

dissolved organic carbon (DOC) 溶解性有機碳 *rongjiexing youjitan*

　　水污染控制中表徵廢水特性的一項綜合指標。在瑞士環境法規中它是一項限制指標。它是可完全溶解（即可通過0.45微米孔徑過濾器）的污染物所含的有機碳的總量。與此不同的是*總有機碳*（TOC）還包括不溶性化合物。

A collective parameter for effluents used in water pollution control, and a limiting parameter in the Swiss environmental legislation. DOC includes the total of organic carbon contained in pollutants which is truly dissolved (i.e. passes a 0.45 micrometer pore filter.) In contrast, the *TOC* value also includes undissolved compounds.

dissolved oxygen (DO) in water 水中溶解氧 *shuizhong rongjieyang*

　　水中溶解氧可以用氧敏感電極或通過與氯化錳發生的標準化學反應來測定。溶解氧是*活性污泥法*運行操作的一項指標；它也是描述水生態系統如湖泊、河流質量的一個參數。溶解氧濃度與溫度有關，其範圍可從大約20 ppm降到*厭氧狀態水體*中的0 ppm。

Dissolved oxygen in water can be measured either by an oxygen sensitive electrode or by a standardized chemical reaction with manganese chloride. The quantity of dissolved oxygen is an indicator in the operation of the *activated sludge process*. It also serves to qualify aquatic ecosystems, e.g. lakes or rivers. The concentration of DO depends on the temperature. It ranges from about 20 ppm down to zero in an *anaerobic waterbody*.

distillation separation 蒸餾分離 *zhengliu fenli*

　　分離不同組分的混合物所廣泛應用的一種技術。它是基於不同組分具有不同蒸氣壓（沸點）來進行分離的，當不同組分的蒸氣通過一個或幾個蒸餾塔時，具有最大蒸氣壓的組分將最先離開塔。蒸餾塔的設計和結構很不相同，有填料塔和盤狀板式塔等。它們的分離效率和特性常常可用*理論板*這一參數來計算。

蒸餾技術的應用非常廣泛，包括應用在污染控制和廢物的循環利用（如廢溶劑的回收）上。在水污染控制中，可用它來分離*揮發性有機化合物*。在許多情況下分離得到的化合物可以被回收和進一步使用。蒸餾過程的設計與被分離的混合物有關。（參見：空氣氣提）

A technology used widely in processing to separate mixtures of different components. Separation is based on different vapor pressures (boiling points) of the components. The vapors pass one or several columns, and the component with the highest vapor pressure leaves the columns first. There are many different designs and constructions, either with a structural filling, or with built in tray-like plates. The parameter *theoretical plates* is used to calculate the separation efficiency and characteristics of such columns.

Distillation technology is used very widely; also in pollution control and recycling, e.g. in the recovery of used solvents. It is used in water pollution control to separate *volatile organic compounds* (*VOC*). In many cases, the separated compounds can be recovered for further use. In detail, the design of the process depends on the mixtures to be separated. (See also *air stripping*)

diversity, biological 生物多樣性 *shengwu duoyangxing*

表示某群落生境中不同生物群體的數量和種類之間關係的術語。如果未受到*人類活動*的干擾，群落生境是數量眾多而且相互作用的各種不同生物群體的天然生境。（參見：捕食者與被捕食者之間的關係、食物鏈、共生現象）。但在許多情況下，人類的活動會影響生態系統的生物多樣性。

A measure and an indicator for the number, kind and magnitude of different populations of organisms occupying a biotope. Biotopes, if not disturbed by *anthropogenic activities*, are a natural habitat for a large number of different populations, interacting with each other. (See also *predator-prey relationship*; *food chain*; *symbiosis*) Activities of man have in many cases reduced the biological diversity of ecosystems.

DO 溶解氧 *rongjieyang*

見：dissolved oxygen in water。

See *dissolved oxygen in water*.

Dobson Units (DU) 多勃遜單位 *duobosun danwei*

表示某一假想空氣柱中臭氧總量的單位。300 DU相當於標準條件下3毫米(未稀釋的)臭氧層。多勃遜單位可用專門的分光光度計測定。(參見:臭氧、紫外輻射)

A unit used to express the total quantity of ozone in a hypothetical cylinder of the atmosphere over a certain location. 300 DU correspond to a 3 millimeter layer of (undiluted) ozone under standard conditions. Dobson Units are measured with a special spectrophotometer. (See also *ozone*; *UV radiation*)

DOC 溶解性有機碳 *rongjiexing youjitan*

見:dissolved organic carbon。

See *dissolved organic carbon*.

DOC/TOC ratio DOC/TOC比 *DOC/TOC bi*

以有機碳進行測定與計算的表示水樣中水溶性污染物與污染物總量的關係的一項指標。(參見:溶解有機碳、總有機碳)

An indicator for the amount of water soluble pollutants in relation to the total amount, measured and calculated on the basis of organic carbon. (See also *dissolved organic carbon*; *TOC*)

double containment 雙重包封 *shuangchong baofeng*

由獨立的兩層屏障構成的包封系統,目的在於萬一發生重大事故時防止危險化學品泄漏或排放到環境中。它被應用於具有高潛在風險的情況。其設計有不同方式,如特殊設計的在底層鋪設有防泄漏*包封池*的倉庫建築,用於貯存桶裝的危險化學品;採用雙壁管道來輸送危險氣體;以及將危險過程的加工設備安置在單獨的屏蔽室內。

A technical containment system of two independent barriers with the objective to prevent leakage or emission of critical components to the environment in case of a major accident. Double containments are applied in cases of elevated risk potential. The design may be in different ways: e.g. a specially designed warehouse building to store critical chemicals in drums with a leakage-tight *containment basin* in the basement, double walled piping to carry critical gases, or an additional housing for the processing equipment of a critical process.

downcycling 下游循環 *xiayou xunhuan*

進行廢物的循環利用以回收低值產品。（參見：循環利用技術）

Recycling of wastes whereby a less valuable product is recovered. (See also *recycling technologies*)

Drager tubes 德拉格管 *delageguan*

見：gas detector tubes。

See *gas detector tubes*.

drop out box 沉降箱 *chenjiangxiang*

見：gravity settling box。

See *gravity settling box*.

droplet separator 液滴分離器 *yedi fenliqi*

從空氣流中去除微細液滴的單元裝置，其基本原理是：廢氣流經該裝置時，氣流中的微細液滴在*撞擊板*上碰撞聚集成較大的液滴，從而被去除。液滴分離器常常用於多級*濕式空氣污染控制單元*中，以減少各級洗滌之間所夾帶的洗滌液。偶而它們也用作為獨立使用的裝置。

A unit to remove fine droplets from an air stream. Basically, the off-gas is streaming against elements which act as *impinger plates*. Fine droplets agglomerate thereby to larger droplets and the resulting liquid is removed. Droplet separators are often used to reduce carry-over of scrubbing liquid between different stages of a multi-stage *wet APC unit*. They are also applied occasionally as stand-alone units.

dry APC-systems 乾式空氣污染控制系統 *ganshi kongqi wuran kongzhi xitong*

見：APC-systems, dry。

See *APC-systems, dry*.

dry dust eliminators 乾式除塵器 *ganshi chuchenqi*

　　非濕法空氣污染控制除塵設施。有許多不同類型的乾式除塵器，它們的除塵效果也很不相同，如*靜電除塵器、過濾器、旋風分離器、重力沉降箱*等。（參見：乾式空氣污染控制系統）

　　Process units for air pollution control which are not based on wet systems. Different types show widely different efficiencies. The group includes *electrostatic separators*, *filters*, *cyclones* and *gravity settling boxes*. (See also *APC-systems, dry*)

dry electrostatic precipitator 乾式靜電除塵器 *ganshi jingdian chuchenqi*

　　見：Cottrell separator, electrostatic。

　　See *Cottrell separator, electrostatic*.

dry flue gas treatment 乾式煙氣處理 *ganshi yanqi chuli*

　　見：flue gas treatment, dry。

　　See *flue gas treatment, dry*.

dual media filtration 雙介質過濾 *shuangjiezhi guolü*

　　見：filtration, dual media。

　　See *filtration, dual media*.

dumps 堆放場 *duifangchang*

　　無控制和無保護的傾倒堆放各種混合廢物的場地，它往往會產生衛生方面和地下水污染的問題，因此往往需要在以後進行代價昂貴的*恢復*和淨化的補救工作。

　　An uncontrolled and unprotected discharge site for all kinds of mixed wastes. Waste dumps often lead to hygienic and groundwater problems. For this reason they often need expensive *remediation* or clean up at a later date.

dust 粉塵 *fenchen*

　　空氣中的顆粒物，其大小很不均一，分佈範圍為10至1000毫微米。對於不同類型的粉塵，其粒子尺度的平均分佈是不同的。可以應用各種乾式和濕式空氣污染控制技術將它們分離。

Particulate matter in an air stream. The size of the particles is not uniform, and its distribution ranges from about 10 to 1000 nm. The average distribution of particle size is different for different types of dust. Dusts can be separated by different dry and wet type APC technologies.

dust monitoring 粉塵監測 *fenchen jiance*

　　見：particulates in off-gas, continuous measurement; plate deposition test。
　　See *particulates in off-gas, continuous measurement*; *plate deposition test*.

dye house effluents 染色廢水 *ranse feishui*

　　紡織印染、染色和前處理操作中產生的各種廢水，其COD濃度隨操作的類型而不同，變化於200–1000 ppm之間。染色廢水中含有許多種污染物質，如殘留染料、鹽類、金屬、螯合劑、鹼、輔助化學品、染色助劑、上漿劑、印染增稠劑、漂白劑等，按照操作過程和所處理的纖維的類型而不同。雖然染料會造成廢水的顏色問題，但其他組分常常會對處理設施造成更多的問題，如有些污染物會抑制*活性污泥處理系統微生物的活性*。

　　染色污水的處理概念必須建立在工廠物料平衡的基礎上，它包括：減少物質的過量使用，選用問題少的化學品和助劑，將殘餘的印染漿料進行焚燒處理，以及選用耗盡程度高的染料。此外，還應在廢水排放到城市污水處理場之前設置*均衡池*。

Different types of effluents from textile printing, dyeing and preparatory operations. The COD concentration depends on the type of operation. Values are between 200 and 1000 ppm. The effluents are contaminated with a wide range of substances such as: residual dyestuffs, salts, metals, chelating agents, alkalis, auxiliary chemicals, auxiliary dyeing assistants, sizing agents, printing thickeners, bleaching agents etc. The contaminants in detail depend on the operation and the type of fiber which is treated. While the actual dyestuffs are visually most apparent, other components often cause more problems in treatment facilities. Some contaminants may inhibit the activity of *activated sludge* treatment systems.

A treatment concept of dye house effluents must be based on a mass balance of all materials entering the factory. It includes steps to reduce excessive quantities of substances; select chemicals and auxiliary agents which are less critical; incinerate excess printing paste; and select dyes with a high degree of exhaustion.

In addition, the installation of *equalizing tanks* may be indicated before the effluent is discharged to a municipal treatment station.

dynamic equilibrium 動態平衡 *dongtai pingheng*

在最適當平衡位置附近周期地震蕩或變動的平衡狀態。生態系統中捕食者與被捕食者之間的平衡是一種動態平衡,其他的自動調節系統中(如商業中)的平衡也屬於這一類。相反,物理系統中大多數的平衡,如天平的平衡,是*靜態平衡*。(參見:捕食者與被捕食者的關係)

An equilibrium state which oscillates or fluctuates periodically around a most favored equilibrium position. Equilibra in ecosystems between populations of predators and prey are dynamic equilibra. Equilibra in other self-regulating systems (e.g. in business) also belong to this group. In contrast, most equilibra in physical systems, the equilibrium of a balance, are *static equilibrium*. (See also *predator-prey relationship*)

E

EC-50 半作用濃度 *banzuoyong nongdu*

會使50%*靶生物*受影響的水中某物質的平均濃度。它可能引起任何失調作用，但不致造成死亡。半作用濃度通常比半致死濃度低。由試驗得出的影響結果（如繁殖率、代謝速率，魚的平衡的喪失和性狀變化等），必須對其接觸時間給與明確說明。（參見：半致死濃度）

The average concentration of a substance in water affecting 50 percent of the *target organisms*. An effect may be any disturbance, but not death. EC-50 is usually lower than LC-50. The effect (e.g. reproduction rate, metabolic rate, loss of equilibrium of fish, behavioral changes, etc.) underlying the test must be clearly specified together with the time of exposure. (See also *LC-50*)

ECD 電子捕獲檢測器 *dianzi buhuo jianceqi*

見：electron capture detector。

See *electron capture detector*.

eco-design 生態設計 *shengtai sheji*

應用於消費產品開發與設計的一種基本觀念。生態設計納入了環境要求，包括諸如選用可再生物質，考慮製品修復的可能性，採用可循環和重複使用的部件，以及最後考慮產品的處置不會產生環境問題。

A design philosophy applied to the development and construction of consumer products. Eco-design integrates environmental needs. Such needs include for instance the selection of renewable materials, the possibility to repair an article, the integration and use of recyclable and reusable parts, and finally the possibility to dispose of a product without problems.

eco-rating 生態評級 *shengtai pingji*

與已有的企業經濟健全性的評級（如穆迪分級法）相類似的環境風險評價評級（正在建議中）。在生態評級中，它根據若干行為參數對企業環境行為和狀況進行評價。目前，這些參數尚未標準化，因此不同的生態

評級是不能直接進行比較的。(參見:購買獲得前審計、環境審計、環境行為評級)

A proposed environmental risk assessment rating similar to the existing rating of the economic soundness of an enterprise. (e.g. the Moody Rating). In an eco-rating, the environmental performance and situation of an enterprise is evaluated on the basis of a number of performance parameters. At present, the parameters are not standardized and different eco-ratings are not directly comparable. (See also *pre-acquisition audits*; *environmental audits*; *environmental performance rating*)

ecobalance 生態平衡 *shengtai pingheng*

評估和分析某一產品在它整個*生命周期*內對環境的影響。生態平衡可用來互相比較兩個類似產品的環境影響。生態平衡的評估以製造某產品的總*物流*(包括能量消耗)為依據,但要準確地比較環境影響是很困難的。對於不同的排放情況,物流與影響之間的關係涉及一些因子,它們具有不同的權重,易受誤差、偏差、個人喜好的影響。此外,當比較兩個產品時,*系統的邊界*與生態平衡的方法必須相同。已經發表了一些生態平衡的方法,並對它們進行了檢驗。(參見:產品的生命周期)

An assessment and analysis of the impact of a product on the environment over its entire *life cycle*. Ecobalances may serve to compare the environmental impact of two similar products with each other. The assessment of an ecobalance is based on the total *mass flow* of manufacturing a product, including also its energy consumption. An exact comparison of the environmental impact is difficult. The relationship between mass flow and impact for different emissions includes factors with different weights which may be subject to errors, bias and personal preferences. Also, when comparing two products, the *system boundaries* and the method of the eco-balance must be identical. Several methods have been published and tested. (See also *life cycle of a product*)

ecochemistry 生態化學 *shengtai huaxue*

化學學科的一個專門領域,它研究生態系統中的化學過程,涉及跨學科研究*生態系統中污染物*的化學相互作用。(參見:環境影響)

A special field of chemistry. Ecochemistry investigates and describes chemical processes in ecosystems. It deals in an interdisciplinary way with chemical inter-actions of pollutants in *ecosystems*. (See also *environmental impact*)

ecoefficient activities or actions 具有生態效益的活動或行動 *juyou shengtai xiaoyide huodong huo xingdong*

既能提高一個國家、一個企業或全體居民的經濟效益，又能改善生態狀況的活動。

Actions which enhance the economic performance (of a nation, an enterprise, the population) as well as the ecological situation at the same time.

ecological assessment (of a project) 建設工程的生態評價 *jianshe gongchengde shengtai pingjia*

建設工程對環境影響的評價，包括鑒別關鍵的因素（排放源）和評價它們的潛在影響。（參見：環境影響評價）

The assessment and evaluation of the impact of a project on its environment. Such an assessment includes the identification of critical factors (emissions) and the evaluation of their potential impact. (See also *environmental impact assessment*)

ecological base line assessment 生態基線評價 *shengtai jixian pingjia*

對未受到任何破壞影響的某一特定的生態系統的狀況進行的評價，這可以在建設工程實施之前進行。

The assessment of the condition of a specific ecosystem without any disturbing effects. Such an assessment may be carried out before the realization of a project.

ecological equilibrium 生態平衡 *shengtai pingheng*

見：biotic equilibrium。

See *biotic equilibrium*.

ecological factors 生態因素 *shengtai yinsu*

決定生態系統中生活條件的因素。有兩類因素：*非生物的生態因子和生物的生態因子*。

The factors which determine the life conditions in an ecosystem. The two groups are: *abiotic ecological factors* and *biotic ecological factors.*

ecological niche 小生境 *xiaoshengjing*

滿足植物和動物群體特殊生活要求的一組獨特的*生態因素*組成的特殊群落生境。這一群體獨特地適應於該群落生境中主導的環境條件。

A particular biotope with a unique combination of *ecological factors* which satisfy the specific life requirements for a plant- or animal population. This population is uniquely adapted to the conditions prevailing in this biotope.

ecological population 生態種群 *shengtai zhongqun*

見：population, ecological。

See *population, ecological.*

ecology 生態學 *shengtaixue*

生物學科學的一個領域，它研究生態圈或其子系統即生態系統中的相互作用，尤其是動物和植物種群之間的關係以及這些種群與它們生活的*群落生境*之間的關係。生態學研究生態系統的平衡和生態破壞的恢復機理。其領域包括*群落生態學*、*個體生態學*和人類生態學。生態學是了解人類排放和環境保護活動的環境影響的重要基礎。（參見：食物鏈、動態平衡、捕食者與被捕食者的關係）

As part of the science of biology, ecology studies the interactions in the ecosphere, or a subsystem thereof, an ecosystem. Of interest are relations between populations of animals and plants themselves, but also of these populations and the *biotope* they live in. Ecology investigates equilibra in ecosystems and also the mechanisms of restoration of disturbances. Subfields of ecology are *synecology*, *autecology* and human ecology. It is an important base for understanding the environmental impact of *anthropogenic emissions* and of environmental protection activities. (See also *food chain*; *dynamic equilibrium*; *predator-prey relationship*)

economic incentives 經濟刺激 *jingji ciji*

激發某個機構或個人去做某件事的經濟因素。在環境立法中已運用經濟刺激和收費一類手段而不是通過*管制法*來減少排放。經濟刺激手段包括：污染者因向環境排放而*付費*，對在環境工作中取得成績者給與*獎勵*，對環境友好的產品給與*補貼*或*減稅*。德國的*排水收費法*就是把經濟刺激手段運用於水污染控制方面的一個例子。

Economic factors to motivate an institution or a person to do something. A group of instruments applied in environmental legislation to lower emission levels through economic motivation and payments, rather than through *police laws*. Economic incentives include *fees* to be paid by the polluter for emissions to the environment, *incentive* reward payments for environmental achievements, *subsidies* or *tax reductions* for environmentally friendly products. The German *law on effluent levies* makes use of economic incentives in the field of water pollution control.

economic instruments in environmental legislation 環境立法中的經濟手段
huanjing lifazhongde jingji shouduan

應用環境立法中的基於經濟刺激的控制手段，指導企業活動沿著*環境可接受的*方向發展。在這種情況下，污染者減少污染的動機不是出於排放限值的考慮，而是基於經濟效益。目前採用的經濟手段有*排污收費*、*排放權交易*和*原料收費*；相關的辦法還有*泡罩原理*和*行業契約*。

The use of control instruments in environmental legislation which are based on economic incentives for directing activities in an *environmentally acceptable* direction. The motivation of the potential polluter for pollution abatement is not based on emission limits but on economic rewards. Different types of mechanism are used: *discharge levy*, *trading of emission rights*, *levies on raw materials*; in a related way also the *bubble principle*, and *branch covenants*.

ecosphere 生態圈 *shengtaiquan*

即地球上一切生命繁衍所在的地球各層的生態系統。生態圈包含許多不同的*生態系統*，由*大氣圈*(包括其中的一切生物體)、*水圈*(包括淡水和海水生態系統，如湖泊、河流、海洋)和*岩石圈*(土壤和地面生物體的*生境*)所組成。(參見：生物圈、群落生境)

The combined ecosystems of the planet earth, the spherical shell housing all life on earth. The ecosphere contains many different *ecosystems*. The ecosphere comprises the *atmosphere* including all organisms therein, the *hydrosphere*, including a variety of sweet- and saltwater ecosystems (lakes, rivers, oceans) and the *lithosphere* (the *habitat* of organisms of soil and ground). (See also *biosphere*; *biotope*)

ecosystem 生態系統 *shengtai xitong*

地球 *生態圈* 的一個組成部分。每一生態系統由某一 *群落生境* 和生活在其中的 *生物群落* 所組成。每一類生態系統可用特定組合的 *生態因子* 來表示其特徵。在 *水圈*、*岩石圈* 和 *大氣圈* 中可以發現一些特定類型的生態系統。

A part or an element of the global *ecosphere*. Each ecosystem is made up of a *biotope* and the community of organisms living in it, the *biocenose*. Each type of ecosystem is characterized by a specific combination of *ecological factors*. Specific groups of ecosystems can be found in the *hydrosphere*, the *lithosphere* and the *atmosphere*.

ecotoxicity 生態毒性 *shengtai duxing*

化學物質或污染物對生態系統中生物體的危害性。生態毒理學主要研究對細胞、生物體、種群以及作為一個整體的生態系統所造成的可逆或不可逆的損害。其 *靶生物* 可能是截然不同的生物體，如在某生態系統中佔主導地位的動物、微生物和植物。

The damaging property of a chemical substance or a pollutant on the organisms in the ecosystem. Focal points of ecotoxicological investigations are reversible or irreversible damages at the levels of cells, organisms, populations and possibly the ecosystems as a whole. *Target organisms* may be distinct organisms, e.g. animals, microorganisms and plants which are predominant in an ecosystem.

effect concentration 作用濃度 *zuoyong nongdu*

見：EC-50。

See *EC-50*.

effluent 廢水 *feishui*

含有各種污染物的液體，通常是水態的廢物流。（參見：污水）

Liquid, generally aqueous waste stream, containing various pollutants. (See also *sewage*)

effluent analysis 廢水分析 *feishui fenxi*

為提供數據以加強管理而進行的廢水的分析。分析得到的數據可用來檢查與法規要求相一致的情況及監控廢水處理的工作效能，可作為數據庫用於處理系統的規劃，以及用於特殊的調查研究中。

廢水分析的系統步驟包括：正確取樣和樣品儲存，用標準方法準確地進行分析，以及對統計數據的解釋。

取樣點可以直接設置在污染產生源，也可以設置在處理系統中各個不同的處理單元之後，以及設置在排放點上。

分析項目包括*綜合指標*和*單一要素指標*，這應根據數據的應用場合及法規的要求來加以選擇。日常分析的項目有：懸浮物、pH、酸度、氧含量、鹽含量(電導率)、總有機碳或化學需氧量、五天生化需氧量、重金屬、各種無機與有機污染物。（參見：廢水取樣、分析的質量標準）

The analysis of effluents with the objective to provide data for further management decisions: to check on compliance with legislation, to monitor the performance of effluent treatment, as a database for planning of treatment systems, and for special investigations.

A systematic approach includes: proper sampling and sample storage, accurate analysis with standardized methods, and statistical data interpretation.

Sampling points may be directly at the source, after different process steps of a treatment system, and at the discharge point.

Parameters include *group parameters* and *single element parameters*. Their selection depends on the use of the data and on legislation. Some often used parameters are: suspended solids, pH, acidity, oxygen, salinity (conductivity), TOC or COD, BOD_5, heavy metals and a wide range of individual inorganic and organic pollutants. (See also *sampling of effluents*; *quality criteria of analysis*)

effluent data sheet 廢水數據卡 *feishui shujuka*

　　水污染控制管理方面的基礎資料，它概括了每一股工藝廢水或母液的數據，包括：管理數據，如廢水編號、產生地點、工廠、建築物、責任單位、費用分攤等；許可的處置方法和設施（如生物降解、焚燒、特殊處理方法）；生產過程及廢水量（如每批生產排出的母液量）；廢水物理性質，如外觀、濁度、顏色、臭味；廢水組成，如五天生物需氧量、化學需氧量、總有機碳、酸度、影響處理的關鍵組分；廢水毒性；日期與簽字。

　　廢水數據卡是針對混合前每一股特定廢水的。如果由於工藝改變使得數據相應發生變化，就應加以修訂。

A document used in connection with *water pollution control* management to summarize data for each separate stream of a process effluent or mother liquor. The document serves as a base for WPC management. It includes: administrative data: effluent code number, site, factory, building, responsibilities, cost allocation, etc.; permitted disposal methods and facilities (e.g. biodegradation, incineration, special processes; the source process and standard quantities of the mother liquor per processing batch; physical properties: appearance, turbidity, color, odor; composition: BOD_5, COD, TOC, acidity, critical components, e.g. which might impede treatment; effluent toxicity; and date and signatures.

The waste data sheet is valid for one specific effluent stream before mixing with others. It should be re-edited if some of the data has changed, e.g. as a result of process changes.

effluent disinfection 廢水消毒 *feishui xiaodu*

　　某些國家的法律規定經過處理後的廢水在排放前需進行消毒，以去除生物處理後仍可能存在的致病微生物或使其失活。

　　廢水消毒可以在生物處理後作為一個最終處理步驟通過氯化來進行。為達到此目的，注入的氯量應使它在處理後仍保持有少量餘氯存在。可以利用彎曲的溝渠系統作為大的連續反應器和利用靜態混合系統。

　　一個應避免的問題是，必須防止通過用氯消毒生成有機氯化合物。姑且不論這一問題和它的處理費用，有人認為這些致病微生物也可以通過正常受納水體中的*原生動物*自然地去除。但目前已很少採用這種方法。

Legislation in certain countries prescribes a disinfection of treated effluents before discharge. The objective is to remove or inactivate pathogenic microorganisms which might still be present after biological treatment.

Disinfection may be carried out by chlorination of the effluent as a last step after biological treatment. To this end, chlorine is introduced in such a quantity that a small excess remains after treatment. A meandering canal system may be used as a large continuous reactor and static mixing system.

The formation of chlorinated organic compounds through this type of disinfection is a problem which has to be avoided. Aside from this problem and the costs of this treatment, it is also thought that such microorganisms are removed naturally by *protozoa* in a healthy *receiving waterbody*. The method is nowadays less used.

effluent inventory 廢水清單 *feishui qingdan*

工廠中從工藝過程末端到最終處理設施排出的全部廢水流的清單或數據庫。其數據包括每一股廢水的組成、數量和連續或間歇的排放方式,從而為制定廢水處理方案提供了依據。

A database or inventory of all effluent streams in a factory. The inventory includes the streams from the end of the process to the final treatment step. The database includes the composition, the volume and the frequency of each stream. An effluent inventory may serve as an information base for the development of a treatment concept.

effluent levy 排水收費 *paishui shoufei*

與水污染控制法規有關的一種經濟手段,它鼓勵企業把廢水的排放減少到低於規定的排放限值。排水費由排放廢水的單位負擔,其費用取決於排放廢水中某些污染物的殘留濃度。(參見:德國排水收費法)

An economic instrument in connection with legislation on water pollution control. This type of legislation is motivating an enterprise to lower its discharges to levels even below the discharge limits. The effluent levy (or fee) is a payment made to authorities for discharging effluents, depending on the residual concentration of certain pollutants in the effluent. (See also *law on effluent levies*)

effluent pretreatment filter 廢水預處理過濾器 *feishui yuchuli guolüqi*

在工藝過程末端進行廢水過濾的預處理裝置。(參見:過濾)

A unit for pretreatment by filtration of effluents at the *end-of-process* level. (See also *filtration*)

effluent toxicity 廢水毒性 *feishui duxing*

説明廢水的毒性,包括廢水中的所有污染物及廢水本身毒性的一項綜合指標。

廢水通常含有(或具有)許多會對水生生態系統的各種生物體產生毒害作用的物質(或性質),如酸度、鹽類、化學物質、懸浮物、缺氧等。被作用的*靶生物*包括魚和作為魚的食物(*食物鏈*)的生物體,如浮游生物、蝦、無脊椎動物、藻類等。因此,廢水毒性是污染控制的一項重要的綜合指標。實際測定時,可以在靜止的*養魚試驗*池中或在流動水槽系統中用魚進行廢水毒性試驗。對於*急性效應*的評估,試驗時間只有幾小時;而對於*長期效應*的研究,試驗可持續幾天。

還有一些用其他生物進行毒性試驗的方法。(參見:對魚的毒性、對蝦的毒性)

A *group parameter* for the toxicity of an effluent including all its contaminants and the water itself.

Effluents in general contain many substances which may exert a toxic effect on different organisms of an aquatic ecosystem: acidity, salts, chemical substances, suspended solids, the absence of oxygen. The affected *target organisms* include fish and organisms which serve as food for fish (*food chain*), e.g. plankton, shrimp, invertebra, algae and others. Effluent toxicity is therefore an important group parameter in the control of pollution. For practical monitoring purposes, testing of effluent toxicity may be carried out with *fish tests* in static testing aquariums or testing in flow-through systems. For the assessment of *acute effects*, testing periods are a few hours; for studies of *chronic effects*, testing may extend to several days.

There are some alternate methods available based on other organisms. (See also *toxicity versus fish*; *toxicity versus shrimp*)

effluent toxicity, luminescence method 廢水毒性，發光法 *feishui duxing, faguangfa*

代替用魚測定廢水毒性的方法。在文獻上已報導了各種不同的方法。市場上現在有成套的測試小箱或系統供應（商品名叫"MICRO-TOX TEST"），它們是基於當微生物被某種物質致毒時微生物會改變其發光活力這一原理而工作的。用這些方法測得的結果可以進行比較，但與魚的毒性測定沒有直接的相關性。

An alternative to the assessment of effluent toxicity with test fish. Various methods are available and have been reported in the literature. Test kits or systems are available commercially (trade name e.g. 'MICRO-TOX TEST'), based on microorganisms that change their luminescence activity when intoxicated by a substance. The methods are valuable for comparisons. They do not correlate directly with fish toxicity measurements.

effluent toxicity with invertebra as test organisms 用無脊椎動物作為試驗生物測定的廢水毒性 *yongwujizhuidongwu zuowei shiyan shengwu cedingde feishui duxing*

某些無脊椎動物如蝦或蝦卵、幼蟲、水蚤等，可以代替魚來測定廢水或某種物質的水生毒性。這種試驗一般是長期的慢性效應試驗，試驗時間要延續到它們的整個生命周期，即從孵化開始直到死亡為止。當用其他替代的靶生物進行試驗時，它們和用魚進行試驗所得到的試驗結果沒有直接的相關性，雖然這些試驗可以用來進行比較。

Certain types of invertebra, e.g. shrimp or shrimp eggs, larvae, daphnia and others may be used as alternatives to fish for the measurement of aquatic toxicity of effluents or of single substances. Such tests are generally longer term tests of chronic effects and they extend over the full life cycle of the invertebra, from hatching to death. As with other alternate target organisms, there is no direct correlation with data from fish, although such tests may be used for comparisons.

effluent treatment 廢水處理 *feishui chuli*

見：water pollution control technology。

See *water pollution control technology.*

effluent treatment plant 廢水處理場 *feishui chulichang*

廢水處理的綜合設施。根據過程單元和處理場設計的不同，它可以處理工業廢水和/或城市污水。(參見：活性污泥技術)

The combined facility for treatment of effluents. Depending on the process units and the design of the plant, it may serve to treat *industrial effluents* and/or *municipal sewage.* (See also *activated sludge technology*)

effluent treatment technologies 廢水處理技術 *feishui chuli jishu*

用於廢水處理使它達到*排放限值*要求的技術。廢水處理技術範圍很廣，在許多情況下，它們與一般化工過程所採用的方法是相同的，但應用的條件不同，即廢水的濃度較稀。

廢水處理的主要一類方法，是將污染物進行分離；這需要進行預處理或化學變化，但所分離得到的物質往往可以回收或再利用。另一類方法是進行礦物化處理，即把有機污染物降解為二氧化碳、水和無機鹽等礦物質，這包括化學法、生物法和熱處理法。

但是必須注意，這些方法只能使廢物組分變得較不令人討厭和易於處理而已。從長期可持續發展觀點看，必須開發與應用*源頭削減污染的*技術。

Technologies to treat effluents containing unavoidable waste components, to meet the requirements of the *discharge limits.* A wide range of technologies are available. In many cases, the same processes are used as in general chemical processing, with conditions applied to the more dilute effluents.

One main group includes treatments to separate pollutants. Pretreatment or chemical changes may be necessary. The separated materials can often be recovered and reused. Another group consists of mineralization processes to degrade organic pollutants to mineral components like CO_2, water and inorganic salts. In this group we find chemical, biological and thermal processes.

It must be kept in mind that these processes only change waste components into components which are less offensive and easier to handle. In the context of long term sustainable development, technologies which aim at *source reduction* must therefore be favored.

EIA 環境影響評價 *huanjing yingxiang pingjia*

見：environmental impact assessment。

See *environmental impact assessment.*

EIR 環境影響報告 *huanjing yingxiang baogao*

見：environmental impact report。

See *environmental impact report.*

electrodialysis 電滲析 *dianshenxi*

一種分離陽離子與陰離子的膜分離方法。它採用的膜具有荷電的官能團，結合在聚合物基體中。這些膜具有*離子交換性質*，可以透過陽離子或陰離子。在電滲析中，採用一組交替排列的陽離子和陰離子滲透膜。滲析過程由該組膜兩端的電位來驅動；液體沿膜的方向，即垂直於電位的方向流動。

電滲析可用來濃縮電解質溶液，如海水濃縮製取食鹽。它還可用在食品和製藥工業、電鍍行業的廢水處理和鹽溶液的化學再生。

A membrane separation process to separate cations from anions. The process uses membranes with charged functional groups bound in a polymer matrix. These membranes show *ion exchange* properties and they are permeable to either cations or anions. In electrodialysis, a stack of alternately cation- and anion permeable membranes is used. The process is driven by an electric potential applied across the stack; the flow of the liquid is along the membranes, i.e. perpendicular to the potential.

Electrodialysis is used to concentrate electrolyte solutions, e.g. of sea water for the production of salt, in the food and pharmaceutical industries, and in electro-plating for effluent treatment and chemical regeneration of salt solutions.

electron capture detector (ECD) 電子捕獲檢測器 *dianzi buhuo jianceqi*

氣相色譜法使用的一種檢測器，它利用放射源來產生連續的電子流，這些電子又由流過檢測池的載氣(氮氣)生成離子，然後用電子學方法測量離子流，產生檢測信號。氣流中的某些微量化合物(所有的鹵素)可能會捕獲原電子，而使檢測器的信號減弱。

電子捕獲檢測器對有機鹵素化合物和某些*光氧化劑*具有很高的靈敏度。它常常用於光化學煙霧的環境研究或飲用水中微量鹵代化合物農藥的測定上。

A detector used in gas chromatography. The detector uses a radioactive source to produce a continuous current of electrons. These electrons in turn produce ions from the nitrogen in the carrier gas, which is flowing through the detector cell. The ion stream is measured electronically to give a detector signal. Certain trace compounds (all halogens) in the gas stream are able to capture the primary electrons. These compounds reduce the signal of the detector.

The electron capture detector is highly sensitive to organic halogen compounds and to certain *photooxydants*. It is often used in connection with environmental investigations of photochemical smog, or in the determination of traces of halogenated pesticides in drinking water.

electrosmog 電煙霧 *dianyanwu*

用於說明電磁場如射頻發射台、高壓輸電線路等對敏感人可能造成的影響和干擾的術語。目前無論對電煙霧或對其影響都還沒有給與明確的解釋。

A term used to summarize effects and possible disturbances to sensitive persons as a result of electromagnetic fields, e.g. around radio frequency transmitting stations, high tension power transmission lines etc. Neither the term nor the effects are as yet clearly defined.

electrostatic precipitators, dry 靜電除塵器(乾式) *jingdian chuchenqi (ganshi)*

見：Cottrell separator, electrostatic。

See *Cottrell separator, electrostatic*.

electrostatic precipitators, wet 靜電除塵器（濕式）*jingdian chuchenqi (shishi)*

見：wet electrostatic preciptator。

See *wet electrostatic preciptator*.

electrostatic separator 靜電分離器 *jingdianfenliqi*

利用靜電吸引力從廢氣中去除塵粒的技術。尾氣中的塵粒被負電極（*冷等離子體電極*），常常是金屬絲電極荷載負電，然後載有負電的塵粒被正電極（*沉積電極*）吸引。兩個電極之間的電位差一般在30千伏左右。通常，靜電除塵器投資大但營運費用低。已有的一些應用是以*濕式靜電除塵分離器*技術為基礎的。

A technology to remove particles from an off-gas stream by means of electrostatic attraction forces. The particles in the off-gas are charged negatively by a negative electrode (*cold plasma electrode*); often, this is a wire electrode. The negatively charged particles are then attracted by a positive *precipitation electrode*. The potential difference between the two electrodes is generally around 30 kV. Electrostatic precipitators usually have high investment costs and low operating costs. There are technical applications available which are based on *wet electrostatic precipitator* technology.

eluate test 洗脫液測試 *xiduoye ceshi*

見：leachate test。

See *leachate test*.

EMAS Directive of the European Union 歐盟《生態管理和審計體系》指令 *oumeng shengtai guanli he shenji tixi zhiling*

1993年歐盟頒佈的有關環境管理和審計的法規（No. 1836/1993），它規定了環境管理標準，要求不斷改善企業的環境狀況，符合法規的要求，強調企業有義務向公眾提供*環境狀況報告*，並且要通過*審計*來證實符合該指令要求的情況。目前，實行歐盟《生態管理和審計體系》雖然是自願的，然而如果某企業的生產現場達到該體系的要求，那麼它就有資格在其商業文件中使用一個特定的標記。

Legislation on environmental management and auditing promulgated by the

European Union in 1993. (No. 1836/1993). The Eco-Management and Audit System (EMAS) Directive specifies environmental management standards. It implies compliance with legislation and it demands constant improvement of the environmental situation of a plant. An *environmental statement* to the general public is compulsory. Compliance with the requirements of the directive is to be verified by an *audit*. Compliance with EMAS at present is voluntary. However, if a site of an enterprise is in compliance, then it is officially entitled to use a specific logo (sign) on its business papers.

emergency holding basins 應急貯存池 *yingji chucunchi*

為了生產廠安全運行而採取的措施，以對嚴重和意料不及排放的廢水（消防水、意料不及的工藝排水、泄漏液體等）起到貯存的作用。它通常設置在廢水處理場廢水入場之前，因此意料不及排放的廢水可以被分流和貯存起來，直到對它們可能進行另外特殊的處理為止。

A measure for the safe operation of a processing plant, functioning as a holding basin for critical unexpected discharges of aqueous effluents (fire fighting water, unexpected discharges from a process, spills of liquids, etc.). The emergency holding basin is generally placed before the inflow of an effluent treatment plant. An unexpected discharge can thus be diverted and stored until special additional treatment is possible.

emergency management 緊急事故管理 *jinji shigu guanli*

見：emergency response management。

See *emergency response management*.

emergency response management 緊急響應管理 *jinji xiangying guanli*

事故及其快速有序響應的管理體系，包括：行政管理措施（警報計劃），以便當發生事故時可自動開始進行正確有效的管理；事故遏制措施，以減小緊急事故的影響。緊急響應管理的基礎在於風險分析，並且要以預防性安全措施為前提。人員的安全教育和安全審計也是緊急響應管理體系的重要基礎。（參見：環境負責組織聯盟原則）

A management system to deal with accidents and emergencies quickly and in

135

an orderly way. It includes administrative measures (*alarm plan*) to automatically initiate a correct and efficient management of the accident and *containment measures* which serve to limit the impact of the emergency. It is based on a *risk analysis* and supported by *preventive safety measures*. Personnel instruction on safety aspects, as well as *safety audits* are also important cornerstones of such a system. (See also *CERES Principles*)

emission 排放 *paifang*

由某種活動產生或釋放出的廢物或污染物。這一術語常常用於指氣態廢物流，但從根本意義上說，它適用於一切物理狀態。排放可來自不同的*排放源*。排放量可以表示為絕對量或*排放因子*。其德語的相關詞為*immission*。

Waste matter or pollutants which are given off or set free by an activity. The term is most often applied to gaseous waste streams; however, basically it applies to all physical states. Emissions may originate from different *emission sources*. The quantity of an emission may be reported as an absolute quantity or as an *emission factor*. For a related term from the German language see *immission*.

emission breakthrough 排放穿透 *paifang chuantou*

污染控制單元所處於的某一操作點，在該操作點系統處於飽和狀態，污染物穿過系統未被去除。(參見：穿透點)

The point in the operation of a pollution control unit at which the system is saturated and the pollutant passes through it without being removed. (See also *breakthrough point*)

emission database 排放數據庫 *paifang shujuku*

見：emission inventory。

See *emission inventory*.

emission factor 排放因子 *paifang yinzi*

單位產品的*排放量*。例如，熱電站生產每千瓦小時能量排出的二氧化硫的公斤數，或生產每噸某化學物質排出的總有機碳的公斤數。

The quantity of an *emission* in relation to a standardized quantity of a product. Example: kg of sulfur dioxide emitted per kWh of energy produced in a thermal power plant; or, kg of TOC discharged per ton of a chemical substance manufactured.

emission inventory 排放清單 *paifang qingdan*

工廠中一切*點源排放*和*擴散排放*的清單，它列出了從工藝過程末端到廢氣處理排出的所有廢氣流。其內容包括每一股廢氣流的組成與數量的有關數據。廢氣排放清單作為基礎數據對於確定處理方案和計算工廠的總排放量有著重要作用。

An inventory of all *point-source emissions* and *diffuse emissions* of a factory. The inventory lists the off-gas streams from the end of the process to the off-gas treatment step. The database includes data on the composition and the volume of each stream. An off-gas inventory may serve as a database for the development of a treatment concept and for calculation of the total emissions of a factory.

emission limit 排放限值 *paifang xianzhi*

在有約束力的文件中規定的限制排入環境中的污染物的數值(包括*法律限值與公司內部限值*)。窄義地說，它僅指廢氣排放。廣義地說，它也包括廢水的*排放限值*。排放限值可以是*濃度限值或總量限值*。

A numerical value which is defined and specified in a binding document (*legal limits, company internal limits*) to limit the discharges of pollutants to the environment. Emission limits apply in the more narrow sense of the term to emissions of off-gases. In a wider sense, emission limits also include *discharge limits* for liquid effluents. Emission limits can be specified as *concentration limits*, or *mass flow limits*.

emission right 排放權 *paifangquan*

環境立法中的經濟手段之一。排放權證由政府主管部門頒發給工廠，允許它在一定時期內排放一定量的污染物。工廠可以決定通過改進工藝程序和採用污染控制技術來減少污染物的排放，於是它可能不再需要排放權，在這種情況下排放權可以進行交易，即可以賣給另一工廠。

因此，排放和排放量就具有經濟價值，從而遵循市場規律。排放權證和排放權交易是一種新的經濟手段，目前只在美國加利福尼亞州個別地方採用。

An element of the group of *economic instruments in environmental legislation*. An emission right certificate is issued by the authorities to a factory and gives permission to emit a certain quantity of pollutant in a given time period. A factory may decide to reduce its emissions through process changes and pollution control technology and it may then not need the emission right any more. Emission rights can under these conditions be traded, i.e. sold to another factory. Emissions and emission quantities are thus equivalent to an economic value, and market forces may apply. Emission right certificates and the trading of emission rights are novel instruments and are up to now used only in few cases. Examples are known from California.

emission right certificates, initial price 排放權證的初始價格 *paifangquanzheng, chushi jiage*

排放權證開始上市時的價格。

The price of an emission right certificate at the time of its initial appearance on the market.

emission right certificates, market price 排放權證的市場價格 *paifangquanzheng, shicang jiage*

排放權證進入市場後按照市場供需關係確定的價格。

The price of an emission right certificate as it develops in the market as a result of supply and demand market forces.

emission sources 排放源 *paifangyuan*

見：point emission source; diffuse source。

See *point emission source*; *diffuse source*.

emulsion 乳液 *ruye*

彼此互不混溶形成真溶液的兩個或多個組分的混合液。通常，乳液含有水相和油相，形成水包油或油包水的乳液。乳化劑會促使乳液形成並使之穩定化。乳化作用在食品、營養、塗料、農業化學品、製藥等行業有著廣泛應用。有些乳液是天然存在的，如牛奶、某些植物的漿汁（膠乳）等。在污染控制中，乳液常常會造成問題，因為對於廢水的處理或預處理來說，將乳液進行分離可能是很困難的，需要採用特殊的技術。例如，機械行業的切削油、加工過程產生的中間相即萃取中不完全的相分離，就屬於這種情況。（參見：乳液分離）

A fluid mixture of two or more components which are not miscible with each other by true solution. Typically, emulsions contain an aqueous phase and an oily phase. This combination can lead to either an oil in water emulsion or a water in oil emulsion. Emulsifying agents promote the formation and the stability of emulsions. Emulsions have many applications in different industries: food, nutrition, paints, agrochemicals, pharmaceuticals, etc. Some emulsions are of natural origin: milk, certain saps of plants like latex. In pollution control, emulsions often cause problems because their separation for effluent treatment or pretreatment may be difficult and may require special techniques. Examples are: cutting oils in the mechanical industry, intermediary phases in processing, e.g. incomplete phase separation in extractions. (See also *separation of emulsions*)

emulsion splitting 破乳 *poru*

乳液的相分離的過程。由於乳液是由多組分組成的，因此在污染控制技術上它常常會造成一些問題。乳液只有在破乳後才能對它的每一相進行特殊處理。

它的分離可通過減弱乳液各相液滴間的穩態力，並同時增大它們之間的吸引力來實現。

採用物理方法如升高溫度，應用電場力或離心的機械應力，通過聚結作用或過濾（微過濾或超濾），可使乳液分離。有時，還可利用重力，通過沉降或氣浮進行緩慢的分離。

有些乳液需要通過化學變化，如 pH 變化或水相離子濃度的變化來進行相分離。在某些情況下，改變乳化液滴的表面性質，如投加混凝劑和絮凝劑，也很有效。另一種破乳方法是對乳化劑進行化學分解破壞。另外，溶劑萃取也可能是有效的。

Processes which lead to a separation of the phases of an emulsion. In pollution control technology, emulsions often cause problems due to their multi-component composition. Breaking such emulsions allows for special treatment of each separate phase.

Emulsions can be separated by weakening the stabilizing forces and at the same time strengthening the forces of attraction between individual droplets of each phase.

Separation can be achieved through physical processes like temperature rise, the application of electric fields, mechanical stress through centrifugation, the use of coalescing units, or filtration (microfiltration or ultrafiltration). Sometimes slow separation through gravity forces by sedimentation or flotation is sufficient.

Some emulsions require chemical changes, like a change of pH, or a change of the ionic strength of the aqueous phase. In some cases a change of the surface properties of the emulsified droplets will be effective; this can be achieved with coagulating and flocculating agents. The chemical destruction of the emulsifier is another possibility. Solvent extraction may also be effective.

end-of-pipe measures 末端處理措施 *moduan chuli cuoshi*

這一術語常常是指在生產過程末端採用的污染控制措施,如工廠的廢水處理場。

A term often used for pollution control measures applied at the end of the manufacturing process, e.g. the effluent treatment plant of a factory.

end-of-process treatment 末端工藝過程處理 *moduan gongyi guocheng chuli*

於工藝過程的末端在把某廢物流與其他廢物流混合之前進行的污染控制處理。例如,對於金屬的處理和回收,末端工藝過程處理是很有效的,因為一旦不同的金屬被混合後,就難以處理和回收了。末端工藝過程處理常常比*末端處理措施*經濟。(參見:末端處理措施)

A pollution control treatment at the end of the process before mixing the waste stream with other waste streams. For example, end-of-process treatment is useful for the treatment and recovery of metals. This cannot be done any more once different metals are mixed. EOP processes are often more economic than *end-of-pipe measures*. (See also *end-of-pipe measures*)

energies 不同形式的能量 *butong xingshide nengliang*

英文energy的複數。在工業術語上指工廠所使用的各種不同形式的能量，即電能、過程熱、室內熱、不同壓力的蒸汽、不同溫度的冷卻鹽水。廣義地說，也包括氣動控制系統和惰化作用使用的壓縮空氣和氮氣。

(Energy, plural). In industrial terminology, the different forms of energy used in a factory: i.e. electrical, process heat, room heat, steam at different pressures, cooling brine at different temperature levels. In a wider sense, compressed air and nitrogen for pneumatic control systems and nitrogen for providing an inert atomsphere are included.

energy 能量 *nengliang*

其物理意義為做功的能力。能量變化總是與物理狀態的變化與運動以及與分子中化學鍵的變化有關。能量可以表現為許多不同方式，如位能、動能、熱能、電能、鍵能、物理吸引或排斥能、電場等等。能量可從一種形式轉化為另一種形式。例如，煤燃燒時，鍵能差可轉化為熱能。這種能量轉換受能量守恆定律支配。愛因斯坦方程式描述了質量與能量之間的關係。

In the physical sense, the capacity for doing work. Energy changes are always associated with changes of the physical status, with motion or with changes of chemical bonds in molecules. Energy manifests itself in many different ways: as potential energy, kinetic energy, thermal energy, electrical energy, bonding energy, energy of physical attraction or repulsion, electrical fields, etc. Energy may be converted from one form into another one. For instance, in the combustion of coal, differences in bonding energy are converted to thermal energy. Such transformations are governed by the law of conservation of energy. The Einstein equation deals with the relationship between mass and energy.

energy, fossil resources 能源，礦物資源 *nengyuan, kuangwu ziyuan*

在煤、原油、天然氣和含焦油的頁岩等礦物資源中所儲存和可獲得的能源。已經定期發表了全世界煤、石油和天然氣的估計用量。按照目前的使用量，估計煤的儲量可使用幾百年，天然氣可使用60年，原油只能使用大約30年。然而必須記住，這種估計常常是變化的，而且似乎是不可靠的。實際上，豐富的礦物資源不是一個絕對的數量問題，而是一個能源經濟學、新儲量開採和提煉技術的問題。

礦物能資源的利用會導致向大氣釋放二氧化碳，它通過*溫室效應*將會增加全球變暖的潛勢。

Energy stored in, and gained from, fossil resources, such as coal, crude oil, natural gas and certain slates containing oil tar. Estimates of the world-wide quantity of coal, oil and natural gas have been published periodically. At the present rate of use, current estimates for reserves of coal are several hundred years. Natural gas has been estimated to be available for 60 years, crude oil for about half of this time. It must be kept in mind, however, that such estimates have often been changed and seem to be unreliable. In reality, the abundance of fossil resources is not an absolute quantity but a question of energy economics, exploration of new reserves and extraction technology.

The use of fossil energy resources leads to the release of carbon dioxide to the atmosphere which will further increase the potential of global warming through the *greenhouse effect.*

energy, grey 灰能 *huineng*

隱藏的能量或灰色的能量，這一術語有時指用來生產某種機械或設備的能量。在討論通過開發特殊技術來採取節能措施時，應將生產設備所需灰能的量包括在內，並且把它與可能被節省的能量相比較。灰能可以看作是投資用於以後節能設備上的初始能源。

Hidden energy, or grey energy, is a term sometimes applied to energy which is used to manufacture a machine or a piece of equipment. In the discussion of energy saving measures through special technical developments, the quantity of 'grey energy' needed to manufacture the equipment should also enter the discussion, and be compared with the quantity of energy which can be saved. Grey energy can be looked at as an initial energy investment into a device used to economize energy later on.

energy, primary sources 一次能源 *yici nengyuan*

由可再生資源或礦物資源如煤、原油、天然氣和水電直接取得的能源。一次能源通常必須轉換為可利用的形態。在此過程中，一般會損失一部分能量成為廢能。

Energy as it is directly available from renewable or from fossil sources: coal, crude oil, natural gas and hydroelectricity. Primary energy usually has to be transformed into a form which can be used. In this process, some of the energy is generally lost as waste energy.

energy, renewable 可再生能源 *kezaisheng nengyuan*

非有限的但可不斷更新的能源，一般認為由風力發電機、太陽能（熱能和光電能）、水電產生的能量以及由可再生生物質（如木材、*沼氣*）產生的能量是可再生的。

Energy from sources which are not finite, but which are renewed constantly. Commonly, energy from wind-powered electric generators, solar energy (heat and photovoltaic), hydroelectricity and energy from renewable biomass (e.g. wood and *biogas*) are considered renewable.

energy recuperation 能量回收 *nengliang huishou*

廢能的回收和再利用。其技術上的可能途徑，可舉例說明如下：（1）回收廢物焚燒爐中廢物的能量來產生生產用水蒸汽、電或供暖用熱能；（2）回收運輸機車制動時的能量來產生電能（現代機車設計有制動動能的回收系統）；（3）優化大型建築的供熱和通風系統，使用*換熱器*；（4）回收大型冷卻系統放出的熱量用於供熱系統；（5）*共生*發電和工藝用水蒸汽。

The recovery and reuse of waste energy. Technical possibilities may be illustrated by examples like: the recovery of the energy content of wastes in waste incinerators, producing process steam, electricity or comfort heat; the recovery of kinetic energy when braking transport vehicles to produce electrical energy (modern designs of locomotives with recuperation braking systems); the optimiza- tion of heating and ventilation systems of large buildings, making use of *heat exchangers*; the recovery of heat given off from large cooling systems for use in heating; the *co-generation* of electricity and process steam.

energy transformation 能量轉換 *nengliang zhuanhuan*

　　一種形式的能量轉換為另一種形式的能量。能量轉換是由熱力學定律支配的。例如，在熱電廠，化學結合能先轉換為熱能，然後又轉換為動能（運動的能量），最後再轉換為電能。完成上述能量轉換的技術設備，其順序包括：蒸汽鍋爐－蒸汽透平機－發電機。由於熱力學原因，每一步的效率平均低於100%，因此導致了能量轉換的損失。

The transformation of one form of energy into another form. Energy trans- formations are governed by the laws of thermodynamics. For example, in a thermal power plant energy of chemical bonding is transformed into heat energy which in turn is transformed into kinetic energy (energy of motion), and this then is transformed into electric energy. The sequence of technical units to accomplish this includes: steam boiler – steam turbine – electrical generator. For thermo- dynamic reasons the efficiency of each step is below 100%, thus leading to transformation losses.

enrichment, sample 樣品富集 *yangpin fuji*

　　見：enrichment step in analysis。

See *enrichment step in analysis*.

enrichment step in analysis 分析中的富集步驟 *fenxizhongde fuji buzhou*

　　在環境痕量分析中，往往必須將污染物從母體中提取出來並加以富集。富集是重要的制備步驟。對水、空氣或浸出液中痕量化合物的富集有若干方法，如用液體吸收，用活性碳、多孔聚合物或硅膠吸附，以及固相提取等。應根據母體與被富集化合物的性質來選擇最佳富集方法。

In environmental trace analysis, pollutants often have to be extracted from the matrix and enriched for the actual analysis. This enrichment step is an important preparative step. Several methods are used for the enrichment of trace compounds in water, air or leachates: absorption in a liquid, adsorption on activated carbon, porous polymers or silica gel, solid phase extraction and more. The best method depends on the matrix and on the nature of the compound to be enriched.

environment 環境 *huanjing*

某一客體周圍(有生命和無生命的)事物。更具體地説,它是生態系統一切組成部分的總和;生態系統可能會對某一客體,如某人、某工廠或全體居民產生影響,並且它也會受到上述客體的作用。客體及其環境之間的相互作用是雙向的:客體會影響環境(或其組成部分),環境也會影響客體。

The surroundings (living and lifeless) of a certain being or object. Specifically, the sum of all components of an ecosystem which may possibly influence an object, e.g. a person, a factory, or the population in general, and which are noticed by said object. The interactions between an object and its environment can take place in both directions: the object can affect the environment, and the environment (or components thereof) can affect the object.

environmental audit 環境審計 *huanjing shenji*

審計方法在環境問題上應用。環境審計的內容一般集中在與環境有關的生產與基礎設施問題上。在特殊情況下也可以審計與產品有關的問題。環境審計是環境綜合管理的一個組成部分。(參見:審計、安全審計、環境管理與審計體系指令、環境管理)

The application of the auditing technique to environmental aspects. The topics covered by the environmental audit generally focus on environmentally relevant aspects of manufacturing and infrastructure. As a special case, the situation regarding the products themselves may also be audited. Environmental auditing is one element of an integral environmental management approach. (See also *audit; safety audit; EMAS Directive; environmental management*)

environmental analysis 環境分析 *huanjing fenxi*

見:analysis, environmental。

See *analysis, environmental.*

environmental compartments 環境的組成部分 *huanjingde zucheng bufen*

相應於把*地球生態圈*劃分為*大氣圈*、*水圈*、*岩石圈*，人類的環境也可以細分為三個部分，即空氣、水和土壤。

Parallel to the segmentation of the global *ecosphere* into the *atmosphere*, the *hydrosphere* and the *lithosphere*, the environment of man may be subdivided into the three compartments air, water and soil or ground.

environmental controlling 環境控制 *huanjing kongzhi*

在環境管理中採用的控制活動。（參見：環境控制）

The activities of controlling, applied in the field of environmental management. (See also *controlling, environmental*)

environmental damage costs 環境損害費用 *huanjing sunhai feiyong*

見：damage costs, environmental。

See *damage costs, environmental*.

environmental data 環境數據 *huanjing shuju*

見：data, environmental。

See *data, environmental*.

environmental economics 環境經濟學 *huanjing jingjixue*

從*宏觀*和*微觀經濟*觀點研究環境保護問題的學科，其研究的重點放在：環境保護費用的付費，確定從經濟角度看也是最適的法律要求水平，以及把污染控制費用作為*內部費用*包括在生產產品中等問題。（參見：污染者付費原則、費用-效益分析、環境費用）

The discipline dealing with the economic side of environmental protection, on the *macroeconomic* as well as on the *microeconomic* level. Questions focus on the payment of environmental protection costs, the finding of a level of legal requirements which is optimal also from the economic side, and questions of inclusion of pollution control costs in the manufactured products, as *internal costs*. (See also *polluter-pays principle*; *cost-benefit analysis*; *environmental costs*)

environmental framework laws 環境框架法 *huanjing kuangjiafa*

見：framework laws, environmental。

See *framework laws, environmental.*

environmental impact assessment (EIA) 環境影響評價 *huanjing yingxiang pingjia*

系統、全面地評價與分析某一擬建項目對環境的潛在影響。它是由有關部門獨立地進行的。環境影響評價以 *環境影響報告* 為依據，它包括評價某一項目潛在影響有關的一切數據和資料。

環境影響評價規範隨各國而不同。在有些國家，如德國，只對大型項目（電廠、機場、高速公路、新建水泥廠等）才強制要求進行環境影響評價。在其他一些國家，如瑞士，即使對較小的項目也要求必須進行環境影響評價。法律規範規定了所需要的數據資料和必須進行評價的項目與設施（類型、規模）。

通過環境影響評價可發現潛在的環境問題，並提出項目修改的建議。（參見：基建項目的環境影響）

EIA, the systematic and overall assessment, analysis and evaluation of the potential impact of a planned project on the environment. The evaluation is carried out independently by the authorities. The EIA is based on an *environmental impact report (EIR)* which includes all relevant information to judge the potential impact of a project.

The application of the EIA procedure depends on national legislation. In some countries, e.g. Germany, it is mandatory for very large projects only (power plants, airports, highways, new cement factories, etc.). In other countries, e.g. Switzerland, the procedure must also be carried out for smaller projects. The legal procedure specifies the information needed and the units (type, magnitude) for which it is mandatory.

The environmental impact assessment procedure serves to identify potential environmental conflicts. It may identify project modifications which may be necessary. (See also *environmental impact of capital projects*)

environmental impact report 環境影響報告 *huanjing yingxiang baogao*

綜述環境影響評價的所有數據與資料以及有關問題的報告，它是大型建設項目計劃管理程序的一個組成部分，由項目建設單位負責編寫。（參見：環境影響評價）

EIR, the report summarizing all data and relevant aspects of an environmental impact assessment. The EIR is to be compiled as part of the planning procedure of a large project by the owner of the project. (See also *environmental impact assessment*)

environmental impact of capital projects 基建項目的環境影響 *jijian xiangmude huanjing yingxiang*

擬建的新裝置可能會對環境造成影響，它的潛在影響隨項目的類型與規模而不同。它可能影響環境的一個或若干個組成部分，可以立即或滯後地、直接地或在過程鏈終端處造成影響。某項目的環境影響是多方面的，必須按照系統的方法來鑒別與評價。對於大型項目，法律規定了必須遵循的影響評價程序（參見：環境影響報告、環境影響評價）；對於較小的項目，應按照公司內部制定的關於影響評價的管理程序（它是綜合環境管理體系的一個組成部分）進行工作。

Projected new installations may have an impact on the environment. The potential impact depends on type and size of the project. It may affect one or several compartments of the environment, and it may manifest itself immediately or after a delay, or directly or at the end of a chain of processes. It follows that the environmental impact of a project is multi-dimensional and its identification and assessment must be based on a systematic approach. For large projects, an impact assessment procedure is often prescribed by law. (See also *EIR*; *EIA*) For smaller projects, a company internal management procedure with the same objectives should be part of an integral environmental management system.

environmental inventory 環境清單 *huanjing qingdan*

見：inventory, environmental。

See *inventory, environmental.*

environmental investments, pay-back time 環境投資，回收期 *huanjing touzi, huishouqi*

見：pay back time; return on investments。

See *pay back time*; *return on investments*.

environmental legislation 環境立法 *huanjing lifa*

為保護人類及其環境所制定的法律、法規和標準，主要包括：排放控制立法(包括空氣、水和廢物管理)、產品和產品使用規範、管理程序規程、自然保護、噪聲和噪聲控制、資源保護、動植物物種保護、污染控制設施貼補等。此外，許多一般性法律如運輸法、景觀保護與區劃法、森林法等，也包括有直接或間接涉及環境問題的條款。

與國家其他立法相同，環境立法也分為不同的結構層次，分別為憲法、基本(框架)法、具體法、細則和條例、技術指南等。環境立法應按照*生態效益*觀點來運用*法律手段*，目前正在朝著運用*經濟手段*的方向發展。

Legislation specifying rules and norms with the objective of protecting human beings in the environment and safeguarding the environment in all its aspects. The main fields of legislation are: emission control legislation (air, water, waste management); specifications for products and product use; specification of management procedures; nature protection; noise and noise abatement; resource conservation; protection of animal and plant species; and subsidies for pollution control facilities. In addition, many general laws include paragraphs which deal directly or indirectly with environmental aspects: transportation; landscape protection and zoning laws; and forestry and others.

Environmental legislation is structured the same as national legislation into constitution, framework laws, specific laws, by-laws and ordinances, and technical guidelines. Legislation should use *legal instruments* which are in line with the idea of *ecoefficiency*. Modern developments go in the direction of applying *economic instruments*.

environmental liability 環境責任 *huanjing zeren*

因環境損害或忽略採取適當防治措施而負有的*責任*。在許多情況

下，土壤或地下水的污染會造成以後代價昂貴的環境責任。(參見：時間範疇上的難題、超級基金計劃)

A business *liability* resulting from environmental damages or from neglecting to take appropriate preventive actions. In many cases, the contamination of ground or groundwater has resulted in costly environmental liabilities at a later time. (See also *time horizon dilemma*; *SUPERFUND program*)

environmental management systems 環境管理體系 *huanjing guanli tixi*

以環境問題為核心的管理體系，它涉及環境的各個組成部分。環境管理體系與一般管理體系沒有什麼根本區別。其管理範圍包括產品價值創造鏈中的所有環節：確定目標、制定實施計劃、實現目標的手段(人力、技術和財力)、提供實施所需的條件(組織機構、職責、人事、培訓)以及確定需要進一步改進的行動。

Management systems focusing on environmental aspects, covering all environmental segments. Environmental management systems do not differ fundamentally from general management systems. The management areas to be addressed throughout all links of the chain of value creation of the products are: defining objectives and goals; planning the implementation; allocating means to reach the goals (personnel, technical, financial); providing frame conditions to facilitate implementation (organization, responsibilities, personnel, training); and controls to identify needs for corrective actions.

environmental officer 環境人員 *huanjing renyuan*

工廠中代表管理部門從事環境保護工作的職工，其任務包括：對環境問題提供諮詢意見、在工廠現場進行日常控制、檢查污染控制設備的效率和正常運行情況、核查符合法律和公司內部標準的情況、對新項目進行評估、以及對已有的和新的生產過程進行評估。在有些國家，對於一定規模的公司，法律規定中要求公司應委任環保人員。

A staff member of a factory, dealing on behalf of management with (all) environmental topics relevant to the plant. The tasks include: consulting on environmental questions; carrying out day-to-day controls at the site; the verification of efficient and correct operation of pollution control equipment; the verification of compliance with laws and company internal norms; the evaluation of new projects; and the evaluation of current and new production processes. In some countries, the assignment of an environmental officer is required by law for companies of a certain size.

environmental performance indicators 環境行為指標 *huanjing xingwei zhibiao*

用於衡量和說明公司的實際狀況與其環境目標或標準相一致情況的數據和信息，它們不僅涉及水、空氣和土壤等環境要素，也還涉及生產過程與產品。

Data elements and information which serve to measure and document the compliance situation with the environmental objectives and norms of a company. The elements cover the environmental segments water, air, and soil or ground. They relate to manufacturing as well as to the products.

environmental performance rating 環境行為評級 *huanjing xingwei pingji*

按某單項指標（或評分等級）對企業的環境行為進行評價的方法。所有這些系統係基於一些不同的參數，如符合法律要求的總體情況、環境管理情況或產品的生態平衡來進行評級。因為這類評級是按某單項指標（或評分等級）來說明總體情況的，因此必須對不同的參數給與不同的權重。但是無論是這些參數或是它們的權重都尚未標準化，所以不同的評級不能直接進行比較。在國際標準組織14000標準系列中，正試圖將這一評級方法加以標準化。（參見：生態評級）

Evaluation procedures which condense the environmental performance of an enterprise in a single index or rating. All such systems are based on a number of different parameters, such as the overall legal compliance, environmental management aspects, or product ecobalances. Since the rating should give an overall picture in a single index (or 'rating'), different parameters have to be given certain weights. As neither the parameters nor their weights are standardized, different ratings cannot be compared directly.

In the context with the *ISO 14000 Norms*, attempts to standardize this procedure are being worked on. (See also *eco-rating*)

environmental performance report 環境工作報告 *huanjing gongzuo baogao*

對與評估某生產廠或企業的環境工作有關的一切因素作出概述的系統和定期報告，它是提供給高層管理人員的公司內部控制文件，是經濟和財務方面的傳統報告的補充。

A systematic and periodical report summarizing all factors which are relevant to the assessment of the environmental performance of a site or an enterprise. The environmental performance report is a company internal controlling document for top management. It complements the traditional reports on economic and financial aspects.

environmental policy 環境政策 *huanjing zhengce*

見：company environmental policy。

See *company environmental policy*.

environmental policy statement 環境政策報告 *huanjing zhengce baogao*

概述企業所遵循的*環境政策*要點的正式文件，它應由企業高層領導正式通過，並成為企業一切環境活動的指導方針。現在，政策報告已愈來愈公開化。

A formal document outlining the key points of the *environmental policy* an enterprise has adopted. The policy statement should be adopted by senior management and form a general guideline for all environmental activities of the enterprise. Policy statements are increasingly made public.

environmental protection (EP) 環境保護 *huanjing baohu*

為保護環境不受*人類活動*的影響而採取的一切行動和措施，重點在保護人類的環境。然而，人類與生態系統在許多方面是有關聯的，因此它還必須包括生態系統的保護。

The total of the activities and measures taken to protect the environment from the impact of *anthropogenic activities*. The focus of the activities centers on protecting the environment of human beings. However, human beings depend in many ways on ecosystems in general, therefore their protection must also be included.

environmental protection (EP) costs 環境保護費用 *huanjing baohu feiyong*

見：costs, environmental protection。

See *costs, environmental protection.*

environmental quality goals 環境質量目標 *huanjing zhiliang mubiao*

為社會所確定的諸如不同水體的環境質量目標或環境空氣污染的長遠控制水平。對於企業來說，目標可以指生產操作的"三廢"排放控制水平。

The goals defined for society specifying for instance environmental quality goals for different types of waterbodies or long term ambient air pollution levels. At the level of an enterprise, goals may specify emission levels for production operations.

environmental report 環境報告 *huanjing baogao*

自願向公眾提供的包括一個公司的某一工廠或若干工廠環境狀況的數據和資料的綜合報告。一些化工公司以及其他行業的公司已發表了此類報告。它們有的是對一個單獨工廠進行說明的，也有的是對一個大型企業進行綜述的。

報告的內容一般包括：政策和計劃；向水體排放的污染物量；空氣污染物的排放量；各種廢物的數量及其處置；安全與事故數據；能耗及能源利用；以及其他適合向公眾報告的關於企業存在的問題和取得的成績的定性和定量資料。

歐洲化學工業聯合中心、國際商會和一些國家的工業組織都已出版了環境報告的編寫指南。

A voluntary and yet comprehensive report to the general public, giving pertinent information and data on the environmental situation of a site, or of a group of sites of an enterprise. Such reports have been issued by chemical companies, and by companies of other industry sectors. They have been published for one single site or as a summary for a large enterprise.

The chapters of the report generally cover: information on policy and planning; discharges to waterbodies; quantities of emissions; quantities of different wastes and their disposal; safety and accident data; energy consumption and use;

and other suitable qualitative or quantitative information to inform the public about problems and achievements of the enterprise.

Guidelines for editing the contents of such a report have been issued, for instance by *CEFIC*, the *ICC* and by national industry organizations.

environmental scientists 環境科技人員 *huanjing keji renyuan*

在與環境問題有關的所有主要領域受過基本訓練的科技人員。在有些大學如蘇黎士的瑞士聯邦工業大學設有相應的培訓課程。在工業部門也可提供在職學習培訓。此外，還設有研究生課程。

Scientists having a basic training in all main fields relating to environmental aspects. A corresponding training curriculum is offered at some learning institutions, e.g. at the Swiss Federal Institute of Technology in Zurich and in other schools. Some study programs can be carried out alongside employment in industry. Postgraduate courses are also offered.

environmental segment 環境的組成部分 *huanjingde zucheng bufen*

環境的一部分，例如水體、大氣、土壤。有時，這一術語也用作為 environmental compartment(*環境的組成部分*)的同義語。

A part of the environment, e.g. all waterbodies, the atmosphere, or ground and soil. The term is sometimes used as a synonym for *environmental compartment*.

environmental statement 環境狀況報告 *huanjing zhuangkuang baogao*

歐盟環境管理與審計體系指令中所要求的向公眾提供的文件資料。這類報告必須提供擬認證為合格工廠現場的環境狀況的有關數據和資料。根據環境管理與審計體系的要求，環境狀況報告必須涉及一個單獨的生產廠，並且必須由一位獨立的審核人對環境狀況報告的正確性進行審計和核實。(參見：環境報告)

A document directed at the general public which is required in the context of the *EMAS Directive of the European Union*. This statement must document the information relevant to the environmental situation at the site which wishes to be certified. According to EMAS, the statement must cover one single site. An independent verifier must audit and attest the correctness of the environmental statement. (See also *environmental report*)

environmental technology 環境技術 *huanjing jishu*

用於保護環境的技術，包括*污染控制技術*和*清潔生產技術*。

Technology which serves to protect the environment. It includes *pollution control technology* as well as elements of *clean production.*

environmentally acceptable 環境可接受的 *huanjing kejieshoude*

用於表示人類活動和工廠的運作或產品的生產合乎環境要求的術語。環境可接受的活動所具有的負面影響，應不超過某一限度，否則管理層應承擔全部的道義責任。因此，必須考慮它的短期和長期的影響。這一術語與*可持續發展*和*殘存影響*有關。環境立法在許多方面對"可接受的"與"不可接受的"之間的界限作出了規定。

A term used as a qualifying adjective for *anthropogenic activities*, the operation of a factory, or products. The negative impact of an environmentally acceptable activity must not exceed a level for which management can assume full moral responsibility. Thereby, short term as well as long term effects must be taken into consideration. The term is related to *sustainable development* and *residual impact.* Environmental legislation in many areas defines the limit between 'acceptable' and 'not acceptable'.

environmentally acceptable products 環境可接受的產品 *huanjing kejieshoude chanpin*

對環境造成的影響被認為是*環境可接受的*工廠的產品。在評價工作中，把*產品生命周期*的各個階段都包括在內，這點是很重要的。

Products of a factory which exert an impact on the environment which can be called *environmentally acceptable.* It is important that in the assessment, all stages of the *life cycle of the product* are included.

EOX 可提取的有機鹵素化合物 *ketiqude youji lusu huahewu*

Extractable organic halogen compounds的縮寫。它是水污染控制中

使用的一項綜合指標，包括可用某種溶劑（如乙酸乙酯）提取的那部分總鹵素化合物。其測定方法與測定*可吸附的有機鹵素化合物*一樣，係將溶劑提取液焚燒，然後測定燃燒氣體中的鹵化物。然而，所使用的溶劑尚未標準化，提取步驟也可能導致錯誤的結果，因此在使用這些結果時必須小心謹慎。

Extractable organic halogen compounds. A group parameter used in water pollution control. EOX includes that fraction of total halogen compounds which can be extracted by a solvent, e.g. ethyl acetate. The solvent extract is incinerated and halide is determined in the combustion gases as with *AOX*. The solvent to be used is not yet standardized and the extraction step may lead to false results. These results must therefore be used with caution.

EP costs 環境保護費用 *huanjing baohu feiyong*

見：costs, environmental protection。

See costs, environmental protection.

equalization 均衡 *junheng*

廢水在進入生物活性污泥池之前均衡流量和組分濃度變化的過程。（參見：均衡池）

The process of equalizing (or buffering) a variable flow rate and composition of an effluent before biological activated sludge treatment. (See also *equalization tanks*)

equalization tanks 均衡池 *junhengchi*

廢水處理進口側體積很大的貯存池（具有緩慢攪拌或不攪拌）。它通常位於中和之後，而在生物處理之前。其體積一般可貯存幾小時至幾天的廢水。均衡池可緩衝進水負荷和濃度的變化，使進入生物處理池的廢水較為均勻，避免由於衝擊負荷造成不利影響。

Large slightly stirred or unstirred holding tanks at the input side of effluent treatment. Equalization tanks are typically located after neutralization and before major treatment steps like biological treatment. A holding volume of a few hours to several days of effluent flow is typical. The equalization tanks serve to buffer variable input loads and input concentrations of effluent in order to give a more even flow to the biological step and to avoid negative effects due to shock loading.

equilibrium 平衡 *pingheng*

見：static equilibrium; dynamic equilibrium。

See *static equilibrium*; *dynamic equilibrium*.

equitable enforcement 公正實施 *gongzheng shishi*

對所有有關方面都同樣嚴格地執行法律，這是理應如此的。

Enforcement of legislation which is carried out with the same strictness for all parties concerned. Equitable enforcement should be the normal case.

error, random 偶然誤差 *ouran wucha*

見：random error in analysis。

See *random error in analysis*.

error, systematic 系統誤差 *xitong wucha*

見：systematic error。

See *systematic error*.

estuarine ecosystem 河口生態系統 *hekou shengtai xitong*

河口的水生生態系統。河口的條件在許多方面係處在河流的淡水入口和開闊的海洋鹽水之間。河口具有淡水淨入流和高低潮汐間水逆流的特點。這會導致特殊的混合條件和鹽濃度的變化。通常，這種流型會在河口形成若干不同的生境。由於所有這些因素的作用，使河口生物群落中物種的多樣性受到了限制；其物種必須能在不斷變化的環境中存活。

地球上許多百萬人口的城市位於河口兩岸，在許多情況下，河口被這些城市的廢物所污染。河口也可能為人類活動造成的大量泥沙和沉積物所危害。因此，污染控制管理是河口管理的一個重要方面。

The aquatic *ecosystem* of an estuary. The conditions in an estuary are in many respects intermediate between the sweet water input from a river and the saltwater open sea. Estuaries are characterized by a net input of sweet water and a reversed flow of water between high tide and low tide. This leads to special mixing conditions and variable salt concentrations. Generally, these flow patterns result in several different habitats in the estuary. As a consequence of all these factors, the diversity of species in the *biocenose* of an estuary is limited; species must be able to survive in the constantly changing environment.

Many mega cities of the earth are situated on the shores of estuaries and in many cases, the estuary is polluted from the wastes of these cities. Estuaries may also be damaged by increased loads of silt and sediments resulting from *anthropogenic activities*. Pollution control management is therefore an important aspect of estuarine management.

eucaryotic cell 真核細胞 *zhenhe xibao*

在所有植物和動物中以及在一些大的*原生物群體如真菌、原生動物和大多數藻類*中的構成單元。

The unit of structure in all plants and animals and in several large groups of *protists*, e.g. *fungi*; *protozoa* and most *algae*.

European Management and Auditing Directive 歐洲管理和審計指令 *ouzhou guanli he shenji zhiling*

見：EMAS Directive of the European Union。

See *EMAS Directive of the European Union*.

eutrophic, eutrophication 富營養的，富營養化 *fuyingyangde, fuyingyanghua*

按字義指提供豐富的食物。它與水生態系統營養充分、過多有關。此術語用以表示水體中藻類生長所需營養物濃度的狀況。綠藻所需的主要營養物是磷和氮的化合物。如果水體中這些化合物過多，則水體處於富營養狀態。有機和無機的氮、磷化合物是尤為重要的富營養化因素。（參見：有營養的、貧營養的）

Literally, providing abundant food. In connection with water ecosystems: well-fertilized, over-fertilized. A term used to indicate the status of a waterbody with regard to the concentration of nutrients for the growth of algae. Critical nutrients for green algae are phosphorous and nitrogen compounds. A waterbody is in an eutrophic state if all these compounds are in over-abundance. Organic or inorganic phosphorous and nitrogen compounds are particularly important eutrophication factors. (See also *trophic*; *oligotrophic*)

exergy 做功的能 *zuogongdeneng*

　　與能量有關並且以其相同單位表示的術語。它是在環境中轉化為功的那部分總能量。在高溫如1000°C下某一定量的能量，與低溫如25°C下同樣量的能量相比，具有較高做功的能。能量與做功的能之間的差異，係基於熱力學第二定律。

A term related to *energy* and expressed in the same units as energy. It is that fraction of the total energy which is available for transformation into work in a certain environment. A given quantity of energy at a high temperature, e.g. at 1000°C has a higher exergy than the same quantity at a lower temperature, e.g. 25°C. The difference between energy and exergy is based on the second law of thermodynamics.

extended aeration systems 延時曝氣系統 *yanshi baoqi xitong*

　　一種廢水生物處理系統，其平均曝氣停留時間很長，因此產生的剩餘污泥量很少。

Biological effluent treatment systems, in which the average *aeration residence time* is high, so that the quantity of excess sludge becomes low.

external costs 外部費用 *waibu feiyong*

　　見：costs, external。

See *costs, external*.

external environmental audit 外部環境審計 *waibu huanjing shenji*

　　對廢物處理的廠外設施進行的*審計*。在這種情況下，應認真確定*審計標準*，而且此標準應與企業利用外部處置設施的*環境政策*相一致。

An *audit* carried out at an off-site facility used for treatment of wastes. The *audit standards* should in such cases be well defined and be compatible with the *environmental policy* of the enterprise using the external disposal facility.

159

extractable organic halogen compounds 可提取的有機鹵素化合物 *ketiqude youji lusu huahewu*

見：EOX。

See *EOX*.

extraction solvent 萃取溶劑 *cuiqu rongji*

用萃取技術將廢水中可溶於溶劑的污染物進行分離所使用的溶劑。

The solvent used in *extraction technology* to separate a solvent-soluble pollutant from an aqueous effluent.

extraction technology 萃取技術 *cuiqu jishu*

用萃取溶劑從廢水中提取某一物質的化學處理方法。利用萃取方法可以將被萃取的物質加以回收。在*水污染控制*中，可以利用這一方法去除廢水中可溶於溶劑的污染物。萃取後應將萃取溶劑進行精餾，以使溶劑重複使用，使污染物得到回收。萃取系統各不相同。其設計涉及適用於稀溶液的一般化工過程原理。

In general, a chemical processing method to extract a substance from a solution with an extracting solvent. The extracted substance can thus be recovered. In *water pollution control*, extraction can be applied to remove a solvent-soluble pollutant from an aqueous effluent. The extraction solvent is rectified and reused and the pollutant can be recovered. Various systems are available. Their design is covered by the principles of general chemical processing, applied to dilute solutions.

F

f/m ratio 食物／生物量比 *shiwu/shengwuliang bi*

見：food to biomass ratio。

See *food to biomass ratio.*

fat soluble 脂溶性的 *zhirongxingde*

可溶於有機溶劑或生物體*脂肪組織*中的化學物質的性質。它在脂肪中的溶解度用*辛醇-水系數*來測定和表示。

The property of a chemical substance which makes it soluble in organic solvents and in *fatty tissue* of organisms. The degree of solubility in fats is measured and reported as the *octanol-water coefficient.*

fatty tissue 脂肪組織 *zhifang zuzhi*

主要由各種不同脂肪（脂類）組成的生物體的組織，它是通過*食物鏈*傳遞並在食物鏈中累積的*親脂性*物質的最終儲存處。（參見：生物富集）

The tissue of an organism composed primarily of different fats (lipids). The fatty tissue is the site of final deposit of *lipophilic* substances transported through, and accumulated in the *food chain.* (See also *bioconcentration*)

fee 費用 *feiyong*

因接受服務或獲得商品而付的款項。付費的多少依所獲得的商品的質量而定。（參見：收費、罰款）

A payment made in return for a service or goods received. The magnitude of the fee depends on the quantity of goods received. (See also *levy*; *fine*)

fee, waste disposal 廢物處置費 *feiwu chuzhifei*

廢物產生者支付的廢物處置費用。它是根據*污染者付費原則*，通常按照廢物的性質和數量來計算的。然而，在許多情況下，廢物處置費是基於污染者付費原則和*社會費用原則*二者並用的方法來計算的，這取決於市政管理當局或有關機構的管理政策。

Fees to be paid by the generator of a waste for its disposal. The fee is normally based on the properties and quantity of the waste, according to the *polluter-pays-principle*. In many cases, however, the fee is calculated on the basis of a mixture of the polluter-pays-principle and the *social cost principle*, depending on the management policy of the municipality or the organization.

fees for effluent treatment 廢水處理費用 *feishui chuli feiyong*

由廢水產生者支付的處理費用。它應盡可能以 *污染者付費原則* 作為實施的基礎。其費用取決廢水的排放量和濃度。這一概念可以應用於企業內，也可應用於社會，涉及城市家庭排出廢水的處理費用問題。在後一種情況下，它是按耗用的水量來收費的，而且假定所耗用的水全部被排放。(參見：污染控制費用的分攤)

Fees to be paid by the generator of an effluent to cover the costs of treatment. Fees should be based as much as possible on the *polluter-pays-principle*. They depend on the volume and concentration of the discharge. The concept of fees can be applied internally within an enterprise, and also externally to cover the treatment costs of effluents from the households of a municipality. In the latter case, they are based on the volume of water consumed. The assumption is that all water consumed is also discharged. (See also *cost allocation of pollution control costs*)

femtogram 毫微微克 *haoweiweike*

10^{-15} 克。

The 10^{-15} part of a gram.

fermentation 發酵 *fajiao*

有機物在缺氧條件下進行分解或變化過程。發酵的產物是低分子量的有機物，例如糖發酵會生成醇。發酵過程總是細菌的微生物代謝過程的一部分。(參見：厭氧過程、好氧過程)

The enzymatic breakdown or modification of organic molecules in the absence of oxygen. Products of fermentation are organic substances with lower molecular weight. The fermentation of sugar, for instance, yields alcohol. Fermentation processes are always part of microbiological metabolic processes of bacteria. (See also *anaerobic processes*; *aerobic processes*)

FID 火焰電離檢測器 *huoyan dianli jianceqi*

見：flame ionization detectors。

See *flame ionization detectors*.

FID analyzer 火焰電離檢測器分析儀 *huoyan dianli jianceqi fenxiyi*

通過在氫火焰中燃燒有機化合物，對電離過程產生的電信號進行檢測、加工處理和記錄，來測定氣流中有機化合物含量的儀器。

An instrument to measure the content of organic compounds in a gas stream by burning its contents in a hydrogen flame. The electric signals generated in the ionization process are detected, processed and registered.

filter aid 助濾劑 *zhulüji*

在過濾前添加到溶液中的高度分散的惰性物質（如硅藻土、活性碳和纖維素纖維等）。助濾劑能吸附微量雜質並且使過濾後的溶液變得潔淨。由於它們在使用後含有微量的各種不同化合物，因此應作為*特殊廢物*進行處置。

Inert, finely dispersed material (e.g. diatomaceous earth, activated carbon, cellulose fibers, etc.) which are added to a solution before filtration. Filter aids adsorb trace impurities and help to get clean and clear solutions after filtration. They then contain traces of different compounds and are therefore to be disposed as *special wastes*.

filter breakthrough 過濾器穿透 *guolüqi chuantou*

過濾器操作中過濾能力被耗盡，從而介質通過過濾器不發生任何變化的操作點。在廢氣過濾時，過濾器穿透會導致突然意外的排放。在嚴重的情況下，需設置第二過濾器作為*控制過濾器*。

The point in the operation of a filter at which the filter capacity is used up and the medium passes through the filter without change. In the case of off-gas filtration, filter breakthrough may lead to sudden unexpected emissions. In critical cases, a second filter is added as a *police filter*.

filter fibers 過濾纖維 *guolü xianwei*

　　一種纖維狀的過濾介質。使用的過濾纖維有毛、棉等天然纖維和合成纖維（聚丙烯腈、聚酯、聚氯乙烯、聚乙烯和其他聚合物）。聚四氟乙烯纖維和無機材料纖維（如碳纖維、玻璃纖維和石英纖維）可耐高溫和耐腐蝕。

　　A filter medium in form of fibers. In use are natural fibers such as wool and cotton, synthetics, (polyacrylonitrile or polyester, PVC and polyethylene). Teflon filters and inorganic materials such as carbon fibers, glass and quartz fibers are able to withstand higher temperatures and corrosive media.

filter medium 過濾介質 *guolü jiezhi*

　　過濾器的關鍵部分，是用於分離的多孔介質。有各種不同的過濾介質：天然和合成紡布、毛氈、礦物纖維、具有不同網目的金屬網、多孔陶瓷材料、砂和有機物質（如泥炭沼）。

　　The active part of a filter, the porous medium used for separation purposes. Filters incorporate a variety of filter media: cloth of natural and synthetic textiles, felts, mineral fibers, metallic grids with various mesh size, porous ceramic materials, sand and organic materials, such as peat moss.

filterpress 壓濾機 *yalüji*

　　化學加工處理中用作間歇過濾的設備，它由若干個鋪放了濾布的板框所構成（通常呈水平排列）。過濾的液體被投加到偶數號的隔間，在那裏把分離得到的殘渣收集起來，然後再排放；而濾液通過濾布進入到奇數號的隔間，同時被收集和導流到排放點。在市場上有許多不同設計形式的可用於不同目的的壓濾機供應。有些壓濾機可自動操作；有些還配有壓濾膜使濾餅的脫水效果更好。

　　A unit used in chemical processing for batchwise filtration. Filterpresses combine (normally in a horizontal array) a number of frames with filter cloth stretched on the frames. The liquid to be filtered is fed into the even numbered compartments where the separated residue is collected and then discharged. The filtrate passes the filter cloth into the odd compartments and is collected and channeled to the discharge point. Many different designs of filterpresses for different purposes are available commercially. Some types allow for an automatic operation; some are equipped with pressing membranes for better dewatering of the filter cake.

filtration 過濾 *guolü*

　　從液相或廢氣中分離懸浮顆粒的過程。由於顆粒大小的不同，比過濾孔徑大的顆粒將被截流，從而可進行分離。小顆粒（小于1000毫微米）的分離還可能受物理力的作用而附著到大表面上。水處理中的*砂濾*或*雙介質過濾*就屬於這種情況。

　　有許多不同設計型式的過濾器，適用於各種不同的情況。可以根據孔的大小來區別傳統過濾與*微過濾*或*超濾*之間的不同。在污染控制技術中，過濾的目的在於：進行母液的*預處理*，用於去除*懸浮物*，進行*廢水的三級處理*，用於*去除磷酸鹽*，以及應用於其他許多方面。

The process of separation of particles of a suspension from a liquid phase or from an off-gas stream. The separation may be the result of a size discrimination, i.e. particles larger than the pores of the filter are held back. Separation of small particles (below 1000 nm) may also be due to adhesion by physical forces to large surfaces. This is the case in *sand filtration* or *dual media filtration* for water treatment.

There are numerous designs of filters for a wide range of applications. Depending on the pore size one distinguishes between conventional filtration, *microfiltration*, *ultrafiltration*. Filtration is applied in pollution control technology with various objectives: *pretreatment* of mother liquors, for removal of *suspended solids*, for *tertiary treatment of effluents* and *phosphate removal* as well as many other applications.

filtration, dual media 雙介質過濾 *shuangjiezhi guolü*

　　用具有兩種或三種不同粒度和不同材料的*砂濾池*進行過濾。這些不同的過濾介質分別鋪設在濾池的各層中。這種系統能克服單層過濾器的堵塞問題。

Filtration through a *sand filter* with two or three distinctly different grain sizes and different materials. These different filter media are kept in separate layers. The system is advantageous in cases where plugging of a single layer filter is a problem.

filtration of off-gas 廢氣過濾 *feiqi guolü*

在產品回收和廢氣處理中應用的一般技術。過濾可以去除*顆粒物*。對於不同大小的粒子和不同的應用情況，市場上有不同結構類型的過濾器供應。（參見：袋式過濾器、袖式過濾器、純粹過濾器）

A common technology applied in the recovery of products and the treatment of off-gases. Filters will remove *particulate matter*. For different size particles and different applications, different construction types of filters are commercially available. (See also *bag filter*; *sleeve filter*; *absolute filter*)

final clarifier 最終澄清池 *zuizhong chengqingchi*

廢水處理場位於活性污泥曝氣池後的澄清池。（參見：二次澄清池）

The clarifier after the activated sludge basin of an effluent plant. (See also *secondary clarifier*)

final sink in the ecosystem 生態系統中的最終去所 *shengtaixitongzhongde zuizhong qusuo*

進入生態系統的某一物質在其遷移的循環過程中被最終結合和除去的場所。"最終"這一術語是相對的，因為物質可在未來通過新的過程再釋放出來。例如，*陸圈*中的石灰石可以認為是碳的最終所在地，*脂肪組織*可以認為是*親脂性物質*的最終去所。

The compartment, specific element or site, in which a substance introduced into an ecosystem is finally bound and removed from the cycle of material transport. The term 'final' is relative, as matter may be released again through new processes at a future time. Example: limestone of the *geosphere* may be regarded as a final sink for carbon, or, *fatty tissue* may be regarded as a final sink for *lipophilic* substances.

fine 罰款 *fakuan*

一種處罰方式。由於不符合法律要求，如不符合水污染控制立法規

定的排放標準，向政府主管部門交付的錢款。在國家法規中對罰款的數額有的規定。（參見：費用、收費）

A form of punishment. A mandated payment to an authority because of non-compliance with legal requirements, e.g. with discharge limits of water pollution control legislation. The magnitude of the fine depends on national legislation. (See also *fee*; *levy*)

finishing pond 最終處理塘 *zuizhong chulitang*

廢水生物處理的最終步驟。將處理後的廢水在排放到受納水體前，在一個池塘（有時在一個天然池塘）中停留較長時間，通過懸浮物沉降、紫外輻射和進一步生物降解等各種過程，使廢水水質得到進一步改善。

A final step in biological effluent treatment. A pond, sometimes a natural pond, in which the treated effluent is kept for a relatively long time before being discharged to the receiving waterbody. In the finishing pond, the quality of the treated effluent may be improved as a result of a variety of processes like sedimentation of suspended matter, oxygen uptake, UV irradiation and additional biodegradation.

fish tests 魚試驗 *yu shiyan*

見：toxicity的不同條目。

See *toxicity* (various entries).

fish-toxic 對魚有毒 *duiyuyoudu*

見：toxicity versus fish。

See *toxicity versus fish*.

fixed bed bioreactor 固定床生物反應器 *gudingchuang shengwu fanyingqi*

利用在反應器填料表面上生長並生成活性生物膜的微生物進行生物化學轉化的反應器。運行時，將廢水用泵打入反應器，在不同的填料類型和反應條件下，通過微生物（*好氧和厭氧微生物*）的作用進行處理。在水污染控制中，其應用包括好氧和厭氧生物降解兩大類處理裝置。

A reactor for biochemical transformations in which the microorganisms grow on the surface of the reactor filling and form an active coating or film. The liquid to be processed is pumped through the reactor. Different microorganisms (*aerobic* as well as *anaerobic*), types of filling, and reaction conditions are used. Applications in water pollution control include units for aerobic as well as for anaerobic bio-degradation.

fixed bed systems 固定床系統 *gudingchuang xitong*

化工工藝中諸如吸附塔的一種處理系統。在該系統的反應器中，固相是靜止、固定的填充層，而使處理的氣體通過該固定床。在固定床活性碳吸附器中，用粒狀活性碳作為吸附劑。（參見：流化床系統）

A processing system in chemical technology, e.g. an adsorption column, in which the solid reactive phase is contained in a reactor as a stationary, fixed filling or bed. The gas or liquid to be processed is passed through the fixed bed. In a fixed bed activated carbon adsorber, granular activated carbon is used as an adsorbent. (See also *fluidized bed systems*)

fixed film biological systems 固定膜生物系統 *gudingmo shengwu xitong*

微生物生長在載體固體表面上的廢水生物處理系統，它與微生物自由地懸浮流動於*混合液*中的系統完全不同。載體表面可以是曝氣塔或*生物轉盤*結構部件的一部分，或者是活性碳顆粒的表面。

Biological effluent treatment systems in which the active microorganisms grow on the solid surface of a support, in contrast to systems in which the micro-organisms are free-flowing in the *mixed liquor*. The supporting surface may be part of a structural element of a column, of a *rotating biological contactor*, or the surface of activated carbon particles or granules.

flame ionization detectors (FID) 火焰電離檢測器 *huoyan dianli jianceqi*

氣相色譜法中通常採用的檢測器，它位於分離混合物的色譜柱之後。其工作原理是：有機化合物在檢測器的氫火焰中被燃燒成為二氧化碳，在電離過程中會產生電信號，然後進行信號加工處理和記錄。在很寬的濃度範圍內，電信號與氣流中有機化合物的濃度成正比。但是，它對鹵代化合物的響應卻是很低的。

A detector commonly used in gas chromatography after separation of a mixture in a column. The organic compounds are burned to carbon dioxide in a hydrogen flame. An electric signal is generated in the ionization process and can be processed and registered. Over a wide concentration range, the signal is proportional to the concentration in the gas stream. The response to halogenated compounds is low.

flame plasma 火焰等離子體 *huoyan dengliziti*

處於高能態的氣體。在等離子體中，物質（分子或原子）的大部分處在極度激活的離子化狀態。等離子體不能用一般熱力學平衡參數來説明。噴入直流電弧中的氣體，是一個典型的等離子體源；處於瞬態放電區的物質同樣也是一種等離子體源。極強烈的火焰（如來自氧氣和氫氣的火焰）可以看作是一種邊界情況。在污染控制方面，它被應用在*等離子體焚燒技術*中。

A gas in a high-energy state. In a plasma, a large fraction of the matter (molecules or atoms) is in a very highly excited ionized state. The plasma defies description by common thermodynamic equilibrium parameters. A gas injected into an electric direct current arc is a typical source of a plasma, as is also the material in a spark channel. An extremely intense flame (e.g. from oxygen and hydrogen) may be looked at as a borderline case. In pollution control, applications of plasma are in the *plasma incineration technique.*

flocculating agent 絮凝劑 *xuningji*

投加到非常微細的懸浮物中能促使小顆粒凝集成較大顆粒，然後可使顆粒物從液相中分離出來的物質。這個過程取決於粒子極性和表面性質（Z–電位）的物理化學變化。市場上有用途不同的各種合成絮凝劑供應。

A substance which, when added to a very fine suspension, will promote the agglomeration of small particles to larger ones which can then be separated from the liquid phase. The process is based on physical chemical changes of particle polarity and surface properties (zeta potential). Various synthetic flocculating agents for different applications are available commercially.

flocculation 絮凝（作用）*xuning (zuoyong)*

懸浮液中小顆粒凝集成較大顆粒，然後可通過沉降、浮選、離心或過濾進行分離的過程。可以投加特定的添加劑來強化絮凝作用。（參見：絮凝劑）

The process of agglomeration of small particles of an aqueous suspension to larger ones which can then be separated by sedimentation, flotation, centrifugation or filtration. Flocculation can be enhanced by the addition of specific additives. (See also *flocculating agents*)

flotation 浮選 *fuxuan*

從液體（如廢水）中分離不溶性物質的過程。如果不溶物比液體輕，它將升起到液體表面而被分離。在技術上可以把空氣通入浮選池的下部來強化浮選作用。這時，微氣泡將附著在不溶顆粒上，使它上升到水面而被去除。投加某些添加劑將會改變懸浮物的濕潤性，從而可提高氣泡的附著能力。

A process for separating insoluble matter from a liquid, e.g. from an effluent. If the insoluble matter is lighter than the liquid it will rise to the surface and can be separated. For technical purposes, flotation can be enhanced by blowing air into the lower part of the basin containing the suspension. Micro air bubbles will attach themselves to the insoluble particles and cause them to rise to the surface from where they can be removed. The attachment of air bubbles can be enhanced by the addition of certain additives which will change the wettability of the suspended matter.

flotation clarifier 浮選澄清池 *fuxuan chengqingchi*

基於*浮選*原理運行操作的澄清池。在廢水處理系統中，有時用它來去除輕的污泥，會產生氣泡的污泥，或用它來濃縮剩餘活性污泥。

A clarifier which operates on the principle of *flotation*. Flotation clarifiers are sometimes used in effluent treatment systems to remove light sludge, sludge which tends to produce gas bubbles, or for concentration of excess biosludge.

flow average sample 流量平均樣 *liuliang pingjunyang*

真正代表了所排放廢水中質量流量的廢水樣。由於流量是變化的，因此這意味著流量大時所取的水樣體積在總水樣體積中所佔的比例必須比流量低時大。如果按照與流量成正比的頻次不斷重複地取小量等體積的水樣，就能實現這一要求。

A sample of an effluent which truly represents the discharged mass flow in an effluent. With a variable flow, this means that at times of high flow volumes, a larger fraction of the total sample volume must be taken than at times of low flow volume. This can be achieved by taking repeated small equal sample volumes at a frequency proportional to the flow rate.

flue gas 煙道氣 *yandaoqi*

焚燒或燃燒過程產生的氣體混合物，它含有物質氧化時所產生的許多化合物。燃燒時有機物被氧化為二氧化碳和水，由於燃燒不完全還可能生成微量一氧化碳，其他元素如硫、磷和氮被轉化為它們的氧化物，鹵素隨燃燒條件的不同則會生成不同的燃燒產物：煙道氣中的氯化物主要是氯化氫，碘化物是游離碘，溴化物是溴化氫和游離溴的混合物。在某些條件下，被部分氧化的有機物分子的碎片會經*重組過程*生成微量的新化合物。煙道氣除含有來自燃燒空氣中的剩餘氧和氮之外，還含有空氣中的氮部分氧化而生成的氮氧化物。這些污染物的濃度主要取決於焚燒物質的化學組成和焚燒時所用的過量空氣的量，但焚燒條件(溫度、時間、焚燒爐類型)也有一定的影響。在報告煙道氣中污染物的濃度時，通常需對它們的濃度通過計算加以調整(即按煙道氣中氧的標準濃度11%進行調整)。

The gas mixture resulting from an *incineration* or *combustion* process. Flue gas contains a number of chemical species which are derived from the substances which are oxidized. Organic components are oxidized to carbon dioxide and water. Trace quantities of carbon monoxide may be present as a result of incomplete combustion. Other elements, like sulfur, phosphorous and nitrogen are transformed into their oxides. Halogens form different products: chlorides are found in the flue gas mainly as hydrogen chloride, iodides as free iodine, and bromides as a mixture of hydrogen bromide and free bromine, depending on the conditions. Under certain conditions, fragments of partially oxidized organic molecules may undergo *recombination processes* to form traces of new compounds. In addition, excess oxygen and nitrogen from the combustion air, as well as oxides of nitrogen from the partial oxidation of air-nitrogen are present. The concentrations of these contaminants depend primarily on the chemical composition of the material which was incinerated, and on the volume of excess air which was used for incineration. The conditions of incineration (temperature, time, type) also play a role. For reporting, the concentrations of the components are generally adjusted to a standard oxygen concentration of 11%.

flue gas treatment 煙氣處理 *yanqi chuli*

消除*煙氣*中污染物以達到法律規定的*排放限值*而採用的聯合處理過程。在*乾式煙氣處理系統*和*濕式煙氣處理系統*中應用了一些處理過程，其單元過程與一般空氣污染控制系統是相似的。在所有的系統中，必須對煙氣處理所產生的廢水（它含有從煙氣中去除的污染物）的處理予以足夠的重視。

A combination of process steps used to eliminate noxious contaminants in *flue gas* to the required legal *emission limit*. A number of process steps are applied in *dry flue gas treatment* systems and *wet flue gas treatment* systems. The unit processes are similar to the ones in general APC-systems. In all systems, care has to be taken to give adequate additional treatment to the aqueous waste streams which contain the pollutants removed from the flue gas.

flue gas treatment, dry 乾式煙氣處理 *ganshi yanqi chuli*

應用乾式單元過程處理煙氣。它可利用固鹼吸著劑如微細粉末氧化鈣，通過化學結合作用去除煙氣中的酸性組分（二氧化硫、有時還有氯化氫）。有些方法係將吸著劑直接投加到燃燒室中，另一些方法則使用一單獨的反應器。所產生的煙塵再用常規方法如*旋風分離器*、*過濾器*和*靜電分離器*去除。在熱電廠採用這種方法時，會產生大量石膏，需要進行處理，如去進行填埋，或者作為二次原材料加以利用。（參見：濕式煙氣處理系統）

The application of dry unit processes to flue gas treatment. Solid alkaline sorbents, e.g. finely powdered calcium oxide, are used to remove acidic components (sulfur dioxide, sometimes also hydrogen chloride gas) from flue gas by chemical binding. In some technologies, the sorbent is added directly into the burning chamber, others use a separate reactor. The resulting dust is then removed by standard techniques. Available are *cyclones*, *filters*, and *electrostatic separators*. Applied to thermal power plants, this absorption and binding process yields a large quantity of gypsum, which needs to be disposed of, either by landfill, or as a secondary raw material. (See also *wet flue gas treatment systems*)

flue gas treatment, wet systems 濕式煙氣處理系統 *shishi yanqi chuli xitong*

利用濕式空氣污染控制原理設計的*煙氣處理系統*，其效率很高，與*除氮氧化物*裝置相結合，代表著現代的工藝技術水平。該系統可應用於處理*廢物焚燒爐*產生的煙氣，其處理系統包括熱回收單元、顆粒物去除單元、煙氣中酸性組分去除單元、氮氧化物去除單元和污染洗滌液處理單元。

此外，由於對排放一氧化碳和揮發性有機化合物也有限制要求，因此必須在最佳條件下進行燃燒。（參見：空氣污染控制過程單元、濕式空氣污染控制系統、乾式空氣污染控制系統）

A custom designed treatment system for *flue gases*, making use of the principles of wet air pollution control steps and units. Wet flue gas treatment systems are highly efficient. Combined with *DENOX* units, they represent the present state-of-the-art. Wet flue gas treatment systems can be applied to the treatment of flue gases from *waste incinerators*. Such a system includes units for heat recovery; removal of particulate matter; removal of acidic components in the flue gas; removal of oxides of nitrogen; and treatment of polluted scrubbing liquors.

In addition, the limitation of the emission of carbon monoxide and of VOC require optimized combustion conditions. (See also *APC-process steps*; *APC-systems, wet*; *APC-systems, dry*)

fluid extraction, supercritical 超臨界流體萃取 *chaolinjie liuti cuiqu*

見：supercritical fluid extraction。

See *supercritical fluid extraction*.

fluid phase 流動相 *liudongxiang*

　與*固相*相反的流動相，它可以是液體中的固體懸浮物或者氣體中的固體物。

The moving phase, in contrast to the *solid phase*. The fluid phase may be a suspension of a solid in a liquid, or a solid in a gas.

fluidized bed biology 流化床生物處理 *liuhuachuang shengwu chuli*

　一種*廢水生物處理*方法。在這種方法中，微生物系生長在活性碳或砂子等顆粒的表面上，通過吸附作用，使微生物所生活的微環境發生變化，從而提高處理效率。在柱式反應器中，係通過上升水流使活性碳或砂子等顆粒保持懸浮狀態。

A type of *biological effluent treatment* in which the microorganisms are growing on the surface of particles, e.g. of activated carbon or sand. Through adsorption processes, the micro-environment of the organisms may also be changed, leading to a higher efficiency. The particles are kept in suspension by the upstream flow of the aqueous effluent in a column reactor.

fluidized bed incineration 流化床焚燒 *liuhuachuang fenshao*

　在*流化床*中進行燃燒的焚燒系統。在該系統中，將空氣通過特殊設計的空氣噴咀噴入系統中，以使細砂在垂直燃燒室的流化床中保持懸浮狀態。焚燒爐襯有特殊的耐火磚。焚燒時把廢物（如廢水處理產生的濃縮污泥）引入到流化床層。用載焰使燃燒溫度保持在650-900°C，焚燒後的殘灰與一部分砂子被從系統中連續取出，並在水中驟冷。有些技術在設計上採用了循環式流化床。

An incineration system in which the actual combustion process is carried out in a *fluidized bed*. In such a unit, fine sand is kept in suspension in the fluidized bed zone in the vertical incinerating chamber by air injected into the system through specially designed air nozzles. The incinerator is lined with special fire brick. Waste, for instance thickened waste sludge from effluent treatment, is introduced into this zone. The combustion temperature is kept at around 650–900°C by support flames. Ashes remaining after incineration are continuously taken out of the system together with part of the sand and quenched in water. Some technical designs use a circulating fluidized bed.

fluidized bed systems 流化床系統 *liuhuachuang xitong*

通過與固相接觸進行液相或氣相處理的一類工藝技術。在該系統的圓柱形垂直反應器中，向上流動的氣體或液體將使固相（中等大小的顆粒）處於恆定懸浮或移動狀態。其升流速度必須大致等於沉降速度。在流化床中，固相反應顆粒不斷地被混合，並且被不斷除去或更新。作為污染控制技術，流化床系統被應用於焚燒爐、吸附系統和生物反應器中。（參見：固定床系統）

A type of processing technology for treatment of a liquid or gaseous phase in contact with a solid phase. In a fluidized bed system, the solid phase in the form of medium size particles is kept in constant suspension or movement in a vertical cylindrical reactor through an upstreaming flow of gas or liquid. The flow rate must be approximately equal to the sedimentation rate. The reactive solid particle phase is constantly mixed and can also be removed and renewed. In pollution control technology, fluidized bed systems are found for example in incinerators, adsorption systems, bioreactors. (See also *fixed bed systems*)

fluorescence spectrophotometry 螢光分光光度法 *yingguang fenguang guangdufa*

一種光致發光法。某些物質能吸收紫外光而發射出可見光波的螢光輻射，它的強度可加以測定。在環境分析中，可利用螢光的性質來鑒別和測定某些物質，廢水或污泥中光增白劑的分析和香煙煙霧中多環芳烴的分析，就是其應用的兩個實例。

A photoluminescence method. Certain substances absorb light in the ultraviolet wavelength range and emit fluorescence radiation in the visible range. The intensity of this fluorescence is measured. Fluorescence properties are used to identify and measure certain substances in environmental analysis. Two examples to show the range of applications are: the analysis of optical whiteners in effluents or sludges and the analysis of polyaromatic hydrocarbons in cigarette smoke.

f/m ratio 食物／生物量比 *shiwu/shengwuliang bi*

見：food to biomass ratio。

See *food to biomass ratio.*

fluorometry 螢光法 *yingguangfa*

基於物質的螢光性原理的分析方法。螢光計是測定溶液螢光的一種簡單儀器，它利用特定的濾光器將激發螢光的汞蒸氣燈所產生的主光譜紫外線分離開，這樣就可以測定處於可見光波長範圍內的螢光。螢光法可用於發螢光化合物如螢光增白劑或多環芳烴的分析。

An analytical method based on fluorescence properties of substances. The fluorometer is a simple instrument to measure the fluorescence of a solution. It uses specific filters to separate the main spectral UV lines of a mercury vapor lamp for excitation of fluorescence. The fluorescence in the visible wavelength range is then measured. The method is used in the analysis of compounds which are fluorescing, for instance optical brighteners or polyaromatic hydrocarbons.

foam control 泡沫控制 *paomo kongzhi*

見：antifoaming agent。

See *antifoaming agent.*

food chain 食物鏈 *shiwulian*

群落生境中一系列的生物體，它們在捕食者與被捕食者關係中是彼此有關的。除了首鏈與末鏈外，食物鏈中的每一生物體既是捕食者又是被捕食者。（參見：營養級、生物富集）

A succession of organisms in a biotope which are related to each other in a predator-prey relationship. With the exception of the first and last links, each organism of the food chain is at the same time a predator and a prey. (See also *trophic level*; *bioconcentration*)

food to biomass ratio (f/m ratio) 食物／生物量比 *shiwu/shengwuliang bi*

廢水生物處理場的一項設計和運行參數。它是以每天投入廢水生物處理場活性污泥曝氣池中的 BOD_5 量與曝氣池混合液中的總生物量之比進行計算的。生物量以 *MLVSS*（*混合液揮發性懸浮物或污泥揮發分濃度*）表示。食物／生物量比的典型值，對BOD去除為0.3，對硝化作用為0.15，對生物降解和污泥穩定化為0.05。

A design and operating parameter of a biological effluent treatment plant. The f/m ratio is calculated as the quantity of daily input BOD_5 into the activated sludge basin of a biological effluent plant in relation to the total biomass in the mixed liquor. The biomass is expressed as *MLVSS*. Typical values for the f/m ratio are 0.3 for full BOD-removal, 0.15 for nitrification, 0.05 for biodegradation and sludge stabilization.

forest damages through air pollution 空氣污染造成的森林破壞 *kongqi wuran zaochengde senlin pohuai*

在工業化國家，由於大氣中污染物的影響，森林已受到不同程度的破壞，表現為針葉樹針葉的落失，及最終造成單株樹木或整片森林死亡。

一方面，對於未經處理的*煙道氣*排出的高濃度酸性污染物（如二氧化碳和氮氧化物；見：*酸雨*）所造成的急性和嚴重的森林破壞，其*因果關係*是比較明顯的。這種後果在工業區附近的森林中可以觀察到。森林破壞已經發覺有幾十年之久。另一方面，有跡象表明，低濃度污染物和*夏季煙霧*的組分也會造成亞急性的破壞，這種破壞不局限於工業區附近的森林。氣候因子如同生物因子（如某些昆蟲）一樣，看來也起著一定的作用。總之，森林破壞的機理現在還沒有被詳細地、確切地解釋清楚。但無論如何，森林是十分重要的，因此應及時採取措施來*控制空氣污染*。

Forests in industrialized countries have suffered damages to a variable degree as a result of pollutants in the atmosphere. The damages manifest themselves in a loss of the needles of conifers, and finally in the death of single trees or whole forests.

On the one hand, the relation between *cause and effect* is relatively clear for acute, severe damages resulting from high concentrations of acidic pollutants (e.g. SO_2 and NO_x; see *acid rain*) from untreated *flue gases*. Such effects are found in forests near heavily industrialized areas. Damages have been observed for decades. On the other hand, there are indications that lower concentrations of pollutants and components of *summer smog* may cause damages on the subacute level. These damages are not limited to forests in the vicinity of heavy industry. Climatic factors also seem to play a role, as do biological factors, such as certain insects. Overall, the mechanism of damage has not in all details been elucidated with certainty. Nevertheless, in view of the great importance of forests it is important that *air pollution control* measures are taken in time.

framework laws, environmental 環境框架法 *huanjing kuangjiafa*

如該名稱所意味的那樣，它是為環境立法中的一類具體法律制定框架的法律，而實施細節在細則和條例中規定。憲法、框架法、具體法和細則是立法體系中的四個層次。（參見：環境立法）

Laws which, as the name implies, set the frame for a group of laws in specific environmental legislative areas. Details for the implementation are then covered in by-laws and ordinances. The constitution, framework laws, specific laws and by-laws represent four levels of a hierarchy in legislation. (See also *environmental legislation*)

free goods 免費物品 *mianfei wupin*

由生產者（或作為整體經濟）使用而不必付費的某種物品或資源。這與企業必須對所使用的物品（原料）付費的正常情況完全不同。例如，利用大氣而排放未處理的廢氣，或者利用水體而排放污染的廢水，就屬這種情況。大氣和水都可以認為是"免費物品"。因為這些資源是免費的，因此使用者節省或限制污染物的排放，不會得到什麼直接的利益，而恢復補救環境污染或它所造成損害的費用將成為*外部費用*。

現在，污染一般已受到法律的管制，生產者必須遵守法律規定的*排放限值*。所以，生產者必須承擔污染預防或污染治理的費用。現在可以說這些費用已成為了*內部費用*。因此，在制定了明確而充分的法規標準並且付諸有效實施的情況下，"免費物品"這一問題已稍有緩解。

A good or a resource, used by a manufacturer (or the economy as a whole) for which the user does not have to pay. This situation is in contrast to the normal case in which an enterprise has to pay for the goods (raw materials) it uses. The use of the atmosphere to discharge of an untreated off-gas stream, or the use of a waterbody to discharge of polluted effluents may be taken as examples. Both, the atmosphere and water, can be considered as 'free goods'. Since these resources are free of charge, the user has no direct interest to economize them and to limit his discharges. The costs to remediate the pollution or the costs of resulting damages will be *external costs*.

Today, pollution is normally limited by legal *discharge limits* which must be observed by the manufacturer. Thus, he has to bear the costs for pollution prevention or pollution abatement. One can say that the costs are now *internal costs*. The problem of 'free goods' therefore is somewhat reduced in a situation with clear and adequate legal limits and efficient enforcement.

freight limits 負載限值 *fuzai xianzhi*

對排放總量的限制。（參見：總量限值）

The limitation of the total mass flow of an emission. (See also *mass flow limits*)

FREON 氟里昂 *fuli'ang*

用作冷卻機中致冷劑一類物質的商品名。（參見：氯氟烴、蒙特利爾議定書）

A trade name for substances used as refrigerants in cooling machines. (See also *chloro fluoro carbons*; *Montreal Protocol*)

fungi 真菌 *zhenjun*

*原生物*王國中的一類生物，其基本結構單元是*真核細胞*。真菌有多種多樣性，主要生活在土壤中，其尺寸小到微生物，大到食用蘑菇。真菌不能同化二氧化碳，因此它們是*異養生物*。它們始終是通過菌絲體來攝取食物的，所以它們需依靠外來的食物源：腐敗的有機物（如腐木），或與其他生物*共生*；有一些真菌是寄生菌，需依賴於活的寄生生物，並且會對它造成損害。許多植物的病害是由真菌造成的，但從食品、健康、生物化學過程、農業、毒理學等觀點來看，真菌是很重要和令人感興趣的。

A group of organisms of the kingdom of *protists*. The basic structural units are *eucaryotic cells*. Fungi show a wide diversity, living mainly in soils, and ranging in size from microorganisms to edible mushrooms. Fungi are not able to assimilate carbon dioxide and are therefore *heterotrophic organisms*. For food, which they always take up through the mycelium, they depend on external sources: on decaying organic material, (e.g. rotten wood), on *symbiosis* with other organisms; some are parasites and depend on a living host organism, which they may damage. Many plant diseases are caused by fungi.

Fungi are interesting from a wide range of viewpoints: food, health, bio-chemical processing, agriculture, toxicology, etc.

G

GACT 粒狀活性碳處理 *lizhuang huoxingtan chuli*

見：granular activated carbon treatment。

See *granular activated carbon treatment.*

galvanic waste 電鍍廢物 *diandu feiwu*

電鍍過程產生的污泥和殘渣。電鍍廢物是一類 *特殊廢物*，常常含有重金屬和氰等有毒物質。對其處置需認真考慮採用特殊的去毒方法。如果將電鍍過程中產生的不同類型的電鍍廢物分離開來，那麼就可以對重金屬進行再生和回收。

Sludges and residues obtained from electroplating processes. Galvanic wastes are a group of *special wastes*, often containing heavy metals from the processes and toxic materials, such as cyanides. Their disposal needs careful consideration and special detoxification processes. If different streams or types of galvanic waste are kept separate at the process level, then a regeneration and recovery of the metal may be feasible.

garbage bag disposal fee 垃圾袋處置費 *lajidai chuzhifei*

見：fee, waste disposal; polluter-pays-principle。

See *fee, waste disposal*; *polluter-pays-principle.*

gas balancing pipes 氣體平衡管 *qiti pingheng guan*

連接運輸槽罐和儲罐的雙管系統，用以減少裝罐時揮發性 *有機化合物* 的排放。當揮發性液體通過一根管子沿一個方向（從罐 A 到罐 B）輸送時，含有飽和揮發性蒸氣的氣相將通過另一根管子沿相反方向（從罐 B 到罐 A）輸送。因此，在裝罐時氣體平衡管可起到減少排放的作用。

A system of double pipes connecting a transport tank with a storage tank, used to reduce emissions of *VOCs* during the filling of a tank. While the volatile liquid is transferred in one direction (from tank A to B) through one pipe, the gas phase, saturated with the volatile vapor, is transferred in the opposite direction (from tank

B to A) through the second pipe. Gas balancing pipes thus serve to reduce emissions during tank filling.

gas chromatography 氣相色譜法 *qixiang sepufa*

分離氣相中混合化學物質的技術。分析時,將混合的揮發性化合物注入到內側塗覆了一薄層固定相的玻璃柱或石英柱中,然後用惰性氣體流將混合物輸送通過柱子,進行*吸附*與*解吸*。由於這些化合物與固定相相互作用程度的不同,而使它們得到分離。分離條件、溫度和塗層類型隨樣品的不同而不同。混合物中的各個物質在分離柱的終端分離後,再去到檢測段進行檢測。市場上有氣相色譜分析所用的各種專門的檢測器和儀器供應。(參見:分析中的分離段、分析中的檢測方法)

A technique for the separation of mixtures of chemical substances in the gas phase.

The mixture of volatile compounds is injected into a glass or quartz column which is coated on the inside with a film of a stationary phase. The mixture is then transported by an inert gas stream through the column. Processes of *adsorption* and *desorption* take place. The compounds are separated as a result of different degrees of interaction with the stationary phase. Separation conditions, temperature and type of coating depend on the sample. The individual substances of the mixture leave the separation column at its end and are detected. A wide variety of specific detectors and instruments for gas chromatographic analysis are available commer- cially. (See also *separation step in analysis*; *detection methods in analysis*)

gas chromatography, capillary 毛細管氣相色譜法 *maoxiguan qixiang sepufa*

用石英毛細管柱*氣相色譜法*分離混合物中的揮發性有機化合物。這一毛細管柱在其管內表面上塗覆了一層穩定的極化硅層。與填充柱標準氣相色譜法比較,毛細管氣相色譜法的分離效果更好。如果樣品含有性質相似的多種不同化合物,則可選用毛細管氣相色譜法進行分析。

The separation of volatile organic compounds in a mixture by *gas chromato-graphy*, using a quartz capillary column. This column is coated on the inner surface with a stable polarized silicon layer. Compared to standard gas chromatography with packed columns, capillary gas chromatography leads to a better separation.

Capillary gas chromatography is the method of choice if the sample contains a large number of different substances with similar properties.

gas chromatography, semi-continuous 半連續氣相色譜法 *banlianxu qixiang sepufa*

標準氣相色譜法基本上是一種間歇的分析方法。在半連續操作中，將未知樣品每隔一定時間重複注射入系統中，然後對注射入的每一樣品只分析其中所含的有限物質。按這種方式，可以利用氣相色譜法進行過程控制或連續監測廢氣。所得到的檢測器的信號可以存貯起來並在以後進行分析，或者可以用於過程控制或在室內監測中對危險狀況直接做出決定。

Basically, standard gas chromatography is a batchwise procedure. In a semi-continuous operation, samples from an unknown sample stream may be injected into the system repeatedly at certain fixed time intervals. Each injected sample is then analyzed for only a limited number of substances. In this way, gas chromatography is available also for process control or for continuous monitoring of an off-gas stream. The detector signals may be stored and analyzed later, or they may be used directly to generate a decision, either in process control, or in risk situations in room monitoring.

gas chromatography-mass spectroscopy (GC-MS) 氣相色譜-質譜分析法 *qixiang sepu-zhipu fenxifa*

把*氣相色譜法*的分離段與質譜分析法的鑒定段結合在一起的一種強有力的分析技術。在一般的色譜法中，被測定的化合物必須用一些獨立分開的方法來鑒別；而在氣相色譜-質譜分析法中，分離後的物質可直接在氣相色譜柱的末端用*質譜分析*來鑒定。

因為質譜分析是在很低的壓力下進行的，而且只需要將少量的未知化合物送入到質譜分析部件，所以現代的氣相色譜-質譜分析系統都使用毛細管柱。但氣相色譜-質譜分析是一種複雜的分析方法，因此它只在特殊的情況下才應用於環境分析中。

A very powerful analytical technique which couples the separation step of *gas chromatography* with the identification step of mass spectrometry. In normal chromatography, the compounds to be measured have to be identified by separate methods. In the GC-MS combination, the separated substances are directly identified at the end of the gas chromatography column by *mass spectrometry.*

All modern GC-MS systems use capillary columns because MS is carried out under a very low pressure and only small amounts of the unknown compounds need to be transferred to the MS part. GC-MS is a sophisticated analytical method which is applied in environmental analysis in special cases only.

gas detector tubes 氣體檢測管 *qiti jianceguan*

市場上出售的小檢測管(商品名為德拉格管)。管中含有吸收物質，並裝填有能與氣流中不同類型的污染物發生反應的特定化合物。監測時，用泵將一定體積的空氣或氣體通過檢測管，使它產生顏色變化，而顏色的變化與樣品中污染物的濃度成正比。德拉格管是一種能快速測定一定濃度範圍的許多特定化合物的簡單設備，但其精確度不高。它們可用於*工作場所測定*和*環境空氣監測*，如*事故釋放*情況下的監測。

Small detection tubes which are available commercially (Trade name: Drger tubes). The tubes contain absorbing materials, loaded with specific compounds, which react with different types of pollutants in the gas stream. A defined volume of air or gas is pumped through the tube, producing a color change. This change is proportional to the concentration of pollutant in the sample. Drger tubes are simple devices for rapid measurement of concentration ranges of a number of specific compounds. They do not produce high precision results. They are used for *workplace measurements* and for *ambient air* monitoring, e.g. in the case of an *accidental release.*

gas monitor 氣體監測器 *qiti jianceqi*

一種連續監測氣體濃度的設備。這種監測儀器是可移動的，而且對某些特定的氣體很靈敏。有些氣體檢測器是以氣流導熱率的變化會導致檢測器元件溫度發生變化這一原理而工作的。它可應用來檢測煉油廠和其他生產裝置的泄漏情況。

A device to continuously monitor gas concentrations. Monitoring instruments are transportable and are sensitive to certain specific gases. Some gas detectors are based on a change of the heat conductivity of the gas stream, resulting in a subsequent change of temperature of the detector element. Applications are for the detection of leaks in refineries and in other production facilities.

gas standards for calibration 校準標準氣 *jiaozhun biaozhunqi*
市場上出售作校準用的裝於小不銹鋼瓶中的標準氣體混合物。

Commercially available standardized gas mixtures in small stainless steel cylinders, used for calibration purposes.

GC-MS 氣相色譜－質譜分析法 *qixiang sepu-zhipu fenxifa*
見：gas chromatography-mass spectrometry。
See *gas chromatography-mass spectrometry.*

GC-MS, capillary technique 氣相色譜－質譜分析，毛細管技術 *qixiang sepu-zhipu fenxi, maoxiguan jishu*
現在，它是*氣相色譜－質譜分析*的標準方法。（參見：*氣相色譜－質譜分析法*）

Today, this is the standard method of *GC-MS.* (See also *gas chromatography-mass spectrometry*)

generator of pollution 污染產生者 *wuran changshengzhe*
污染源一方。

The party which is, or was, the source of a pollution.

geosphere 陸圈 *luquan*
見：lithosphere。
See *lithosphere.*

global atmospheric carbon dioxide (CO₂) 全球大氣的二氧化碳 *quanqiu daqide eryanghuatan*
由煤、木材和烴類的燃燒過程產生的二氧化碳。它也可通過微生物好氧氧化而產生。大氣中的二氧化碳是一個重要的組分，它與引起全球氣候變化的溫室效應有關。現在已獲得了許多有關二氧化碳濃度的資料，它們是基於在夏威夷的羅阿（Loa）山上進行的測定，以及對冰川和極地冰冠的冰中長期儲存的二氧化碳進行測定而得到的。在1800年二氧化碳的濃度約為280 ppm，而在1990年已增加到360 ppm左右。測定結果表明，目前大氣中的二氧化碳每年以1-2 ppm的速度在明顯加速地增加。

Carbon dioxide is resulting from all combustion processes of coal, wood and hydrocarbons. It is also produced through microbiological aerobic oxidation. CO_2 in the atmosphere is a critical component. It is of interest in the context of the *greenhouse effect*, which may be linked to global *climatic changes*. A long series of concentration values is today available. The values are based on measurements carried out on Mount Loa in Hawaii, as well as on measurements of CO_2 stored over long periods of time in the ice of glaciers and the polar ice caps. While in the year 1800, the concentration was about 280 ppm, in 1990 it had increased to about 360 ppm by volume. The measurements indicate at present a significant and accelerated increase of about 1 to 2 ppm each year.

grab sample 瞬時樣 *shunshiyang*

任意取的某個樣品（如廢水樣）。這種樣品不一定能代表某一時間內廢物流的平均組成。由瞬時樣得到的結果，其統計意義是有限的。

A single sample (e.g. of an effluent) taken at random. The sample does not necessarily represent the average composition of a waste stream over a specified period. The statistical significance of results from grab samples is limited.

gradient elution 梯度洗脫 *tidu xiduo*

色譜法中採用的一項特殊技術，其洗脫是在一定的變化條件下進行的。在氣相色譜法中，係按溫度梯度進行梯度洗脫，即慢慢提高分離柱的溫度，使得保留時間逐漸縮短。在逆相高壓液相色譜法中，通過形成溶劑和水的混合物來產生洗脫劑極性梯度而改變洗脫條件，從而導致系統的洗脫過程發生變化。

A special technique used in chromatography in which the elution is carried out under conditions which are changing in a defined way. In gas chromatography, gradient elution is carried out in a temperature gradient. That is, the temperature of the separation column increases slowly, resulting in a gradual reduction of the retention time. In reversed phase HPLC, the elution conditions are modified by generating mixtures of the solvent with water, resulting in a polarity gradient of the eluting agent. This again results in a modified elution behavior of the system.

granular activated carbon filter 粒狀活性碳過濾器 *lizhuang huoxingtan guolüqi*

用活性碳吸附氣態污染物(溶劑)的過濾器。(參見:活性碳、吸附塔)

A filter used for the adsorption of gaseous pollutants (solvents) on granular activated carbon. (See also *activated carbon*; *adsorption columns*)

granular activated carbon treatment (GACT) 粒狀活性碳處理 *lizhuang huoxingtan chuli*

用活性碳吸附去除廢水中物質的處理方法。處理時,將廢水通過吸附塔中的粒狀活性碳,使可被吸附物質得到去除。從經濟上考慮,粒狀活性碳必須進行再生以恢復其吸附性。(參見:穿透點、粉末活性碳處理)

A treatment for aqueous effluents containing mainly substances which can be adsorbed on activated carbon. The aqueous effluent is passed through granular activated carbon contained in an adsorption column, thereby adsorbable substances are removed. For economic reasons, the granular activated carbon must be re-generated to regain its adsorption activity. (See also *breakthrough point*; *powdered activated carbon treatment*)

grate incinerator 爐篦式焚燒爐 *lubishi fenshaolu*

一般用於焚燒城市垃圾的焚燒爐。廢物由鑄鐵或鋼製成的移動機械爐篦推進通過爐子。爐篦可由一個單獨的單元或由若干部件組成,朝燃燒室向下傾斜。它們有不同的設計型式可供選用。爐灰和燒渣通過爐篦落下而被除去。焚燒溫度約為700–800°C。現代的爐篦式焚燒爐具有完全燃燒的二次燃燒室、熱回收產生蒸汽的廢熱鍋爐和煙氣處理系統。但爐篦式焚燒爐不適用於焚燒*特殊廢物*,因為在爐灰中會偶然發現有未燃物質。

An incinerator design which is generally used for the incineration of municipal trash. The waste is propelled through the furnace by a moving mechanical grate made of cast iron or steel. The grate may consist of one single unit or several parts, all slanting downwards towards the combustion section. Various designs are used. Ashes and slag fall through the grate and are removed. The incineration temperature is about 700–800°C. Modern grate incinerators are equipped with a second combustion chamber for complete combustion, steam production for heat recovery and flue gas treatment systems. Grate incinerators are not suitable for the incineration of *special wastes* because some unburnt material is occasionally found in the ashes.

gravity settling box 重力沉降箱 *zhongli chenjiangxiang*

　　分離塵粒的最簡單的設備。當氣流通過沉降箱中低氣速區，重的塵粒可通過重力沉降去除。這種方法的處理效率不很高，只適用於處理大于100微米的粒子。

The simplest way to separate dust particles. The gas stream is passed into a zone of low gas velocity in a settling box. Heavy particles are removed by gravitational forces and settled out. The system is not very efficient and it is only applicable to particles larger than ca. 100 micrometers.

green products 綠色產品 *lüse chanpin*

　　用於指環境可接受的產品和消費品的一般性術語。它包括從工業產品到蔬菜範圍廣泛的不同產品。目前，對綠色產品還沒有給出明確的定義，也未標準化，因此消費者可能被誤導。在一些國家以及在國際上，正試圖對它給出明確的定義。

A general term used for the designation of products and consumer goods which are *environmentally acceptable*. The term is used for different kinds of products, ranging from industrial products to vegetables. The term is not well defined and not standardized. Consumers may therefore be misled. There are endeavors on national and international levels to issue clear definitions.

greenhouse effect 溫室效應 *wenshi xiaoying*

用以說明由大氣中微量污染物(*溫室氣體*)引起潛在氣候變化的一般術語。溫室氣體主要是二氧化碳和氯氟烴(氟里昂),它們會導致地球溫度升高,其作用被認為與花匠溫室中玻璃板或塑料膜的功能相似,即玻璃板或塑料膜可使光和紫外輻射透過進入到溫室,同時減少紅外熱輻射的逸散。地球大氣具有與溫室玻璃板或塑料膜類似的作用。雖然在過去幾十年裏大氣中二氧化碳的濃度有了明顯增加,但是它對地球變暖有多大影響的問題仍在爭論之中。有些簡化數學模式的計算結果預測在下一世紀地球將變暖大約1-2°C。然而,對這個問題在科學上還沒有取得一致的意見。

A general term used to describe potential *climatic changes* as a result of trace pollutants (*greenhouse gases*) in the *atmosphere*. These gases, mainly CO_2 and chloro fluoro carbons (FREONs), may lead to an increase of the global temperature. Their effect is thought to be similar to the function of glass plates or plastic sheeting in a gardener's greenhouse: light and UV radiation is permitted to enter the greenhouse, while the dissipation of infrared heat radiation is reduced. The atmosphere of the earth has a similar effect. While the atmospheric carbon dioxide concentration has clearly increased over the past decades, the question of the magnitude of global warming is still under debate. Some simplified mathematical model calculations predict a warming of ca. 1-2°C during the next century. Scientific consensus has, however, not been reached on this question.

greenhouse gases 溫室氣體 *wenshi qiti*

大氣中能讓短波輻射(可見光、紫外光)到達地面又能減少紅外長波輻射(熱輻射)逸散的微量氣體。這些氣體被認為是造成大氣潛在變暖(*溫室效應*)的原因。主要的溫室氣體是二氧化碳和氯氟烴(氟里昂),其次是甲烷和氮氧化物。但溫室效應的討論主要集中在二氧化碳和氟里昂上。大氣變暖的相對潛勢按下列順序增加:二氧化碳(0.00076),氟里昂22(0.36),氟里昂11(1.0),氟里昂12(3.1)。雖然二氧化碳的潛勢較小,但由於大量釋放造成大氣中二氧化碳濃度明顯增高,因此它對地球變暖的貢獻卻最大。

Trace gases in the atmosphere which let short wave length radiation (light, UV light) reach the surface of the earth, while the dissipation of longer wave length infrared radiation (heat radiation) is reduced. These gases are thought to be responsible for a potential warming of the atmosphere. (*greenhouse effect*) The main greenhouse gases are: carbon dioxide, fluoro chloro carbons (FREONs) and, to a lesser extent, methane and oxides of nitrogen. The discussion on greenhouse effect centers mainly on carbon dioxide and the FREONs. The relative potential for atmospheric warming is increasing from CO_2 (0.00076); to FREON 22 (0.36), FREON 11 (1.0), and FREON 12 (3.1). Despite the relatively low potential of carbon dioxide, its contribution to global warming is the greatest as a result of its concentration and the large quantities being set free.

grey energy 灰能 *huineng*

見：energy, grey。

See *energy, grey.*

grit 砂礫 *shali*

城市污水中攜帶的比水重的粗粒（砂、礫石和碎屑等），它們來源於街道等的排水。

Coarse material (sand, pebbles, debris, etc.) which is heavier than water and is carried along with the general sewage of a municipality. Grit originates from the drain-off from streets, etc.

grit removal 除砂礫 *chushali*

*廢水生物處理場*中最初段的處理過程。砂礫可以用較小的具有漏斗狀截面的沉降池去除。除砂池要設在廢水處理場廢水的入流處。

An initial process step in a *biological effluent treatment plant*. Grit can be removed by a relatively small sedimentation basin with a funnel-like cross section. This unit is placed at the inflow of the effluent plant.

groundwater 地下水 *dixiashui*

位於地表面下不同深度處於沙、砂礫或喀斯特石灰石滲透層中的水體。地下水體可以像地下河流那樣流動，也可能是相對靜止的。如石灰石岩洞中的地下水（稱為*喀斯特水*）就是可自由流動的。地下水體通常與地表水流是相連通的。在某些地方，也可能存在若干互相連通的地下水層。地下水是灌溉用水、飲用水和工業用水的極為重要的水源，但地下水體通過與之相連通的地面水而可能被地面污染物所污染，而且污染可在不同地下水層之間轉移，因此保持地下水不受污染是非常重要的。地下水的*恢復補救*行動，目的在於淨化污染源。（參見：超級基金）

Waterbodies which are located below the surface of the ground, at different levels, in a permeable stratum of sand, gravel or karstic limestone formations. Groundwater bodies may flow like subterranean streams, or they may be relatively still. Groundwater contained in cavities of rocky limestone may be free flowing; it is called *karstic water*. A groundwater body is generally connected to a surface water stream. At a given location, several groundwater bearing strata at different levels may exist. These strata may be connected with each other. Groundwater reserves are extremely important sources of water for irrigation, drinking and industrial use. Through connection with surface water, a groundwater body may become contaminated by pollutants from the surface. Contamination may be transmitted between different levels. It is of utmost importance to keep groundwater reserves uncontaminated. Groundwater *remediation* activities serve to decontaminate polluted sources. (See also *SUPERFUND*)

groundwater contamination 地下水污染 *dixiashui wuran*

地下水污染是人為污染物使地下水不適於使用要求而造成的。有許多不同的污染源和污染物類型，如混凝土污水管道的泄漏，地下貯槽的泄漏，廢水處理混凝土池的泄漏，廢物無保護填埋和廢物堆放產生的滲流水，來自污染生產場地的滲流水，以及過量施用可溶肥料等等。危險物品運輸或其他事故造成的*事故釋放*，往往也會造成地下水污染。（參見：超級基金、恢復補救）

The pollution of groundwater reserves through *anthropogenic* pollutants which make the groundwater unfit for the planned use. There are many sources and types of different pollutants: leaking concrete sewage pipes, leaking underground storage tanks, leaking concrete basins in effluent treatment, seepage water from unprotected landfills and waste dumps, seepage from contaminated industrial sites, excessive use of soluble fertilizers and many more. Groundwater contamination is often also the result of *accidental releases* from transports of critical goods or from other accidents. (See also *SUPERFUND*; *remediation*)

groundwater monitoring 地下水監測 *dixiashui jiance*

見：piezometer。

See *piezometer*.

groundwater protection zones 地下水保護區 *dixiashui baohuqu*

*城市區劃法*中規定的保護區，其根本目的是保護地下水水質以提供飲用水。通常有若干不同的保護區。其限制是比較嚴格的，涉及可能造成地下水污染的一些活動，如貯罐區、地下管道、危險物品的運輸以及其他類似的潛在污染源。

Zones defined in the context of *municipal zoning laws*. The basic objective is to protect groundwater quality to allow for the production of drinking water. Generally there are several different zones. Restrictions are more or less strict and cover those activities which have a potential for groundwater pollution, like tank storage, underground pipes, transport of critical goods and similar potential sources of pollution.

groundwater quality data 地下水水質數據 *dixiashui shuizhi shuju*

説明地下水水質的數據元。水質參數的選擇取決於*地下水*的類別以及它的使用。其數據還可用作為評估是否需採取*恢復補救*行動的決策依據。由於地下水的背景值會隨自然和人為的影響而有很大變化，因此確定在所有情況下都適用的地下水的總體質量目標是不可能的。

就用作飲用水來說，需優先考慮細菌學方面的數據。其他重要的參數包括外來物質（*異生化合物*）的類型與濃度、BOD_5、銨鹽、硝酸鹽、磷酸鹽、pH和電導率。世界衛生組織和歐盟已公佈了許多這類化學參數的限值。（參見：荷蘭限值表、干預值、恢復補救項目的管理）

Data elements which describe the quality of *groundwater*. The selection of quality parameters depends on the type of groundwater as well as on its planned use. Data is also used as a decision base to evaluate the need for *remediation* activities. General quality goals for groundwater covering all cases are not possible because background levels depend on widely varying natural and *anthropogenic* influences.

For use as drinking water, bacteriological data has priority. Important para-meters then include type and concentration of foreign substances (*xenobiotica*), BOD_5, ammonium salts, nitrate, phosphate, pH and conductivity. Limiting values for a number of these chemical parameters have been published e.g. by the World Health Organization and the European Union. (See also *Holland List*; *intervention values*; *remediation project management*)

groundwater remediation 地下水補救恢復措施 *dixiashui bujiuhuifu cuoshi*

消除地下水體污染的行動和措施，如果可能，則恢復到它原來的水質狀況。地下水的補救恢復包括現狀評價、防止進一步污染的技術與行政管理措施、減少或消除已有污染的技術措施。（參見：補救／恢復）

Activities and measures carried out to eliminate the pollution of a groundwater body and if possible to restore its original quality. Groundwater remediation includes an adequate assessment of the situation, technical and administrative measures to prevent further contamination, technical measures to reduce or remove the existing contamination. (See also *remediation*)

group parameter 綜合指標（參數）*zonghe zhibiao (canshu)*

在污染控制技術和分析中以及作為法律中規定的*限值*所使用的一類指標。它與一些不同類的化學物質的總量有關，可用來表徵廢水、特殊廢物或廢氣的性質，如化學需氧量、生物需氧量、總有機碳、揮發性有機碳、懸浮物、粉塵等。（參見：單一要素指標）

A type of parameter used in pollution control technology, analysis and as *limiting value* in legislation. Group parameters relate to the total quantity of several different chemical species. Group parameters are used to characterize effluents, special wastes, or air streams. Examples of group parameters are: COD, BOD_5, TOC, VOC, SS, dust. (See also *single element parameter*)

H

habitat 生境 *shengjing*

生態系統中種群生活的空間。（參見：群落生境）

The space for living of populations in ecosystems. (See also *biotope*)

Halon 哈龍 *halong*

用於消防滅火的一類 *氯氟烴物質* 的商品名。控制臭氧耗竭物質的 *蒙特利爾議定書* 對哈龍的使用規定了嚴格的限制。

A trade name for a group of chloro fluoro carbons (*CFCs)* used in fire fighting. The use of Halons is restricted by the *Montreal Protocol* on ozone depleting substances.

hazardous waste 危險廢物 *weixian feiwu*

在裝卸、運輸和處置中具有一定危害性的固體或濃稠廢物（污泥、泥漿、液體），但最好把它稱之為 *特殊廢物*。

Waste of any consistency (solid, sludge, slurry, liquid) which is associated with certain hazards in handling, transport and disposal. A preferable term is *special waste*.

headspace analysis 液上氣體分析 *yeshang qiti fenxi*

廢水或固體樣品中 *揮發性有機化合物*（VOC）的間接分析。在進行液上氣體分析時，將與液體或固體樣品處於平衡狀態的氣相的樣品體積（樣品瓶中的頂空）輸送到氣相色譜儀中，進行 *氣相色譜分析*。氣相中揮發性有機化合物的濃度與液體或固體中的濃度有關。用已知濃度進行校正，就能確定它們的確切關係。

The indirect analysis of volatile organic compounds (*VOC*) in an effluent or a solid sample. In headspace analysis, a sample volume of the gas phase in equilibrium with the liquid or solid sample (the 'head space' in a bottle containing the sample) is transferred into a gas chromatograph and subjected to *gas chromatography*. The concentration of VOC in the gas phase is related to the concentration in the liquid or the solid. Calibration procedures with known concentrations serve to define the exact relationship.

heat exchanger 熱交換器 *rejiaohuanqi*

將系統中熱能從熱態轉變為冷態的設備。在許多情況下，可將熱液流或氣流的廢熱加以回收，用以加熱待進一步加工處理的冷物料。例如，在高溫*濕式空氣氧化過程*和*選擇性催化還原去除氮氧化物過程*中，都要使用換熱器來回收反應熱。發電廠的鍋爐，如大建築物中熱能回收和再利用的設備一樣，本身就是一種換熱器。換熱器的設計在技術、介質和應用溫度上是很不相同的。鍋爐、蒸發器、節熱器、過熱器、冷凝器和冷卻器等，都是特殊設計的換熱器。

A unit which serves to transfer thermal energy in a system from a hot phase to a cold one. In many applications, waste heat from a stream of hot liquid or gas is recovered and used to heat a cold stream of material for further processing. As examples, heat exchangers are used in the high temperature *wet air oxidation* process and in the *SCR DENOX process* to recover the heat of reaction. Boilers in a power plant are heat exchangers, as are units applied in the recovery and reuse of heating energy in a large building. There is considerable variety in the design of heat exchangers for different technologies, media and temperature levels. Terms like boiler, evaporator, economizer, superheater, condenser, cooler and others are used and are all special designs of heat exchangers.

heavy metals 重金屬 *zhongjinshu*

在環境問題討論中常常使用而無明確、統一意義的術語。其定義尚未標準化，所涉及的金屬元素的範圍隨所討論的問題而定。在沒有進一步給與詳細說明的情況下，不應使用"重金屬"這一術語。一般往往將密度大於4.5的金屬包括在重金屬的範圍內，有汞、鎘、銅、鉻、鉛、釩、鋅和許多其他金屬。但是重金屬的毒性與密度並沒有明顯關係。另外，許多情況也表明，某些所謂的"重金屬"是生物系統中的重要元素。

A term often used in environmental discussions without a clear and uniform meaning. The definition is not standardized and the range of metallic elements depends on the question under discussion. The term 'heavy metals' should not be used without further specification. Often, metals with a density greater than 4.5 are included. The list of these metals includes mercury, cadmium, copper, chromium, lead, vanadium, zinc and many others. In the investigation of toxicity of heavy metals, no relation with the density is apparent. Furthermore, in many cases it is evident that certain so-called 'heavy metals' are essential elements in biological systems.

hedge hog filling for packed columns 用於填料塔的刺猬型填料 *yongyu tianliaotade ciweixing tianliao*

在*填料洗滌塔*中使用的一種特殊類型的填料。它是經特殊設計由聚乙烯材料製成的,大小像乒乓球或網球。這種填料為氣體與洗滌液之間的接觸提供了巨大的比表面,而*壓力降*卻很小。它在市場上有供應。

A special type of column filling for *packed column scrubbers*. Specially designed elements with the size of a ping-pong ball or a tennis ball are made of polyethylene. The hedge hog filling provides a high specific contact surface between the gas and the scrubbing liquid. The *pressure drop* with this type of column filling is low. Hedge hog filling is available commercially.

herbivore 食草動物 *shicao dongwu*

僅僅吃植物的動物。(參見:食肉動物)

A zoological species feeding exclusively on plants. (See also *carnivore*)

heterotrophic organisms 異養生物 *yiyang shengwu*

依靠有機化合物和營養物來生活和生存的一切生物,如人和動物。(參見:自養生物)

All organisms which depend on organic compounds and nutrients for living and sustainment e.g. man, and animals. (See also *autotrophic organisms*)

high load biology 高負荷生物處理系統 *gaofuhe shengwu chuli xitong*

負荷(*食物與生物量比*)等於或大於0.3的*活性污泥法*廢水處理系統。高負荷操作條件對廢水多級生物處理是有利的。

An *activated sludge* effluent treatment system with a *food to biomass ratio* (f/m ratio of about 0.3 or higher). High load operating conditions are advantageous in multi-stage biological treatment.

high pressure liquid chromatography (HPLC) 高壓液相色譜法 *gaoya yexiang sepufa*

在高壓下分離非揮發可溶性化合物的一種技術。這種分離是在色譜系統中進行的。現代的高壓液相色譜法係基於逆相系統進行工作,即首先塗覆固定相並用石蠟樣的物質來改性,以在它的表面形成一層具有憎水性的有機層。然後用含水的溶劑混合物進行分離和洗脫。利用壓力泵使樣品由洗脫溶劑輸送通過一個小直徑的柱子,由於物質具有不同的吸附性質,因而得到分離。

高壓液相色譜法的檢測段依分離物質的化學結構而不同。通常採用的方法是測定波長254毫微米的紫外光的吸收能力。

高壓液相色譜法可用於分析非揮發水溶性混合物中的化學物質;作為一種特殊的應用,它也可用於環境分析中作為富集段以定量分析地下水中的除草劑。此外,它還可應用於生物化學或人體健康調查研究所進行的某些分析中。

A separation technique under high pressure for non-volatile soluble compounds. The separation is carried out in a chromatographic system. Modern HPLC works with reversed phase systems. That is, the stationary phase is first coated and modified with a paraffin-like substance to provide an organic layer on its surface, which has hydrophobic properties. Separation and elution takes place with an aqueous solvent mixture. The sample is transported through a small diameter column by the eluting solvent, using a pressure pump. The substances are separated because of different adsorption properties.

The detection step in HPLC depends on the chemical structure of the separated substances. Absorption of UV light of a wavelength of 254 nm is frequently used.

HPLC is used for example to analyze chemical substances in mixtures of water soluble, non-volatile compounds. As a special application, HPLC is also used in environmental analysis in the enrichment step to quantify herbicides in groundwater. Applications are also in certain analyses for biochemical or human health investigations.

high pressure oxidation chamber 高壓氧化室 *gaoya yanghuashi*

濕式空氣氧化處理設施中的主要工藝設備，它具有泡罩塔結構，一般用耐腐蝕材質構成。

A processing unit, for instance the central processing unit in a *wet air oxidation* facility. The high pressure oxidation chamber may be constructed as a bubble column and generally is made of corrosion resistant materials.

high pressure wet air oxidation 高壓濕式空氣氧化 *gaoya shishi kongqi yanghua*

見：wet air oxidation, high pressure。

See *wet air oxidation, high pressure.*

high temperature combustion 高溫燃燒 *gaowen ranshao*

見：high temperature incineration。

See *high temperature incineration.*

high temperature incineration 高溫焚燒 *gaowen fenshao*

通過在高溫下焚燒來破壞有機污染物和廢物。這種方法可用於範圍廣泛的各種廢物流，包括有毒廢物。（參見：焚燒）

The destruction of organic contaminants and wastes through incineration at high temperature. This method is used for a wide range of waste streams, including toxic substances. (See also *incineration*)

Holland List 荷蘭限值表 *helan xianzhibiao*

表中列出了在土壤和地下水中一般會發現的許多污染物，同時還列出了對*恢復補救項目*管理的決策具有重要作用的*干預值*。此限值表最初是由荷蘭政府主管部門建議制定的。

A list of a number of commonly found contaminants in soil and groundwater, together with *intervention values* which are important in decision making in the management of *remediation projects*. The list was initially proposed by the Dutch authorities.

hollow fiber membrane 中空纖維膜 *zhongkong qianweimo*

　　用於逆滲透和氣體分離的中空纖維狀*半滲透膜*，其外徑約為100微米，材料為聚碸、聚醯胺和其他材料。中空纖維膜可用於脫鹽過程、氧和氮的分離。雖然它的價格較便宜，但膜的中空纖維對雜質很敏感。

A *semi-permeable membrane* in the form of a hollow fiber, used for reversed osmosis and also for gas separation. The outer diameter of such a fiber membrane is about 100 micrometers. Polysulfone, polyamide and other materials are used. Hollow fiber membranes are used for desalination processes and for the separation of oxygen and nitrogen. The price is relatively low, but membranes in hollow fiber geometry are sensitive to impurities.

HPLC 高壓液相色譜法 *gaoya yexiang sepufa*

　　見：high pressure liquid chromatography。

See *high pressure liquid chromatography*.

Husmann Test 哈斯曼試驗 *hasiman shiyan*

　　見：continuous measurement of biodegradability。

See *continuous measurement of biodegradability*.

hybrid technology incinerator 復合技術焚燒爐 *fuhe jishu fenshaolu*

　　通常由兩個或多個不同焚燒系統復合組成的焚燒爐，特別是在燃燒段由平爐與流化床組合而成的焚燒爐。廢物如污泥首先在焚燒爐上部的爐膛系統中進行乾燥與部分焚燒，然後掉入焚燒爐下部以砂子為介質的流化床中。起流化作用的空氣和輔助燃料通過設在下部的噴咀徑向噴入爐內。爐灰與餘砂一起被取出。另外，它還包括有熱回收和煙氣處理的輔助設備。這種類型的焚燒爐具有多爐膛焚燒爐(乾燥段)和高效流化床焚燒爐二者的優點，但必須小心避免污泥中存在無機鹽，因為它會導致形成低共熔鹽的混合物，從而會使流化床中的砂粒熔融成團塊。除了這一可能產生的問題之外，它是處理廢水污泥的一種有效的焚燒爐。

In general, an incinerator design combining two or more different incineration systems. In particular, an incinerator combining in the combustion section an open hearth part with a fluidized bed part. Waste, e.g. waste sludge, is first dried and partially incinerated in the hearth system in the upper part of the incinerator. It then falls into the lower part of the incinerator. This section is designed as a fluidized bed with sand. Air for fluidization and additional support fuel is introduced radially through nozzles at this level. Ash is taken out with the excess sand. Additional process units are included for heat recovery and flue gas treatment. This incinerator design combines the advantage of the multiple hearth incinerator (drying section) with the higher efficiency of a fluidized bed system. Care must be taken to avoid inorganic salts in the sludge which may lead to eutectic mixtures of salts and, as a result thereof, to a melting-together of the individual sand particles in the fluidized bed. Apart from this potential problem, this is an efficient incinerator for effluent sludges.

hydrosphere 水圈 *shuiquan*

地球*生態圈*的三個主要組成部分之一。水大約佔地球表面的74%，主要以海洋、河流、淡水湖、地下水和極地冰的形式存在。水圈包括全部水生*生態系統*，它又可進一步分類為*湖沼生態系統*（淡水湖、池沼與河流的生態系統）和*海洋生態系統*。

One of the three main compartments of the global *ecosphere*. Water covers about 74% of the surface of the earth. It is present mainly in the oceans, rivers, sweetwater lakes, groundwater and as polar ice. The hydrosphere includes all of the aquatic *ecosystems*. These in turn can be grouped into *limnic ecosystems* (ecosystems of sweetwater lakes, ponds and rivers) and *marine ecosystems*.

hygienization 衛生化處理 *weishenghua chuli*

將物品如食品、衣服或生活居室進行衛生處理以供人們安全使用，但衛生處理並不一定意味是滅菌消毒。（參見：滅菌消毒）

The process of rendering a material safe for human use, e.g. food, clothing or living quarters. Hygienic does not necessarily mean sterile. (See also *sterilization*)

hygienization, thermal, of activated sludge 活性污泥的熱衛生處理 *huoxing wunide reweisheng chuli*

對於在農田上進行處置的剩餘活性污泥，有關法規可能要求對它進

行衛生處理。這種處理過程有時也被錯誤地稱為"滅菌消毒"。對於衛生
處理來說,如果將污泥加熱到70–80°C維持至少15–20分種,一般就足夠
了。衛生處理的設施通常是連續操作並具有熱交換設備的裝置。所需投
入的熱能,往往是由現場污泥發酵產生的沼氣產生的。

Hygienization of excess activated sludge may be demanded by legislation for activated sludge which is to be disposed on agricultural land. The process is sometimes also wrongly called 'sterilization'. Hygienization is generally sufficient if the sludge is heated to 70–80°C for at least 15–20 minutes. Facilities for hygienization generally are continuous flow through systems using *heat exchanger* units. Often, the necessary heat input is generated on site from biogas from sludge fermentation.

I

IC-50 半抑制濃度 *banyizhi nongdu*

見：inhibitory concentration (activated sludge)。

See *inhibitory concentration (activated sludge)*.

ICC 國際商會 *guoji shanghui*

International Chamber of Commerce的縮寫。其總部設在巴黎。國際商會已對工業界頒佈了許多政策文件、環境管理和審計規範。

International Chamber of Commerce, with administrative headquarters in Paris. The ICC has issued a number of policy documents and environmental management and auditing specifications for industrial application.

ICP-MS 感應耦合等離子體質譜分析 *ganying ouhe dengliziti zhipu fenxi*

見：inductively coupled plasma mass spectrometry。

See *inductively coupled plasma mass spectrometry*.

identification step in analysis 分析中的鑒定段 *fenxizhongde jiandingduan*

未知混合物的分析通常包括四個主要階段：將樣品注入儀器中，用若干分離技術中的一種方法進行分離(分離段)、信號的產生和檢測、信號的解析、以及未知化合物的鑒別。鑒別隨化合物的類型而不同，一般以標準的物理和化學方法為基礎。

在氣相色譜法和液相色譜法中，分離後的大部分組分可以通過將其保留時間和洗脫特徵與已知物質進行比較來加以鑒別。在較複雜的*氣相色譜-質譜法*中，檢測段是直接與鑒定段(如與*質譜*)結合在一起的。在*紅外分析*中，是基於與已知物質的光譜(可由某些數據庫中獲得)相比較來鑒別的。

The analysis of mixtures of unknown substances usually includes four main steps: sample introduction into the instrument; separation step, carried out by one of several separation techniques; production and detection of a signal; and interpretation of the signal and identification of the unknown compound. The identification depends on the type of compound. It is based generally on standard physical and chemical methods.

In gas chromatography and liquid chromatography, most of the components after separation are identified by comparing the retention times and the elution behavior with known substances. In more sophisticated methods, such as *GC-MS*, the detection step is coupled directly with the identification step, e.g. with *MS*. In *infrared analysis*, the identification is based on a comparison with spectra of known substances, available in certain databases.

immaterial goods 無形商品 *wuxing shangpin*

無形價值的同義詞。

A synonym for *immaterial values*.

immaterial values 無形價值 *wuxing jiazhi*

一般不能作為商品來生產和貿易的東西和價值，但是它對於定義和評估*生活質量*是很重要的，如文化、潔淨的大自然、美學、沒有緊張壓力、無噪聲、社會關係等。無形商品在環境*費用效益*的考慮中具有重要作用。然而，對無形商品的賦價在很大程度上取決於個人的喜愛和文化差異。

Goods and values which generally can not be manufactured and traded as merchandise, but which are important in the definition and appreciation of the term *quality of life*. Culture, clean nature, esthetics, absence of stress, absence of noise, social relations, etc. can be taken as examples. Immaterial goods play an important role in environmental *cost-benefit* considerations. However, the assignments of costs to immaterial values is strongly based on personal preferences and on cultural differences.

immission 排放 *paifang*

德語中有時使用的與*排放*(emission)一詞有關的術語。它係指作用於實驗對象的某種污染物的量，例如：會影響人類健康的環境空氣中*臭氧*的含量。"排放濃度"一詞比較正確的英語用語應為"環境空氣濃度"。

A term sometimes adopted from the German language related to *emission*.

Immission is specifying the quantity of a pollutant which acts upon a subject, e.g. the quantity of *ozone* in the ambient air which has an impact on human health. The more correct English term for 'immission concentration' is 'ambient air concentration'.

immission limits 排放限值 *paifang xianzhi*

有時用於*環境空氣濃度限值*或*環境空氣目標*的一個不正確的術語。

An incorrect term used sometimes for *ambient air concentration* limits or *ambient air goals*.

immobilization of wastes 廢物固定化 *feiwu gudinghua*

見：wastes, immobilized。

See *wastes, immobilized.*

immobilized wastes 固定化廢物 *gudinghua feiwu*

見：wastes, immobilized。

See *wastes, immobilized.*

immunoassay analysis 免疫測定法分析 *mianyi cedingfa fenxi*

應用免疫化學反應和原理的一類分析方法。它已在醫學分析中應用了相當時間。它在環境領域中的應用是基於某些生物活性污染物，如土壤中的除草劑，會干擾免疫系統的反應這一事實。已建議將免疫測定法分析作為一種現場測定分析方法應用在土壤恢復補救項目或水污染控制中。據說這種方法比傳統的取樣和分析方法快，但是對其結果的解釋需要有豐富的經驗。在市場上已有專門成套的測定箱供應。

A group of analytical methods which are based on the application of immuno-chemical reactions and principles. Immunoassays have been applied in medical analyses for some time. Applications in the environmental field are based on the fact that certain biologically active pollutants, e.g. herbicides in soils, interfere with reactions of an immune system. The application of immunoassay analysis has been proposed as a field method in soil remediation projects or in water pollution control. The method is said to be faster than conventional sampling and analysis but the interpretation of the results needs much experience. Specific kits are available commercially.

impact assessment and analysis 影響評價和分析 *yingxiang pingjia he fenxi*

見：environmental impact assessment。

See *environmental impact assessment.*

impact of an accident 事故影響 *shigu yingxiang*

某一事故對直接當事人、一般居民和環境造成的負面影響。通過*事故遏制措施*可使潛在嚴重的影響減少到較可接受的程度。

The negative influence of an accident on the persons directly involved, the population in general, and the environment. A potentially large impact can be reduced to a more acceptable level through *containment measures.*

impact of sound 聲音的影響 *shengyinde yingxiang*

見：sound, impact of。

See *sound, impact of.*

impact on ecosystems 對生態系統的影響 *duishengtai xitongde yingxiang*

各種類型廢物的排放對生態系統不同組成部分的影響。這種影響是由若干因子造成的，其中包括：排放的類型、*靶生物*的類別、地點和時間。在最簡單的情況下，這種影響是排放點和影響點之間直接相互作用的結果，在時間上無任何滯後。例如，有毒物質的排放對河流中魚產生的即刻影響。相反，有的情況表明，排放的影響只有在遲滯相當的時間後，並在離排放點一定距離的地方才會顯現出來。以森林破壞為例，早期排出的不同類型的排放物會使遠離排放源的森林遭受毀壞，但它還會受氣候和風向等複雜因子的影響。

The influence of emissions of all types upon different components of an ecosystem. This impact is the result of several factors, amongst them: type of emission, type of *target organism*, location and time. In the simplest case, the impact is the result of a direct interaction between point of emission and point of impact and without time delay. Example: the immediate impact of a toxic discharge on the fish

205

in a river. On the other end of the scale, there is the case of an emission which shows an effect only after a considerable delay and at a distance from the emission point. As an example we may look at the damages to forests, resulting from different types of emissions at an earlier time, at a distance away from the source, and influenced by complicating factors such as the climate and wind directions.

impingement separator 撞擊式分離器 *zhuangjishi fenliqi*

用於廢氣處理去除液滴、煙霧和塵粒的一種分離器。處理時，將污染廢氣射向撞擊板，由於氣流轉向而使液滴或塵粒凝結和分離。它單獨的分離效率是較低的，因此通常需與其他的空氣污染控制設備一起使用。(參見：空氣污染控制設備)

A type of separator used in off-gas treatment to remove droplets, mist and dust particles. The polluted off-gas stream is directed against impingement plates. It is then deflected and droplets or particles are aggregated and separated. The separation efficiency of this element alone is relatively low. It is generally used in connection with other APC process steps. (See also *air pollution control equipment*)

in-plant measures 廠內措施 *changnei cuoshi*

在工廠內和在工廠管理部門負責下實施的措施。

Measures carried out within a factory and under the responsibility of the factory management.

in-process measures 工藝措施 *gongyi cuoshi*

污染控制措施上通常應用的術語，即作為生產過程的一個部分，通過工藝過程的優化來控制污染。(參見：清潔工藝)

A term generally applied to pollution control measures applied as part of the production process, by an optimization of the process. (See also *clean technologies*)

in-site remediation 現場恢復(補救) *xianchang huifu (bujiu)*

此術語主要用於對被污染的土壤和地下水所進行的 *恢復補救*。它意味著在現場進行處理，而不是把土壤挖掘走或用泵把地下水打到別處進行處理。

A term mainly used in *remediation* of contaminated soil and groundwater. It implies in-place treatment, without excavation of the soil, or pumping of the groundwater for external treatment.

incentive systems 刺激 (獎勵) 制度 *ciji (jiangli) zhidu*

利用獎勵來激發職工做出工作成績的管理制度。它可以獎勵個人，也可以獎勵集體；可以發給獎金，也可以其他形式進行獎勵。

Management systems using incentives to motivate employees to achieve the desired results. Incentives may be made to individual persons or to organizational groups. Incentives may be in form of payments or in any other form.

incentives in environmental management 環境管理中的刺激手段 *huanjin guanlizhongde ciji shouduan*

潛在的污染者因避免、減輕或完全消除了污染而獲得的經濟回報 (收益、利潤)，其形式可能包括減少*收費*，減稅，有時也可給與*補貼*。

在*綜合環境管理*中，對於個人在環境事務中取得的特殊成績，也可對職工個人予以獎勵。(參見：人事激勵制度)

Economic rewards (gains, profits) accruing to a potential polluter for avoided, reduced, or totally eliminated pollution. Such rewards may include reduced *levies*, tax reductions and in some cases payment of *subsidies*.

In the context of an *integral environmental management* approach, incentives can also be given to individual employees for special personal achievements in environmental matters. (See also *personnel incentive systems*)

incineration 焚燒 *fenshao*

按照字義，指化為灰。焚燒是在空氣中燃燒有機廢物的過程。焚燒過程產生的揮發性產物存在於*煙氣*中，一些無機元素如汞、鋅和鎘是部分揮發性的，其他的無機鹽和元素則成為燒渣和灰。因此，特殊廢物焚燒產生的煙氣，在排入大氣前必須進行有效處理。(參見：燃燒)

Literally, to reduce to cinders or ashes. The process of burning organic wastes in air. The volatile products of the incineration process are found in the *flue gas*. Some inorganic elements, like e.g. mercury, zinc, and cadmium are partially volatilized. Other inorganic salts and elements remain as slag and ashes. The flue gas from the incineration of special wastes therefore has to undergo an efficient flue gas treatment before discharge into the atmosphere. For the more general term see *combustion*.

incineration chamber 焚燒室 *fenshaoshi*

焚燒爐的一個組成部分，焚燒過程就是在其中進行的。不同類型的焚燒爐有著不同的構造原理。（參見：焚燒爐）

The unit of an *incinerator*, in which the *incineration* process takes place. Depending on the type of incinerator, various construction principles are used. (See also *incinerator*)

incinerator 焚燒爐 *fenshaolu*

工業規模物料焚燒所採用的技術設備。在污染控制中，它被用於焚燒廢物。其主要部件包括：廢物進料系統、焚燒室、輔助燃料點火設備、節熱設備（如鍋爐）、煙氣處理系統（包括洗滌液處理）、除無機燒渣或灰的機械設備、以及必要的過程控制系統和設備（包括煙氣監測設備）。焚燒爐有各種不同型式。（參見：冷壁式焚燒爐、流化床焚燒爐、爐箅式焚燒爐、複合技術焚燒爐、平爐式焚燒爐、回轉窯焚燒爐、固定爐室焚燒爐）

A technical facility used for large scale incineration of materials. In pollution control, incinerators are used for the incineration of wastes. The main parts are: a waste feeding system; an incineration chamber; support fuel firing; heat economizing units, e.g. boilers; flue gas treatment systems with associated treatment of the scrubbing liquors; mechanisms for removing inorganic slag and ashes; disposal facilities for slag and ashes; the necessary process control system and units, including monitoring equipment for flue gas.

For individual types see *cold wall incinerator*; *fluidized bed incinerator*; *grate incinerator*; *hybrid technology incinerator*; *open hearth incinerator*; *rotary kiln incinerator*; *static chamber incinerator*.

incinerator, cold wall 冷壁式焚燒爐 *lengbishi fenshaolu*

　　一種在設計上直接進行熱交換而具有最佳熱交換性能的焚燒爐，其燃燒室不襯耐火磚，而是鋼製的雙壁結構，可直接把熱量傳遞給水或熱傳導油等熱傳導液體。通常蒸汽鍋爐就是以這種方式設計的，但它也可應用於處理廢溶劑的焚燒爐中，以*回收能量*。這種焚燒爐的材料必須耐腐蝕，並且還必須對要進行焚燒處理的廢物加以選擇以避免腐蝕問題。

A type of incinerator designed directly for optimal heat exchange. The combustion chamber is not lined with fire brick. The double walled construction made from steel leads to a direct heat transfer to the heat transfer liquid which is either water or a heat transfer oil. Boilers for steam generation are generally designed in this way, but the design has also been applied to incinerators for waste solvents which then serve as an energy source (*energy recuperation*). The material of the incineration chamber has to be resistant to corrosion and any wastes incinerated must be selected with this problem in mind.

indirect costs 間接費用 *jianjie feiyong*

　　與產量不直接有關的產品生產費用，包括諸如產品的研究開發、生產管理、生產設備和基礎設施的借貸基建費用、房屋建築、人工或部分人工、質量控制、產品特定的環境控制和其他的費用要素。間接費用可按照某種分配原則分攤到*產品生產總成本*中。（參見：成本計算方法）

Those manufacturing costs of a product which are not directly related to the quantity of production. Indirect costs include for instance research and development of products, management of production, cost of borrowing capital for production equipment and infrastructure, general services of a building, manpower or a fraction thereof, quality control, product specific environmental controls and other cost elements. Indirect costs can be allocated to the *true total production costs* on the basis of a distribution key. (See also *cost calculation procedure*)

indirect pollution control costs 間接的污染控制費用 *jianjiede wuran kongzhi feiyong*

　　工廠中污染控制的費用要素，它與產品的生產產量或銷售量無直接關係。例如，環境管理、環境服務及其實驗室的運作、對環境法規要求

相符合情況的監控、環境數據管理等方面的費用。污染控制設施的運行
費用,部分為直接費用,部分為間接費用。

Elements of pollution control costs in a factory which do not depend directly
on the sales- or production volume. Examples are: environmental management, the
operation of the environmental service and its laboratories, monitoring of
environmental compliance, management of environmental data, etc. The operation
of pollution control facilities is in part direct and in part indirect.

inductively coupled plasma mass spectrometry (ICP-MS) 感應耦合等離子體質
譜分析 *ganying ouhe dengliziti zhipu fenxi*

用於水樣中多元素分析的一種儀器方法。在感應耦合等離子體質譜
法中,氬等離子體在溫度9000–10000°K下生成質譜系統所需的離子。這
種技術使得可以測定的元素超過90%,而且在濃度極低(1–100毫微克/
升)的情況下不需進行富集。

An instrumental method which was developed for multi-element analysis of
aqueous samples. In ICP-MS, an argon plasma at temperatures between 9,000–
10,000° Kelvin serves to produce the ions for the MS system. This technique makes
possible the determination of over 90% of the elements at extremely low concentrations without any enrichment step, in certain cases as low as 1 to 100 nanogram/
liter.

industrial effluent 工業廢水 *gongye feishui*

含有從工業生產過程中產生的 *污染物* 的廢水。一般來說,工業廢水
的處理方法與 *城市污水* 不同。

An aqueous effluent characterized to a large fraction by *pollutants* resulting
from industrial production processes. Industrial effluents normally need different
treatment technologies than *municipal sewage*.

industrial effluent treatment plant 工業廢水處理場 *gongye feishui chulichang*

特別為工業廢水末端處理而設計的廢水處理場。在某些情況下,對
於特定的污染物也可採用 *末端工藝過程* 處理。如果採用 *活性污泥法* 處
理,則必須考慮適應於廢水生物降解性較差的條件,因此廢水需要在曝

氣池中停留較長時間。另外，由於廢水流量變化不定，所以還需考慮設置均衡池以提高處理效率。

An effluent treatment plant designed specifically for the *end-of-pipe* treatment of industrial effluents. In some cases, pollutant-specific *end-of-process* treatments are used. If the treatment is based on *activated sludge technology*, then the plant must be adapted to the generally lower degree of *biodegradability* of the effluent. For this, longer aeration times are needed in the *aeration basins*. Because of the variable flow rates, *equalization basins* may also enhance the efficiency.

industrial hygiene 工業衛生 *gongye weisheng*

見：occupational health。

See *occupational health*.

infrared (IR) analysis, non-dispersive 非色散紅外分析 *feisesan hongwai fenxi*

基於氣流吸收特定紅外波長的分析，它所選用的波長依測定的化合物而定。非色散紅外分析可用於連續定量測定煙氣中的二氧化碳、一氧化碳和氮氧化物。與紅外光譜分析不同，它不測定光譜。

Analysis based on the absorption of specific wavelengths of IR by a gas stream. The wavelength depends on the compound to be measured. Non-dispersive IR analysis is applied in the quantitative continuous determination of CO_2, CO and NO_X in flue gas. In contrast to IR spectral analysis, no spectrum is measured.

infrared (IR) spectral analysis 紅外光譜分析 *hongwai guangpu fenxi*

由於紅外輻射與分子振動相互作用，有機物分子對紅外光有一定的吸收譜帶。實際上，一個吸收光譜包含有許多不同的譜帶和在不同波長處的峰值。紅外光譜學是通過紅外區的各個吸收光譜來區別物質的儀器分析方法。紅外分析係基於紅外光譜作出的定性和半定量的描述。

為便於比較和鑒別，可利用已有的紅外光譜數據庫。在環境領域，可應用紅外光譜法來分析許多不同物質。對於汽油或礦物油造成的污染，採用紅外光譜分析是非常有效的。這類混合碳氫化合物具有典型的紅外吸收光譜。（參見：分析中的鑒定段）

Organic molecules show well-defined absorption bands for infrared light as a

211

result of the interaction of IR radiation with molecular vibrations. In practice, an absorption spectrum contains a large number of different bands and peaks at different wavelengths. Infrared spectroscopy is an instrumental technique which is capable of differentiating substances by their individual absorption spectra in the infrared range. IR analysis is based on the qualitative or semi-quantitative interpretation of IR spectra.

For comparison, there are large data bases of IR spectra to facilitate identification. In the environmental field, infrared spectroscopy is used in the analysis of many different substances. It may help in cases of pollution with gasoline or mineral oil as contaminants. Such hydrocarbon mixtures have typical IR absorption spectra. (See also *identification step in analysis*)

inhibition of biodegradation 生物降解的抑制 *shengwu jiangjiede yizhi*

廢水處理中某些生物毒性物質對微生物生物降解過程產生的不良作用。某物質的抑制作用可以通過在不同濃度條件下進行的 BOD(生物需氧量)測試來評估,如果某物質在高濃度時生物降解速率較低,則該物質具有抑制作用。

The negative effect of certain biotoxic substances on the microbiological biodegradation processes in effluent treatment. The inhibition effect of a substance can be evaluated and assessed with a BOD test carried out at different concentrations. A substance has an inhibitory effect if its rate of biodegradation is lower at high concentrations.

inhibitory concentration (activated sludge), IC-50 抑制濃度(活性污泥) *yizhi nongdu (huoxing wuni)*

某一物質使活性污泥中微生物的活性50%受到抑制時的濃度。在好氧活性污泥系統和污泥厭氧發酵過程中都可能產生抑制作用。抑制濃度可在添加和不添加某試驗物質時測定污泥的氧耗量來確定。

The concentration of a substance which inhibits the activity of microorganisms in activated sludge by 50%. Inhibition may take place in the aerobic *activated sludge* system and also in the anaerobic *fermentation* process of sludge. The inhibitory concentration is determined by measuring the oxygen consumption of sludge with and without added test substance.

injector 噴射器 *pensheqi*

曝氣系統的一個組成部分，用以將空氣或氧在水面下噴入污泥混合液中。

Element of an aeration system to inject air or oxygen into the mixed liquor below the surface.

inoculation 接種 *jiezhong*

用微生物在生物反應器中進行初始引種，然後微生物再增長和繁殖到所需要的平衡濃度。

The initial seeding of a bioreactor with active microorganisms which then grow and multiply to the desired equilibrium density or concentration.

instrumental methods of analysis 儀器分析方法 *yiqi fenxi fangfa*

主要基於應用較複雜的儀器進行分析的方法。儀器分析是在1850年進行首次光譜分析後才逐漸發展起來的。現在，大多數分析方法都以不同方式應用儀器來進行分析，傳統的*化學濕分析法*已逐漸被儀器分析所代替。有各種不同的儀器分析方法，但所有的方法都包括了四個基本單元：首先，進行樣品處理以產生能被檢測的信號（如*質譜法*中的離子化和裂解作用）；其次，用分離單元進行信號處理（如光譜法中波長的分離，質譜法中根據其質量進行離子的分離等等）；再次，用專門的設備對信號進行檢測（如質譜法中碎片質量的鑒別）；最後，進行信號的加工處理以得到可理解的信號。

在以上這一基本概念內，有許多方法可供採用，並且正在不斷地發展新的方法。目前它正朝著更準確、更靈敏的方向發展。儀器分析是一種自動和快速的分析方法，大多包括有數據處理單元。它可以測定一種以上的元素。但是，現代化的儀器需要高水平的培訓，而且其價格可能是很昂貴的。

Methods of analysis which are based mainly on the use of more or less complex instruments. Instrumental analysis has developed gradually since 1850, when the first spectral analyses were carried out. Today, most analytical methods include instruments in one way or another. Traditional *chemical wet analyses* are gradually replaced by instrumental methods.

Instrumental methods are very diverse. All methods, however include four basic modules. First, the treatment of the sample in such a way that a signal is produced which can be detected and measured (e.g. ionization and fragmentation in *mass spectrometry*). Second, the working up of the signal, e.g. by a dispersive element (such as separation of wavelengths in spectroscopy, separation of ions according to their mass in mass spectrometry, etc.). Third, the detection of the signal by a specific device (in mass spectrometry, the identification of the mass of a fragment) and lastly, the processing of the signal to an understandable information. Within this basic concept, many methods are used and new ones are constantly being developed.

Instrumental analysis today allows for more accurate and more sensitive analyses. The methods are automatic and more rapid and many have integrated data processing units. The measurement of more than one element is possible. The other side of this development is the fact that modern instruments require high level training and may be quite expensive.

integral environmental management 綜合環境管理 *zonghe huanjing guanli*

把有關環境保護的所有重要問題納入到管理體系中的管理方法。其基本概念包括：污染預防、節約資源、有效的污染控制技術和有效的管理。(參見：環境保護的管理手段)

A management approach integrating all significant aspects relating to environmental protection in the management system. The cornerstones of the underlying concept include: prevention, resource economy, efficient pollution control technology and efficient management. (See also *management instruments in environmental protection*)

integral process development 完整的工藝過程開發 *wanzhengde gongyi guocheng kaifa*

把環境和能源問題與產品質量、生產效率和價格等因素一起納入到生產過程的開發和改進中。其目的是在源頭消除不要的副產品或廢物（因此也稱之為"源削減"），或者把它們變為問題較少的產品。完整的工藝過程開發是可持續發展的最重要內容之一，它要求樹立總體觀念，防止解決一個問題又產生一個新的問題。它是*清潔工藝*的一個重要組成部分。

An activity which serves to integrate environmental and energy aspects in the development and improvement of production processes together with quality, efficiency and cost optimization. The objective is to eliminate unwanted side products or wastes as they are formed at the source (therefore also called 'source reduction') or to change them to less problematic products. Integral process development is one of the most important aspects to gain sustainability. It demands an overall perspective and prevents from solving a problem by generating a new one. It is an important element of *clean technologies*.

inter-laboratory tests 實驗室間的相互測試 *shiyansijiande xianghu ceshi*

在環境分析中往往很難相互比較分析結果，除了分析方法本身的誤差外，測定誤差可能來自樣品本身以及樣品的處理和貯存。

實驗室間的相互比較，可以起到檢驗方法和實驗室工作可靠性的作用。在進行比較時，需製備相同的樣品，分發給參與比較的實驗室，並用統一的標準方法進行分析。在測試開始時，必須向參與的實驗室分發明確的工作指南，通常由一個委員會負責監督各個階段的測試工作，包括樣品的製備、所用方法的等效性以及結果的評價和解釋。

準備在環境分析方面進行合格認證的分析實驗室，必須參加一次或多次實驗室間的相互測試。

In environmental analysis, it is often difficult to compare analytical results with each other. Errors of measurements may be due to the samples themselves, to sample handling and storage, in addition to errors of the analytical method itself.

Inter-laboratory comparisons serve to check on the reliability of a method and of a laboratory. In such a comparison, identical samples are prepared and given to participating laboratories. The samples are analyzed by the same standardized method. The participating laboratories have to be clearly instructed at the beginning of such a test. A committee will normally be responsible for the supervision of all steps of the test, beginning with the sample preparation, the equivalence of the methods used, and the evaluation and interpretation of the results.

Analytical laboratories who intend to be certified for environmental analyses are obliged to participate in one or more inter-laboratory tests.

interconnected databases 互聯數據庫 *hulian shujuku*

　　互聯數據庫是涉及不同問題和專題的幾個數據庫，其設計方式使得一個數據庫的數據可以直接與其他數據庫的數據相聯。例如，將企業數據庫與排放數據庫以及與生態系統地理分佈數據庫互聯，可以提供某一地區污染負荷的信息。（參見：數據管理）

Databases, covering different aspects and topics, which are designed in such a way that data from one database can be correlated or linked directly with data from other databases. Examples: the interconnection of a database on enterprises with a database on emissions with a database on the geographical distribution of eco-systems may provide information of the pollution load in a certain area. (See also *data management*)

internal costs 內部費用 *neibu feiyong*

　　見：costs, internal。

See *costs, internal*.

internal polluter-pays-principle 內部的污染者付費原則 *neibude wuranzhe fufei yuanze*

　　將與產品生產有關的污染控制費用分攤到各個產品的生產成本中的基本管理原則。採用這一原則，必須確定生產過程產生的污染控制費用，並把它納入內部成本計算體系中。

The basic management principle of allocating product specific pollution control costs to the production costs of individual products. For the application of this principle, pollution control costs arising in the manufacturing process must be identified and measured and be integrated in the internal cost calculation system.

internalizing external costs 內部化的外部費用 *neibuhuade waibu feiyong*

　　一種費用分攤機制，其目的在於將*外部費用*納入到污染源產品的成本結構中。通過內部化，對所謂的"*免費物品*"賦以價格，將使經濟因素和機制在減少污染中發揮更大的作用。因此，產品價格應包括這些外部費用要素，並真實地反映全部的成本。（參見：污染控制費用的分攤）

A mechanism of cost allocation with the objective of integrating *external costs* into the cost structure of the products which are the sources of pollution. Through this internalization, so called '*free goods*' will be given a price, and economic factors and mechanisms will play a greater role in the reduction of pollution. Product prices should then include these external cost elements and truly reflect all costs. (See also *allocation of pollution control costs*)

International Standards Organization (ISO) 國際標準組織 *guoji biaozhun zuzhi*

　　總部設在日內瓦的一個國際組織，其任務是制定和公佈各行業的技術標準。國際標準組織通過它的委員會、小組委員會和工作組，並且與其他機構和各國標準組織密切合作與磋商而進行工作。（參見：國際標準組織14000標準系列）

An international organization with administrative headquarters in Geneva, Switzerland. ISO is active in the field of developing and publishing norms for many technical fields and applications. ISO works through its committees, subgroups and task forces in close cooperation and consultation with other bodies and with similar normative organizations on the national level. (See also *ISO 14000 Norm series*)

international water management 國際水域管理 *guoji shuiyu guanli*

　　見：water management basin。

See *water management basin*.

intervention values 干預值 *ganyuzhi*

　　地下水和土壤中污染物的濃度值，它們作為限值用在*恢復補救管理*中。當發現污染物的濃度超過這些限值時，必須採取恢復補救行動。例如，在*荷蘭限值表*中就列有干預值。

Concentration values for contaminants in groundwater and in soil. The values are used in *remediation* management as limiting values. When concentrations exceeding these limits are found, remediation activities have to be undertaken. Intervention values are published for instance in the *Holland List*.

inventory, environmental 環境清單 *huanjing qingdan*

在環境保護方面對與污染有關的因素，如污染源、影響、已污染的生態系統、已污染的老填埋場、擬保護的景觀等列出的綜合與全面的數據表。

In the context of environmental protection, any comprehensive and complete listing of data elements related to pollution: e.g. pollution sources, effects, polluted ecosystems, old contaminated landfills, landscapes to be protected, etc.

inventory of emissions of a factory 工廠排放清單 *gongchang paifang qingdan*

包括工廠排放污染物的定性和定量數據在內的所有的排放源的資料或數據。（參見：點源、擴散源）

A documentation or database including all sources of emissions of the site, including qualitative and quantitative data on emitted pollutants. (See also *point sources*; *diffuse sources*)

investment costs 投資費用 *touzi feiyong*

提供經營條件所需的開支，它可能包括房屋建築和基礎設施、新設備、已有設備更新所需的加工設備以及污染控制設施的投資。

The expenditures needed for providing the means to conduct business. Investments may include buildings and infrastructure, new equipment, processing units for the renewal of existing equipment, and also pollution control facilities.

investment costs, environmental protection 環境保護投資費用 *huanjing baohu touzi feiyong*

投資在污染控制技術上的費用。除了*末端處理設備*外，還包括用在工藝改革設備的投資，但有時很難定義它們的界限，因為現代化的生產設備往往同時會減少污染。所以，它同時可起到雙重目的的作用。

The costs for investments into pollution control technology. In addition to units for *end-of-pipe treatment*, investments into process dedicated equipment are also included. The definition of borderline cases is sometimes difficult, because modern production equipment frequently leads to lower pollution at the same time. It may therefore serve two purposes at the same time.

ion analysis 離子分析 *lizi fenxi*

　　與水體中含鹽量問題有關的無機離子的分析。它可用*離子選擇(特效性)電極*通過直接電位測定進行分析，如對氟化物的測定。另一種可能是應用諸如*離子色譜法*的分離技術。相反，*電導率*測定只能得到總離子濃度的數據。

Analyses of inorganic ions is of interest in connection with problems of the *salinity* of waterbodies. Ions can be analyzed by direct potentiometric measurement with *ion specific electrodes*, as e.g. for fluoride. Another possibility is to use a separating technique such as *ion chromatography*. In contrast, *conductivity* measurements only provide data on the overall ionic concentration.

ion chromatography 離子色譜法 *lizi sepufa*

　　用一種儀器可測定許多共離子如溴化物、氯化物、氟化物、硝酸鹽、亞硝酸鹽、磷酸鹽和硫酸鹽的方法。它在水污染控制中是很有用的。測定時，首先將水樣稀釋到適合的濃度，然後將它注射入流經*離子交換柱*的碳酸鹽－碳酸氫鹽洗脫溶液中，測定洗脫液的電導率就可檢定被分離的離子，將其*保留時間*與參比物質或標準溶液進行比較則可加以鑑別。離子色譜法可進行濃度為幾ppm的痕量分析。

The determination of common ions, such as bromide, chloride, fluoride, nitrate, nitrite, phosphate and sulfate with one single instrument. This determination is of interest in water pollution control. The water sample is diluted to a suitable concentration. It is then injected into a stream of carbonate-bicarbonate eluant solution which is passed through an *ion exchange column*. The separated ions are detected by measuring the conductivity of the eluate stream. For identification, the *retention times* are compared with reference substances or standard solutions. The method is capable of analyzing trace concentrations in the low ppm range.

ion chromatography, chemical suppression 離子色譜法，化學抑制 *lizi sepufa, huaxue yizhi*

"化學抑制"這一術語是指離子色譜法分析的預處理。其目的是通過減少樣品溶液的總離子強度來提高離子色譜法的靈敏度。化學抑制預處理係在一裝填了具有高交換能力的陽離子交換樹脂的柱子中進行。所有的陽離子被氫離子交換，因此溶液變成酸性。然後，將該溶液用通常的*離子色譜法*進行處理。

The term 'chemical suppression' is used to describe a pretreatment to ion chromatographic analysis. The objective is to increase the sensitivity of IC by reducing the total ionic strength of the sample solution. This will result in a better signal to noise ratio. Modules for chemical suppression use a column, filled with a high capacity cation exchange resin. All cations are exchanged for hydrogen ion and the solution becomes acidic. This solution is then treated by normal *ion chromatography*.

ion exchange 離子交換 *lizi jiaohuan*

結合在離子交換基體中的陽離子(或陰離子)被水溶液中的陽離子(或陰離子)取代的化學過程。因此，水溶液中的陽離子(或陰離子)可通過與結合在基體中的陽離子(或陰離子)進行交換而去除。這一過程是可逆的，可通過再生來恢復基體的初始狀態。離子交換法可用於污染控制，如去除廢水中的重金屬離子(汞離子)，也可用於水的淨化，以及作為分析化學中的一種分析方法。(參見：離子交換樹脂、陰離子交換、陽離子交換)

The chemical process in which either cations (or anions), bound in an ion exchange matrix, are replaced by cations (or anions) of a solution. These cations (or anions) are thereby removed from the solution and exchanged by the cation (or anions) initially bound on the matrix. The process is reversible and the initial form of the matrix can be regenerated. Ion exchange processes can be used in pollution control to remove e.g. heavy metal ions (mercuric ion) from an effluent. Ion exchange is also used for water purification and as a method in analytical chemistry. (See also *ion exchange resin*; *anion exchange*; *cation exchange*)

ion exchange columns 離子交換柱 *lizi jiaohuanzhu*

裝填一種或幾種離子交換樹脂的柱型反應器。當擬淨化的水溶液通過離子交換柱時，其中的某些離子態污染物（如 Hg^{2+}）可被去除。在離子交換樹脂被耗盡後，可對它進行再生。

A column-type reactor containing one or several ion exchange resins. The aqueous solution to be purified is passed through this column and certain ionic pollutants (e.g. Hg^{2+}) are removed. After exhaustion, the ion exchange resin is regenerated.

ion exchange resins 離子交換樹脂 *lizi jiaohuan shuzhi*

作為離子交換基體的一種特殊類型的合成聚合物，通常呈小珠狀。離子交換樹脂在其聚合物骨架上含有特定的反應基，它們具有離子交換的屬性。某些天然礦物也具有離子交換性能，其基體由礦物的晶體結構所構成。（參見：沸石）

A special type of synthetic polymer, usually in form of small beads, serving as a matrix with ion exchange properties. Ion exchange resins contain specific reactive groups bound to the polymer skeleton. These groups with binding activities for cations or anions are responsible for the ion exchange properties. Certain natural minerals also show ion exchange properties. In these cases the matrix is formed by the crystal structure of the mineral. (See also *zeolites*)

ion exchange separation 離子交換分離 *lizi jiaohuan fenli*

見：ion chromatography。

See *ion chromatography*.

ion specific electrode 離子選擇（特效性）電極 *lizi xuanze (texiaoxing) dianji*

對某種離子如氟化物特別敏感的電極。可用 pH 測定所使用的標準電位計來測定電極電位。為了得到可靠的結果，需根據被測定離子的不同濃度，來確定緩衝溶液中總離子的最適濃度。測定時，需用已知的標準濃度作為參比。

An electrode with a specific sensitivity for one single type of ion, as e.g. for fluoride. Measurement of the electric electrode potential is carried out with a standard potentiometer, e.g. as used in pH measurement. To obtain reliable results, one needs to optimize the total ionic strength in a buffered solution at different concentrations of the ion to be analyzed. Known concentrations of standards are used for reference.

IR analysis 紅外分析 *hongwai fenxi*

見：infrared analysis。

See *infrared analysis.*

ISO 國際標準組織 *guoji biaozhuan zuzhi*

見：International Standards Organization。

See *International Standards Organization.*

ISO 14000 Norm series 國際標準組織14000標準系列 *guoji biaozhun zuzhi 14000 biaozhun xilie*

國際標準組織頒佈的有關環境管理和環境審計的一組標準。它是基于自願採用的原則進行環境管理的標準規範。符合該規範要求的公司，可獲得國家授權機構頒發的證書。

A set of norms issued by the International Standards Organization specifying environmental management and environmental auditing. The norm is based on a voluntary application of environmental management instruments. Companies fulfilling the requirements of the specifications may receive a certificate from a national authorized certification institution.

J

jet aeration system 射流曝氣系統 *sheliu baoqi xitong*

廢水生物處理的一種曝氣設備，其特點在於將水和空氣混合引入水面下混合液中的噴咀系統。射流曝氣系統比*擴散器*不易發生堵塞。

A type of aeration equipment for biological effluent treatment. It is characterized by a system of nozzles through which a mixture of water and air is introduced into the mixed liquor below the surface. Jet aeration systems have a lower tendency for plugging than *diffusors*.

jet scrubber 噴射洗滌器 *penshe xidiqi*

一種簡單的用於廢氣處理的濕式洗滌器，它是基於吸氣器原理工作的。噴射洗滌器具有*文丘里管道*的形狀，氣體通過水的噴射引入裝置，使污染物從氣相轉入液相。其洗滌液可以循環使用。

A simple wet scrubber for off-gas treatment. It is based on the principle of an aspirator. The jet scrubber has the shape of a *venturi canal*. The gas is transported through the unit by a jet of water and pollutants are transferred from the gas phase to the liquid phase. The scrubbing liquid may be circulated.

job description 工作職責説明 *gongzuo zhize shuoming*

目標管理的一個組成部分。它涉及到職工個人或者一個機構（單位），確定了某一工作職責的目標、責任和權力。在*綜合環境管理*中，工作職責説明還必須包括與環境保護有關的責任和權力。

An element of management by objectives. A job description refers to an employee or to an organizational unit. It defines the objectives, the responsibilities and the authorities of a job. In an *integral environmental management* approach, job descriptions must also include responsibilities and authorities relating to environmental protection.

K

karst water 喀斯特水 *kasite shui*

在岩石地下層，通常為石灰石地下層喀斯特型空間所含的地下水。喀斯特水是一種特殊的地下水。

Underground water which is contained in karst-type open spaces in a rocky underground, usually of limestone. Karst water is a special type of groundwater.

kiln, rotary 回轉窯 *huizhuanyao*

焚燒爐中緩慢旋轉的水平燃燒室，其旋轉軸微微向下傾斜若干度，窯中通常襯有耐火材料層。（參見：回轉窯焚燒爐）

The slowly rotating horizontal combustion chamber of an incinerator. The axis of rotation is slanting downward by a few degrees. The kiln is usually lined with a refractory lining. (See also *rotary kiln incinerator*)

Kjeldahl nitrogen analysis 克氏氮分析 *keshidan fenxi*

測定胺中氮的克氏法。測定時，將樣品置於克氏燒瓶中於沸騰的硫酸中進行消解，並用硫酸汞作催化劑加速反應，使各類胺氮變為硫酸銨。然後投加鹼使混合液呈鹼性，將氨蒸餾出來，並用容量滴定法定量測定氨。克氏法不能分析硝酸鹽、亞硝酸鹽、腈以及一些其他類別的氮。目前，對傳統克氏測定法已作了改進，採用了不同的消解介質和其他的催化劑。從環境上考慮，分析中應避免使用汞和硒。

The Kjeldahl method serves to determine nitrogen in the trinegative state, i.e. amin nitrogen. The sample is digested in a Kjeldahl flask in boiling sulfuric acid. The reaction is catalyzed by mercuric sulfate. Different types of amine nitrogen are changed into ammonium sulfate. In a subsequent step, the mixture is made alkaline and ammonia is set free by distillation. Ammonia is then quantized by volumetric titration. The Kjeldahl method fails in analyzing several forms of nitrogen such as nitrate, nitrite, nitriles and others. Modern versions of the traditional Kjeldahl determination use different media for digestion and other catalysts. Mercury and selenium should be avoided for environmental reasons.

L

laboratory effluents 實驗室廢水 *shiyanshi feishui*

從實驗室或中試裝置運行中排出的廢水。雖然它們的水量較少，但由於含有實驗室中所涉及的各種各樣的化合物，如溶劑、混合化學物質或有毒催化劑（如化學需氧量分析中使用的催化劑），而會導致產生廢水管理中的問題。因此實驗室廢水在排放之前應對其組成進行測定。如果需要，應進行專門的預處理，或者應將它們與工藝廢水一起排放和處理。

Effluents from laboratories or from pilot plant operations. Although their volume is relatively low, these effluents may lead to problems in effluent management as a result of the diversity of compounds handled in a laboratory: solvents, mixtures of chemicals or toxic catalysts (e.g. in COD analysis). The composition of laboratory effluents should therefore be evaluated before discharge. If necessary, specific pretreatment should be applied, or they should be discharged and treated together with process effluents.

lambda sensor 蘭達傳感器 *landa chuanganqi*

汽車催化轉化器的控制回路中使用的一種特殊傳感器。它的一個特殊的含有二氧化鋯的陶瓷傳感器元件，可依據尾氣中氧的分壓發出電信號，來控制汽油燃爆馬達中汽油的噴射，從而使催化轉化器有效地運作。

A special sensor used in the control loop of an *automotive catalytic converter*. A special ceramic sensor element containing zirconium dioxide furnishes an electric signal which depends on the partial pressure of oxygen in the exhaust gas. This signal is used to control the gasoline injection in the gasoline explosion motor in such a way that the operation of the catalytic converter is most efficient.

landfill, secure 安全填埋 *anquan tianmai*

見：secure landfill。

See *secure landfill*.

landfill gas 填埋氣體 *tianmai qiti*

它是大約含70-80%甲烷和10%二氧化碳的混合氣體(沼氣),是處置有機廢物的填埋場在厭氧微生物的作用下產生的。所生成的沼氣量取決於填埋的年代和填埋的廢物量。在一些改進的老填埋場,沼氣被收集起來用火炬燒掉;而在現代化的填埋處置場,它被收集起來用於沼氣發動機發電或用於供熱。

A gas mixture containing about 70-80% of methane and about 10% of carbon dioxide. Landfill gas is produced by anaerobic microbiological reaction in a disposal site containing reactive organic material. The quantity depends on the age of a landfill and its contents. In some upgraded older sites, this gas is collected and flared off. In modern concepts of disposal sites, the gas is collected and used for electricity generation with gas motors, or for heating purposes.

landscape protection 景觀保護 *jingguan baohu*

環境保護觀念的一個組成部分。其目的在於保護風景異常優美和具有特殊價值的景觀,以防止利用它作為工業或居住用地。為此,可採取與一般環境保護相類似的措施:(1)確定擬保護景觀的定義;(2)制定擬保護景觀的國家名錄;(3)制定法律來控制敏感地區的開發;(4)把保護區設立作為國家或地區公園。

An element of an encompassing concept of environmental protection. Its objectives are to protect landscapes of exceptional beauty and value from construction and use as industrial or residential land. The instruments to achieve these objectives are similar to general environmental protection instruments: a definition of landscapes to be protected; a national inventory of such landscapes; legislation to control development of sensitive zones and areas; and the setting away of areas to be protected as national or regional parks.

laser in analysis 激光在分析中的應用 *jiguan zaifenxizhongde yingyong*

見:LIDAR。

See *LIDAR*.

law on effluent levies 排水收費法 *paishui shoufeifa*

德國和瑞士頒佈的法律，採用*排水收費*的辦法來指導水污染控制。排水收費以一些指標和排水量為依據；排放每單位污染物的收費標準，在預先規定的期間內將定期提高。

Laws promulgated in the Federal Republic of Germany and in the Netherlands, directing water pollution control by the application of *effluent levies*. Effluent levies are based on several parameters and the discharged effluent volume; the levy per unit of discharged pollutant increases periodically over a predetermined time period.

law on the conservation of material 物質守恆定律 *wuzhi shouheng dinglü*

自然科學的一個基本定律，即物質不會失去或消失。物質守恆定律與環境保護有關，它表明進入某系統的所有物質必須全部離開該系統。這意味著投入到某加工裝置的全部物料必須產出作為產品及成為各種廢物和排放物。

A basic law of natural sciences. It implies that no material gets lost or disappears. It is of relevance in environmental protection because it makes clear that all matter entering a system must also leave the system. In a processing plant this means that all input quantities must be accounted for as outputs, either products or wastes and emissions of different kinds.

LC 液相色譜法 *yexiang sepufa*

見：liquid chromatography。

See *liquid chromatography*.

LC-50 半致死濃度 *banzhisi nongdu*

某一毒物會導致與其接觸的50%*靶生物*死亡的濃度。（參見：半作用濃度）

The average lethal concentration of a toxicant which leads to a 50% mortality rate, i.e. when 50% of the exposed *target organisms* die. (See also *EC-50*)

leachate 浸出液 *jinchuye*

與有害（或特殊）廢物的特性及其處置有關的術語。浸出液是接觸了廢物並含有來自廢物的微量污染物的廢水。例如，在未加保護的填埋場，於雨季期間，當有害（或特殊）廢物與水接觸時，廢物中的化合物會溶於水中。（參見：浸出液測試、洗脫液測試）

A term used in connection with characterization and disposal of (hazardous or special) wastes. Leachate is the aqueous solution which has been in contact with a waste and which contains traces of pollutants from the waste. Compounds contained in such wastes may be dissolved in water when these wastes are brought into contact with water, as for instance in an unprotected landfill during rainy weather. (See also *leachate test*; *eluate test*)

leachate control and treatment in landfill 填埋場中浸出液的控制和處理 *tianmaichangzhong jinchuyede kongzhi he chuli*

在含可溶性組分廢物的填埋處置中所產生的滲流水，它可能會導致地下水或地面水的污染。因此，現代化的填埋處置場在概念上應包括採取三個層次的措施：（1）避免填埋含有可溶性組分的廢物，以防止滲流水的污染（見：*浸出液測試*）；（2）盡可能在處置場設立防雨屏障，以減少滲流量；（3）用排水管系統收集污染的滲流水，並建立處理設施以便在排放到河流或其他地面水體前進行適當的處理。

Seepage water generated in a disposal site containing wastes with soluble components may contribute to a contamination of groundwater or surface water. Modern concepts of disposal sites therefore include measures at three levels: prevention of the contamination of seepage by avoiding wastes with soluble components (see *leachate test*); reduction of the seepage volume by shielding the disposal site from precipitation as much as possible; collecting contaminated seepage water by a system of drainage pipes and installing facilities to give appropriate treatment before its discharge to a river or to another surface waterbody.

leachate test 浸出液測試 *jinchuye ceshi*

模擬填埋場中所處置的某種廢物的洗脫性質而進行的測試過程。由於它與地下水的潛在污染有關，因此是廢物填埋可接受性判據之一。測試時，將廢物與水一起振搖進行浸出，然後測定洗脫液中廣譜微量雜質。在有關法規中對測試方法、應評價的化合物和它們的限值，都作出了明確規定。（參見：層次分析計劃）

A test for the simulation of the elution properties of a waste for disposal in a landfill. The eluate test is relevant in connection with a potential contamination of groundwater and it is therefore one of the acceptance criteria for a waste deposit. Leaching is carried out by shaking a waste with water. Subsequently, a broad spectrum of trace impurities is determined in the aqueous eluate. The test method, the range of compounds to be assessed and their limiting values are defined by legislation. (See also *analysis plan in tiers*)

legal compliance 符合法律要求 *fuhe falü yaoqiu*

符合法律上對工廠規定的所有要求。

The conformity with all legal requirements of a factory.

legal instruments 法律手段 *falü shouduan*

在立法中應用的管理和控制手段，主要包括：制定標準（限值）；規定"怎麼做"和"不做什麼"；為實現目標而運用的經濟機制和經濟刺激等手段；以及對不符合法規要求而課以的處罰。

各個國家所實施的立法體系具有不同的基礎，因此它們側重的方面有所不同，採用的機制也不同。

Management and control instruments used in legislation. The main legal instruments include: the setting of limits, the prescription of procedures 'how to do something' and 'what not to do', economic instruments to achieve certain goals through economic mechanisms and incentives, and the penalties for not complying with legislation.

The legal systems which are in force in different countries have developed from different roots. Consequently, they put emphasis on somewhat different aspects and mechanisms.

legal instruments using market mechanisms 運用市場機制的法律手段 *yunyong shichang jizhide falü shouduan*

見：economic instruments in environmental legislation。

See *economic instruments in environmental legislation.*

legal responsibility 法律責任 *falü zeren*

自然人或法人對應遵守的法律所負的責任。

The responsibility of a natural person, or management of a legal entity, to abide by the applicable legislation.

legislation 法規 *fagui*

社會通過其政府制定的規章和標準。為了全社會的利益，它規定了哪些行動是容許的，哪些是不容許的。為此，需運用各種*法律手段*。

Rules and norms specified by a society through its government. Legislation defines actions which are allowed and actions which are not allowed, all in the interest of all groups of the society. To achieve this, legislation makes use of different *legal instruments*.

legislation on products and product use 有關產品和產品使用的法規 *youguan chanpin he chanpin shiyongde fagui*

當按照供應商的說明使用所銷售的產品或活性組分時，對它們在環境中的影響進行管理的法規。它對產品的生態和毒性試驗規定了要求。對某些類產品，如農藥和農業化學品，它還規定了產品登記的程序和條件，只有在得到政府主管部門許可後才可投放市場。

Legislation covering aspects of the impact of sales products or of active ingredients in the environment when used according to the instruction of the supplier. It may specify requirements for ecological and toxicological testing of products. For certain types of products, e.g. for pesticides and agricultural chemicals, this legislation may also include procedures and conditions for registering the products with the national authorities to obtain permission to put them on the market.

lethal effect 致死效應 *zhisi xiaoying*

會導致 *靶生物* 死亡的效應。

An effect resulting in the death of the *target organism*.

levy 收費 *shoufei*

向政府主管部門繳費以換取進行某種行動的權利。收費的多少並不與政府主管部門賦與的經濟方面的權利直接有關。收費通常會對某些行動產生導向作用。它也被認為是一種解除動機以避免某些行動或狀況的有效措施。收費的多少及其運作是由政治上的因素決定的。例如,聯邦德國的 *排水收費法* 就是水污染控制的一個具體例子。(參見:費用、稅收)

A payment made to an authority in return for the right to carry out a certain action. The magnitude of the levy is not directly related to economic aspects of the right granted by the authority. A levy generally has a steering effect on certain actions. It may also be regarded as a demotivating factor with the objective to avoid certain actions or situations. The magnitude of a levy, and the application of it are the result of a political decision.

For a specific example in water pollution control, see *law on effluent levies* of the Federal Republic of Germany. Related topics are *fee*; *taxes*.

levy on raw materials 原料收費 *yuanliao shoufei*

環境立法中一種可能的經濟刺激手段。由於工廠購買的大部分溶劑最終將作為 *揮發性有機化合物* 損失掉而進入環境或廢水中,因此通過經濟手段對工廠購買的溶劑進行收費,將促使工廠減少溶劑的消耗,從而減少排放。對溶劑的收費已在實行,而對於其他原料的收費則正在討論之中。(參見:環境立法中的經濟手段、收費)

A possible economic incentive instrument in environmental legislation. Solvents may be taken as an example. As a large fraction of solvents bought by a factory is also lost to the environment as *VOC* or in the effluent, levies on solvents purchased for use in a plant will tend to reduce the consumption of solvents through economic mechanisms. The reduction of the consumption will also automatically lower the emissions. Such levies are applied or under discussion for other raw materials. (See also *economic instruments in environmental legislation*; *levy*)

levy on the discharge of CO_2 二氧化碳排放收費 *eryanghuatan paifang shoufei*

> 見：discharge levy for CO_2。
>
> See *discharge levy for CO_2*.

levy on VOC input 揮發性有機化合物的收費 *huifaxing youji huahewude shoufei*

> 見：levy on raw materials。
>
> See *levy on raw materials*.

liability 責任 *zeren*

會導致在將來需要支付費用的潛在的商業或個人責任。（參見：環境責任）

A potential business or personal obligation, which may lead to expenditures in the future. (See also *environmental liability*)

LIDAR 光達 *guangda*

測定大氣中污染物的遙感方法。光達是"激光探測和測距（Light Detection and Ranging）"的縮寫。它利用能發射某些特定波長的激光，將激光束沿研究對象所在的方向發射出去，光將被大氣中的氣溶膠散射回來，而部分被污染物所吸收。散射光的強度可用安設在光發射點處的望遠鏡來測量，利用電子儀器對信號進行檢測和加工處理，得到沿激光束軌線的污染物濃度的數據。由於條件的限制和干擾作用，這種方法只能應用於一些特殊情況。在這方面已發表了有關某電廠煙道氣煙羽中汞的數據，以及在整個城市上空二氧化硫、氮氧化物、臭氧和煙霧成分等一些污染物濃度的資料。

A *remote sensing* method proposed for the measurement of pollutants in the atmosphere. LIDAR stands for 'Light Detection and Ranging'. The method makes use of a laser, which can be tuned to specific wavelengths. The laser beam is emitted in the direction to be investigated. Light is back-scattered by atmospheric aerosols and is partially absorbed by pollutants. The intensity of back-scattered light is measured with an optical telescope at the point of light emission. The signal is detected and processed electronically, to give data on the concentration of pollutants along the path of the laser beam. As a result of limitations and inter- ferences, the applications of LIDAR cover specific cases only. Data have been published for mercury in the flue gas plume of a power plant and for concentrations of a number of pollutants distributed over whole cities, e.g. sulfur dioxide, nitrogen oxides, ozone and smog components.

life cycle analysis 生命周期分析 *shengming zhouqi fenxi*

見：ecobalance。

See *ecobalance.*

life cycle of a product 產品的生命周期 *chanpinde shengming zhouqi*

某個產品的生命的各個階段。產品的生命始於初始原料的開採，包括生產過程的各個階段和產品的銷售，並以在使用後進行最終處置而告終。為了評估產品的全部環境影響，必須將它生命周期的各個階段都考慮在內。（參見：生態平衡）

All steps of the life of a particular product. The life of a product starts with mining of the primary raw materials. It includes the different steps of the production process itself, the distribution to the customers and it ends with the final disposal after use. For estimation of the total environmental impact of a product, all steps of the life cycle need to be considered. (See also *ecobalance*)

life sciences 生命科學 *shengming kexue*

此術語用於概括與人、動物和植物生命有關的科學，它主要包括醫學、生物學、營養學和食物生產以及許多有關的分支領域。

A term used to summarize sciences which are related to human, animal and plant life. The main elements are medicine, biology, nutrition and food production as well as the large number of associated sub-fields.

limit of detection 檢測極限 *jiance jixian*

見：detection limit。

See *detection limit.*

limnic ecosystems 湖沼生態系統 *huzhao shengtai xitong*

表徵淡水水體如湖泊、池塘或緩慢流動的河流的生態系統的一般術語。湖沼生態系統的鹽含量低，它可用礦物質的量來測定。此外，湖沼具有明顯不同的分層，即湖上層和湖下層，而表徵這兩層特性的參數會在這兩層之間的中間溫變層中過渡變化。在此系統中相互作用的種群的詳細情況，取決於礦物質組成、氧含量、污染程度、太陽輻射、光透射、溫度、風的影響、不同層的循環、湖泊類型（淺湖、深湖、沼澤）和季節變化等因素。（參見：富營養化）

A general term summarizing ecosystems in a freshwater body, e.g. a lake, a pond or a slow moving river. The salinity of all limnic ecosystems is low; it is determined by mineral factors. Distinct zones at different levels are the epilimnon (upper zone) and the hypolimnon (bottom zone). The parameters characterizing these two zones change in the intermediate thermocline zone. Details of the populations interacting in such systems depend on factors such as mineral composition, oxygen content, degree of pollution, solar irradiation, light transmission, temperature, wind effects, turnover of different layers (strata), the type of lake (flat, deep, swampy) and seasonal changes. (See also *eutrophication*)

LINDOX system 林多克斯系統 *linduokesi xitong*

應用純氧技術的一種特殊的廢水生物處理系統，它由林德公司市場化。該技術已有幾項很成功的專門應用，如煉油廠廢水處理。（參見：純氧曝氣系統、優諾克斯系統）

A special application of biological effluent treatment with technically pure oxygen. LINDOX is marketed by LINDE. Several highly specialized successful applications of this technology are in operation, e.g. for treatment of effluents from rendering plants. (See also *aeration systems with oxygen; UNOX*)

line function 指揮職能 *zhihui zhineng*

工廠的指揮機構。（參見：參謀職能）

An organizational unit of the line organization of a plant. (See also *staff function*)

line organization of the plant 工廠的指揮機構 *gongchangde zhihui jigou*

工廠中直接負責生產運行的綜合機構及其下屬的單位。生產部門就是一個典型的負有指揮職能的機構。

The network of those units of the overall organization of a factory which contribute directly to the manufacturing operations. The production department is a typical line function.

line responsibility 指揮職責 *zhihui zhize*

負有指揮職能的經理為實現其工作目標而承擔的直接責任。責任與指揮行動的權力是連在一起的。在*職責說明*中應對責任和權力加以明確規定。在理想的情況下，應只在費用開支限額和人事管理政策上對其權力有所限制。

The direct responsibility of the manager of a line function to reach the objectives which have been assigned to the line function. The responsibility goes hand in hand with the authority to mandate the necessary actions. Responsibilities and authorities should be defined in a *job description*. In the ideal case, the authority to mandate is only limited by a limit on expenditures and policies on personnel management.

linear activity 線性活動 *xianxing huodong*

某一活動的控制論條件。它是在無反饋，即不分辨其副作用下進行的活動。（參見：控制論管理方法、閉環活動）

A cybernetic qualification of an activity. An activity which is carried out without feedback, that is, without recognizing the (side) effects of it. (See also *cybernetic management approach*; *closed cycle activity*)

lipophilic 親油（脂）的 *qingyou(zhi)de*

可溶於脂肪介質或有機介質中。親油（脂）性是評價物質在食物鏈中行為的重要因素。化學物質的親油（脂）性可以用某物質在水與正辛醇之間的分配系數表示。（參見：辛醇系數）

Soluble in a fatty medium or in an organic medium. Lipophilicity is an important factor in the assessment of the behavior of a substance in the food chain.

Lipophilic properties of chemicals can be characterized by the partition of a substance between water and n-octanol. (See also *octanol coefficient*)

liquefaction 液化 *yehua*

將氣體變為液體的過程；它是汽化的逆過程。液化可通過降低蒸氣溫度或增加蒸氣壓力或同時通過二者來實現。它在加工過程和分離技術中應用非常廣泛，並具有各種各樣的設備。在污染控制中，如果廢氣中揮發性有機化合物的濃度較高，可利用液化技術從廢氣中回收揮發性溶劑。

The process of changing a gas into a liquid; it is the reverse process of vaporization. Liquefaction may be achieved either by lowering the temperature of a vapor or by increasing its pressure, or both. Liquefaction is used very widely in processing and separation technology and a wide range of equipment is available. In pollution control, the process has been applied to recover volatile solvents from off-gas streams if VOCs are present in relatively high concentration.

liquid chromatography 液相色譜法 *yexiang sepufa*

於正常壓力下在分離柱中進行色譜分離的技術。它所用的吸附劑和洗脫溶劑可能是很不相同的，其選擇取決於被分離的混合化合物和它們的*吸附*性質。在環境分析中，這種分離技術主要應用於分析前的淨化、*母體*分離或*富集*等步驟。例如，在收集了焚燒裝置的*煙氣*樣品後，進行煙氣中二惡英的分析，就需應用液相色譜法進行前處理。

A chromatographic separation technique in a separation column under normal pressure. The adsorbent and the solvent for elution may vary considerably. The selection of solvent and adsorbent depends on the mixture of compounds to be separated and their *adsorption* properties. In environmental analysis this separation technique is mainly used in clean-up, *matrix* separation or *enrichment* steps prior to analysis. The analysis of dioxins after collection of the samples from *flue gas* of an incineration plant may serve as an example.

liquid/liquid extraction 液液萃取 *yeye cuiqu*

應用於特殊廢水或母液處理或預處理的*萃取技術*。

The application of *extraction technology* to the treatment or pretreatment of special effluents or mother liquors.

lithosphere 岩石圈 *yanshiquan*

*生態圈*的三個主要組成部分之一。岩石圈是地球的硬殼；但此術語在使用上常常具有不同的含義。從根本上說，它包含了幾乎所有的地球的物質。岩石圈與水圈在不斷地相互作用，並在一定程度上它通過風化過程與大氣在不斷地相互作用。

One of the three main compartments of the *ecosphere*. The lithosphere is the hard crust of the earth; the term is often used with a variable meaning. Basically, the lithosphere contains almost all of the minerals of the earth. It is in constant interaction with the hydrosphere and to a lesser degree with the atmosphere through erosion processes.

load limits 負荷限值 *fuhe xianzhi*

見：mass flow limits; concentration limits。

See *mass flow limits*; *concentration limits*.

LOPROX technology (Bayer) 洛普羅克斯技術(拜耳公司) *luopuluokesi jishu (bai'er gongsi)*

見：low pressure effluent oxidation。

See *low pressure effluent oxidation*.

low pressure effluent oxidation 廢水低壓氧化 *feishui diya yanghua*

由拜耳公司開發的*廢水預處理*技術，其商業名稱叫做洛普羅克斯技術。它是在壓力5至20巴、溫度低于200°C下，利用二價鐵離子作為*催化劑*，於*泡罩塔反應器*中用空氣中氧進行廢水的氧化預處理。該技術可應用於處理含高濃度有機物的廢水。在上述反應條件下，某些*難降解污染物*被部分氧化為可*生物降解*的物質。另外，它也可用來處理剩餘活性污泥。其工藝設備還包括熱交換器和*降壓段*。(參見：高壓濕式空氣氧化)

A technology developed by Bayer under the trade name LOPROX for *pretreatment of effluents* with oxygen from air in a *bubble column* reactor at pressures between 5 and 20 bar and at temperatures below 200°C. Iron-2 ion is used as a *catalyst*. The technology is applicable to highly concentrated effluents containing mainly organic wastes. Under these reaction conditions, some *refractory contaminants* are partially oxidized to render them more suitable for *biodegradation*. The technology can also be applied to the treatment of excess biosludge. The process units include also heat exchangers and a *decompression stage*. (See also *wet air oxidation, high pressure*)

luminescence effluent toxicity 發光法廢水毒性 *faguangfa feishui duxing*

見：effluent toxicity, luminescence method。

See *effluent toxicity, luminescence method.*

M

macroeconomic aspects 宏觀經濟問題 *hongguan jingji wenti*

會對一個國家的經濟產生影響的問題。

Aspects and influences on the national economy of a country.

maintenance and repair costs 維護維修費用 *weihu weixiu feiyong*

工廠中設備維護維修的費用,它是工廠運行費用的一部分。(參見:維修與維護)

The expenses resulting from maintenance and repair activities for equipment in a factory. Maintenance and repair costs are part of the operating costs of a factory. (See also *repair and maintenance*)

MAK 最大容許濃度 *zuida rongxu nongdu*

工作場所中某物質的最大容許濃度。其值由德國MAK委員會發佈。它類似於*閾限值*,也是對工作場所中某物質的濃度作出規定。它是基於如下假定,即一個健康工人在室內於該物質的濃度下工作不超過8小時而得到的。短期接觸的限值可按最大容許濃度表中給出的公式進行計算。在德國,最大容許濃度是具有約束力的。它與美國的閾限值以及與英國的EH40/96職業接觸限值相似,但不相同。(參見:閾限值)

Maximum allowable concentration of a substance at a workplace. The MAK values are published by the German MAK-Committee. Similar to the *TLV*s, MAK values also specify workplace concentrations. The limits are based on the assumption that a healthy worker spends not more than 8 hours in a room with a substance at this concentration. Limits for short term exposure are calculated according to a formula given in the MAK list. MAK values are binding in the Federal Republic of Germany. They are similar — but not identical — to the American TLV values and the British EH40/96 occupational exposure limits. (See also *TLV*)

malodors 惡臭 *echou*

令人討厭的強烈臭味。

Bad odors, an odor nuisance.

man monitoring 隨身監測 *suishen jiance*

用一種小型便攜式取樣裝置監測工人在工作場所接觸空氣中污染物的情況。這種取樣裝置可在工人工作時隨身攜帶，然後對所吸附的組分在實驗室進行分析測定。隨身監測是職業衛生監測的一種特殊技術。（參見：職業衛生）

Monitoring of the exposure of a worker to pollutants in the workplace atmosphere by a small portable sampling and collecting device which is carried by the worker during his assigned work. The actual analytical determinations of the adsorbed components are carried out in the laboratory. Man monitoring is a special technique of occupational health monitoring. (See also *occupational health*)

management, emergency 緊急事故管理 *jinji shigu guanli*

見：emergency response management。

See *emergency response management.*

management by objectives 目標管理 *mubiao guanli*

在確定具體的目標、賦予職工明確的責任和權力的基礎上進行管理的方式。目標和相應的責任和權利，可按機構、業務部門、工作小組、領導和職工分別確定。

A management style which is based on the philosophy of defining specific objectives and goals, assigning responsibilities, and assigning authority to act to employees. The objectives and corresponding responsibilities and authorities can be specified for organizations, business groups, business teams, leaders and employees.

management control instruments 管理控制手段 *guanli kongzhi shouduan*

為進行控制所採取的滿足其基本要求的管理手段，包括數據收集、數據管理和數據評估等環節。（參見：環境審計、控制、工作報告）

Management instruments which fulfill the basic need to carry out controls. Control instruments include the element of data collection and data management and the evaluation of relevant data. (See also *environmental audit*; *controlling*; *performance report*)

management instruments 管理手段 *guanli shouduan*

制定一個或若干個*管理規程*，以保證對某一特定方面的管理達到所需要的廣度和深度。

One or several *management procedures* which together make certain that the treatment of a specific area of management receives the necessary coverage and depth.

management instruments in environmental protection 環境保護的管理手段 *huanjing baohude guanli shouduan*

為執行*公司的環境政策*和使工廠的環境保護工作取得良好成績而採用的管理手段（體系和規程），主要包括：（1）確定目標（見：立法、公司環境政策）；（2）制定規程（見：環境影響）；（3）優化實施措施，包括組織機構、職責、人事管理、培訓（見：人事管理）；（4）技術管理（見：技術的管理）；（5）費用管理（見：費用/成本）；（6）控制措施，即在生產現場的控制措施，包括環境服務、審計、工作報告（見：審計、環境工作報告、環境人員）；（7）與政府主管部門以及與公眾的溝通與交流。

Management instruments (systems and procedures) which serve to implement a *company environmental policy* and to achieve and maintain a high performance in environmental protection of a factory. The main groups of instruments are: setting of goals and objectives (see *legislation*; *company environmental policy*); planning procedures (see *environmental impact*); instruments serving to optimize the implementation such as organization, responsibilities, accountabilities, personnel management and training (see *personnel*); technology management (see *management of technology*); management of costs (see *costs*); control instruments at the production level, including environmental service, auditing, performance reporting (see *audit*; *environmental performance report*; *environmental officer*); and communication with the authorities and the public.

management of environmental costs 環境費用的管理 *huanjing feiyongde guanli*

涉及與環境保護活動有關的經濟問題的管理體系與方法，主要包括：評審*環境保護*全部*費用*的方法、程序和有關資料；以及把環境費用納入到產品價格中所採用的方法。（參見：環境費用的分攤、內部費用）

241

Management systems and procedures which deal with economic aspects related to environmental protection activities. The main elements are: instruments, procedures and documents to give an overview over total *costs of environmental protection*, and procedures ('tools') used for the integration of environmental costs into the product prices. (See *allocation of environmental costs*; *internal costs*)

management of off-gas emissions 廢氣排放的管理 *feiqi paifangde guanli*

工廠中空氣污染控制的系統管理方法,其主要內容包括:確定空氣污染控制的目標,建立作為基礎數據的全部廢氣的清單(包括所有排放數據),只要有可能則通過工藝改革在源頭解決污染問題,根據充分的技術資料選擇污染控制設備,測定設備的效率,檢查與標準相符合的情況,在設計新設備時應用和反饋有關的數據資料與經驗。

A systematic approach to air pollution control in a factory. The main elements are: clear objectives in APC; an inventory of all off-gas streams as a database, including data on all emissions; whenever possible, problem solutions at the source, by process modifications; the selection of pollution control equipment based on adequate technical data; measurements of the efficiency of the equipment, checks on compliance with norms; and the application and feed back of data and experiences in the design of new equipment.

management of special waste disposal 特殊廢物處置的管理 *teshu feiwu chuzhide guanli*

為保證*特殊廢物*以環境可接受的正確方式進行處置所需採取的全部管理措施。特殊廢物的管理必須遵循*"從搖籃到墳墓"的原則*。除了在*源頭減少廢物*之外,其管理還包括:鑒別和分析特殊廢物,選擇可靠的廢物處置技術,通過*廢物多聯單*來跟蹤廢物,以及將處置費用分攤到源頭產品上。在此管理體系中,必須對以上所有的責任予以明確規定,並且必須對排放特殊廢物加以控制。

The total of the management activities needed to make sure that *special wastes* are disposed of in a proper and environmentally acceptable manner. Management of special wastes must be based on the *cradle-to-grave principle*. In addition to activities for *waste reduction at the source*, it includes procedures to characterize and analyze special wastes, to select adequate waste disposal techniques, to apply a *waste tracking procedure* by a *waste manifest*, and to allocate the disposal costs to the source products. In such a system, all responsibilities must be clearly defined and their discharge must be controlled.

management of special wastes 特殊廢物的管理 *teshu feiwude guanli*

控制與解決 *特殊廢物* 問題的管理體系。其主要內容包括開發環境友好的產品(*綠色產品*);在生產方面,應通過 *再使用、循環利用、完整的工藝過程開發* 來防止產生廢物;最後,對於不可避免產生的廢物,必須按環境可接受的原則進行適當的 *廢物處置*。(參見:特殊廢物處置的管理)

A management system to control and solve the problem of *special wastes*. The main elements include on the one side the development of environmentally friendly products (*green products*). On the manufacturing side, important aspects are the prevention of the generation of wastes through *reuse, recycling, integral process development*. Finally, unavoidable wastes must undergo environmentally acceptable *waste disposal*. (See also *management of special waste disposal*)

management of technology 技術的管理 *jishude guanli*

進行技術特別是技術裝備投資方面的管理所需採取的管理措施,其主要內容包括:確定明確的目標和預期結果,具有計劃工作所需的完善的數據庫,確定工作的重點,提供選擇最佳項目方案的決策方法,嚴格的項目管理,財務管理,人事管理,效果的最終評價,以及把獲得的新知識納入到新項目中。

Management instruments necessary to manage technical aspects, specifically investments in technical installations. Main elements are: the specification of clear goals and expectations; the provision of an adequate planning database; priority setting; decision techniques for selecting the best project option; proper project management; management of financial aspects; personnel aspects related to management of technology; and final evaluation of the efficiency and integration of gained new knowledge in new projects.

management principles in general 一般管理原則 *yiban guanli yuanze*

　　工廠管理所遵循的原則和概念。重要的是要規定明確的目標和責任。另外，人事管理應把各層次的人員納入到公司的各類運作和過程中。賦予權力（即有權做出決定）能激發人員的積極性，而人員的參與也能把全體職工的智慧與能力匯入公司的運作中。

　　Principles and concepts by which a factory is managed. Clear specifications of objectives and responsibilities are important. Personnel management should integrate personnel at all levels into all types of operations and processes. Empowerment, i.e. the delegation of power to decide, serves to motivate personnel. Participation of personnel also serves to integrate the full capabilities and capacities of all employees in the company operation.

management procedure 管理規程 *guanli guicheng*

　　進行某一方面（如有害廢物處置）管理所需要採取的一系列步驟和詳盡行動。

　　A defined set of steps and detail actions necessary to manage a certain aspect, e.g. the disposal of hazardous wastes.

management system 管理體系 *guanli tixi*

　　由一些具體內容構成的系統而一致的管理方法。（參見：各項管理工作的管理手段）

　　A systematic and consistent approach to management, including specific elements. (See also *management instruments* for individual management tasks)

manufacturing costs 生產費用 *shengchan feiyong*

　　見：product manufacturing costs。

　　See *product manufacturing costs*.

marine ecosystems 海洋生態系統 *haiyang shengtai xitong*

　　大海或大洋的生態系統，即世界上最大的水生生態系統。它基本上是一個單一的互相聯系的生態系統。雖然所有海洋的化學組成是很相似的，但是由於溫度、深度、太陽輻射、複雜的物理化學的相互作用和兩極之間的循環等因素的影響，它們也有很大的不同。

海洋生物群落包括有多種多樣的生物體，如各種浮游生物、自由游動的生物(魚、海洋哺乳動物和烏賊等等)，甚至還有海龜和海蛇。

海洋可以看作是一切進入物質的主要儲存地或*最終去所*。雖然海洋非常廣闊，但是人類向海洋排放污染物、人為的變化和事故性排放，在許多情況下已對海洋生態系統造成重大損害。現在，這類排放已受到國際條約的控制。

The ecosystems of the seas or oceans, the largest of all aquatic ecosystems. Basically, the marine ecosystem is one single interconnected ecosystem. While the chemical composition of all oceans is very similar, there are nevertheless significant differences due to other factors like, for instance, temperature, depth, solar irradiation, complex physical chemical interactions and circulations between poles.

The marine biocenose includes a high diversity of organisms. Populations include many different organisms of plankton as well as free swimming organisms (fish, oceanic mammals, squid and many others) and even turtles and sea snakes.

Oceans may be regarded as a main sump or *final sink* of all matter introduced. Despite the vastness of the oceans, man-made discharges of pollutants into the seas, artificial changes, and accidental releases have in many cases led to significant damages in marine ecosystems. Today, these discharges are controlled by international treaties.

market instruments in pollution control 污染控制的市場手段 *wuran kongzhide shichang shouduan*

見：economic incentives in environmental protection。

See *economic incentives in environmental protection.*

mass flow 物流 *wuliu*

通過某工藝裝置、某生產過程或某工廠的物料流。根據物質守恆定律，進入某過程的所有物質必須全部離開該過程。在環境管理上，可以利用物流數據來確定所有的廢物流。

The flow of material through a process unit, a production process, or a factory.

Basically, the law of the conservation of matter demands that all materials entering a process must also leave it. In environmental management, the application of mass flow data may serve to identify all waste streams.

mass flow cycles in ecosystems 生態系統中的物流循環 *shengtai xitongzhongde wuliu xunhuan*

通過生態系統各不同組成部分的物流。一般它具有如下特點，即物質不會流失，而是往往被另一種生物再利用。許多過程和局部循環通常是互相關聯的，而且它們組成一複雜的網絡。同時，生態系統的某些組成部分可起到作為某種物質的*最終去所*的作用。（參見：碳循環、氮循環、氧循環）

The flow of material or mass through different parts of an ecosystem. Eco-systems are generally characterized by the fact that no mass is lost and materials are often reutilized by another organism. Many processes and partial cycles are generally interlinked in a complex network. Certain compartments of an ecosystem may act as a *final sink* for a substance. For three examples see *carbon cycle*; *nitrogen cycle*; *oxygen cycle*.

mass flow limits 總量限值 *zongliang xianzhi*

對廢物流中某特定污染物允許的排放總量限值，而不是其濃度限值。限定排放總量相當於限定了該污染物對環境的影響，因為影響與總量而不是與濃度有關。進行總量控制的優點在於，它使工廠有較大的自主性來滿足限制標準的要求。（參見：濃度限值）

The limitation of the permitted discharge of a specific pollutant in a waste stream by specifying the total quantity (the mass flow or load) of the pollutant and not its concentration. Limiting the mass flow is equivalent to limiting the impact of said pollutant on the environment, since the impact is related to the total mass flow and not to the concentration. As an advantage, mass flow limits leave more flexibility to a factory for meeting the restrictions. (See also *concentration limits*)

mass flow meter, ultrasonic 超聲波質量流量計 *chaoshengbo zhiliang liuliangji*

用電子學方法測量質量流量（如廢水流量）的儀器。它用超聲波設備

來測量通過文丘里管道液體的液面；由於液面與流速成正比，因此可測
得質量流量。

A device to measure mass flow, e.g. the flow of an effluent, electronically. Ultrasonic devices are used to measure the level of liquid passing a *venturi canal*. This level is proportional to the flow rate.

mass spectrometry 質譜分析 *zhipu fenxi*

在研究和分析中用來鑒別未知化合物的一種非常重要的儀器分析手
段。在質譜分析中，把分子注入真空離子化室，用不同技術使它們離子
化，得到荷電粒子、離子或離子碎片。再將這些荷電粒子流通過一強磁
場，根據這些碎片的相對豐度和最大的質子數，就可以鑒別出未知的化
合物。

現在，只要將所記錄得到的質子碎片與已知物質的碎片相比較，就
能鑒別許多化合物質。為此，有一些大型的光譜數據庫可供利用。此
外，質譜儀也可用作為一種檢測器，將它與其他分析儀器聯用，可用來
鑒別和定量未知物質。（參見：氣相色譜－質譜分析、感應耦合等離子體
－質譜分析）

A very powerful instrumental tool, used in research and analysis, for the identification of unknown compounds. In mass spectrometry, molecules are introduced into an evacuated ionization chamber where they are ionized by different techniques. Charged particles, ions or ionic fragments are obtained. By directing the stream of these charged particles through a strong magnetic field, they are separated according to their mass to charge ratio. A detecting device serves to record mass and quantity of the ionic charged particles. The relative abundance of these fragments and the highest mass serve to identify the compound under question.

Today one can identify a large number of chemical substances by comparing the recorded mass fragments with those of known substances. For this, large spectral data libraries are available. MS is also used as a detector to identify and quantify unknown substances in connection with other instrumental analyses. (See also *GC-MS*; *ICP-MS*)

material balance 物料平衡 *wuliao pingheng*

用以表徵物料在過程中發生變化的一組數據，包括投入的全部物料、產出的全部產品和所有的廢物流。按照*物質守恒定律*，對某一過程而言，全部投入的總和應等於全部產出的總和。

A set of data characterizing a process in which materials undergo changes. It contains all input quantities of materials as well as all outputs, including outputs of the manufactured products and outputs of all waste streams. The physical *law on the conservation of matter* demands that the sum of all inputs to a process equals the sum of all outputs.

mathematical dispersion models 數學離散模式 *shuxue lisan moshi*

見：dispersion models。

See *dispersion models*.

matrix of a sample 試樣母體 *shiyang muti*

母體常常被認為是試樣中含有痕量污染物的主要組成部分。作為某些分析方法的第一步，必須將污染物從母體中分離出來。母體在性質上與被分析的化合物可能是相同的，也可能是不相同的。根據其相似性的不同，必須採用不同的試樣制備與分析方法。因此，痕量污染物含量的分析測定，將受環繞痕量化合物的母體的性質所影響。所採用的分析方法必須使這種干擾不致產生錯誤的結果。

Often, the matrix is the main part of a sample in which trace quantities of pollutants are suspected. As a first step of some methods of analysis, the pollutant has to be separated from the matrix. The matrix may be similar in nature to the compound to be analyzed, or it may be dissimilar. Depending on this similarity, different methods of sample preparation and analysis must be used. The analytical determination of the quantity of a trace pollutant is therefore influenced by the nature of the matrix which surrounds the trace compound. The method of analysis has to be such that this interference does not lead to false results.

measurement ports 測量口 *celiang kou*

見：sampling and measuring ports。

See *sampling and measuring ports*.

membrane 膜 *mo*

　　由有機高分子材料制成的用來使兩相分離的柔韌薄膜。依膜材料和厚度而異，它們可以或多或少地滲透過不同大小的分子或離子，但卻不能滲透過粒子。膜在微生物過程、*滲透*、*超濾*中具有重要應用，它還在*逆滲透技術*中用作為*半透膜*。

　　A thin sheet or foil, often flexible, made of an organic polymer, used to keep two phases separate. Depending on the membrane material and the thickness, membranes may be more or less permeable to molecules or ions of different size. They are not permeable to particles. Membranes are important in microbiological processes, in *osmosis*, in *ultrafiltration* and as *semi-permeable membranes* in the *reversed osmosis technology*.

metabolism 新陳代謝 *xinchendaixie*

　　細胞中通過生物化學變化將複雜分子分解為較小分子的所有化學反應。在進行新陳代謝時，細胞吸收營養物，把它們轉換為細胞成分，從而消除了廢物，並為細胞提供了能量。酶作為*生物催化劑*，會對新陳代謝過程起到催化作用。新陳代謝反應是一切生命的基礎，它同樣也是廢水生物處理過程的基礎。在新陳代謝過程中，上一代謝階段的產物常常是下一代謝階段的營養物。

　　All chemical reactions that are carried out in cells by biochemical transformations, including the breaking down of complex molecules to smaller ones. During metabolism, cells take in nutrients and convert them into cell components. Waste products are eliminated. The metabolic process provides energy for the cell and is catalyzed by enzymes. Enzymes are biological *catalysts*. Metabolic reactions are fundamental for all forms of life. They are also responsible for the processes in biological effluent treatment. The products of one metabolic step most often serve as input to the next step.

meteor water 雨水 *yushui*

　　見：precipitation water。

　　See *precipitation water*.

microbiological mineralization 微生物礦質化 *weishengwu kuangzhihua*

見：mineralization; mineralization by microorganisms。

See *mineralization; mineralization by microorganisms.*

microbiological treatment 微生物處理 *weishengwu chuli*

在一般意義上指為改變物質的化學結構利用微生物對物質所進行的任何處理。

In the very general sense, any treatment of a substance by microorganisms in order to change its chemical structure.

microbiology 微生物學 *weishengwuxue*

研究微生物及其作用過程的學科，它是生物學的一個分支學科。

The science dealing with microorganisms and the processes carried out by them. A subdiscipline of biology.

microcenose 微生物群落 *weishengwu qunluo*

某群落生境中微生物的不同種群。（參見：生物群落）

The different populations of microorganisms of a given biotope. (See also *biocenose*)

microeconomic aspects 微觀經濟問題 *weiguan jingji wenti*

與某個企業經濟有關的問題。

Aspects related to the economy at the level of a single enterprise.

microfiltration 微過濾 *weiguolü*

過濾器的孔徑為20-5000毫微米的過濾過程。其過濾器可由陶瓷材料、高分子材料和金屬製成。操作時，將懸浮液在大約1-5巴的壓力下通過過濾器。微過濾可應用於處理含懸浮物的廢水，以及從廢水中回收產品或重金屬。（參見：過濾、超濾、逆滲透）

A filtration process in which the pore size of the filter unit ranges from 20 to 5000 nm. The filter unit can be made of ceramic materials, polymers and also of metals. The suspension is pressed through the unit at a pressure of about 1 to 5 bar.

Microfiltration is applied in the treatment of effluents e.g. in the removal of particulate matter (suspended solids), and in the recovery of products or heavy metals. (See *filtration*; *ultrafiltration*; *reversed osmosis*)

microgram 微克 *weike*

　　百萬分之一克；環境痕量分析中使用的單位。它大約是一顆直徑為 0.1–0.05毫米的塵粒的重量。

　　The one millionth part of one gram; a unit in environmental trace analysis. As an illustrative example, it represents the weight of one single dust particle of a diameter of about 1/10 to 1/20 mm.

micrometer 微米 *weimi*

　　1米的百萬分之一，或1毫米的千分之一，或1000毫微米。

　　The one millionth part of one meter; the one thousandth part of one millimeter; or one thousand *nanometers*.

milligram 毫克 *haoke*

　　千分之一克；環境痕量分析中使用的單位。它大約是一根20厘米長 的細頭髮的重量。

　　The one thousandth part of one gram; a unit in environmental trace analysis. Approximately the weight of about 20 cm of one fine human hair.

milling 粉碎 *fensui*

　　把粗顆粒破碎為細顆粒的化工單元過程。在工業上有許多不同的粉 碎設備。乾式粉碎常常是產生噪聲、粉塵與顆粒物的污染源。

　　A unit process in chemical technology; the breaking down of coarse particles into fine ones. A number of different designs of milling equipment are available commercially. Dry mills are often a source of emissions of noise, dusts and particulates.

mineralization 礦質化 *kuangzhihua*

有機物分解為小分子無機物（如二氧化碳、水和其他物質）的過程。燃燒就是一種礦質化過程。

The general process of breaking down organic material into small inorganic molecules like carbon dioxide, water and others. Combustion is a mineralization process.

mineralization by microorganisms 由微生物礦質化 *you weishengwu kuangzhihua*

由微生物進行的礦質化過程，即有機物通過生物降解完全分解為簡單分子如二氧化碳、水、硫化氫、氮的過程。

The general mineralization process carried out by microorganisms. The complete biodegradation of an organic compound to simple molecules like e.g. carbon dioxide, water, hydrogen sulfide, and nitrogen.

mixed liquor 混合液 *hunheye*

廢水生物處理場曝氣池中的*活性污泥懸浮液*。乾活性污泥的濃度在2-10克/升範圍內。（參見：污泥濃度、污泥揮發分濃度）

The aqueous suspension of *activated sludge* in the aeration step of a biological effluent treatment station. The concentration of activated (dry) sludge is in the range of 2 to 10 g/l. (See also *MLSS*; *MLVSS*)

mixed liquor suspended solids (MLSS) 污泥濃度（混合液懸浮固體）*wuni nongdu (hunheye xuanfuguti)*

見：MLSS。

See *MLSS*.

mixed liquor volatile suspended solids (MLVSS) 污泥揮發分濃度（混合液揮發性懸浮固體）*wuni huifafen nongdu (hunheye huifaxing xuanfuguti)*

見：MLVSS。

See *MLVSS*.

MLSS 污泥濃度（混合液懸浮固體）*wuni nongdu (hunheye xuanfuguti)*

Mixed liquor suspended solids的縮寫。它是廢水生物處理使用的一項綜合指標，即曝氣池中乾固體物質的總濃度。它包括活性生物污泥、無活性的無機物和有機物。

Mixed liquor suspended solids. A group parameter applied in biological effluent treatment; it specifies the total of the concentration of dry solid matter in the aeration basin. The MLSS value includes the active biological sludge and inactive, inorganic and organic matter.

MLVSS 污泥揮發分濃度（混合液揮發性懸浮固體）*wuni huifafen nongdu (hunheye huifaxing xuanfuguti)*

Mixed liquor volatile suspended solids的縮寫。它是廢水生物處理使用的一項綜合指標，是將乾污泥樣品在600°C下灼燒，污泥濃度中揮發消失的部分。它包括活性生物污泥和任何其他的有機物。

Mixed liquor volatile suspended solids. A group parameter specifying that fraction of the total MLSS which is eliminated from the dry sample by ignition at a temperature of 600°C. The MLVSS value includes the active biological sludge components as well as any other organic material.

Montreal Protocol 蒙特利爾議定書 *mengteli'er yidingshu*

1989年1月生效的關於臭氧耗竭物質的使用和排放的協議，1990年在倫敦又通過了該協議的修正案。它要求分階段地大量削減使用臭氧耗竭物質，主要是氯氟烴、哈龍、四氯化碳、三氯乙烷和溴代甲烷。蒙特利爾議定書已被大約50個國家批准。

An agreement on the use and emission of ozone-depleting substances, effective as of January 1989. An amendment was passed in London in 1990. The Montreal Protocol calls for a stepwise and significant reduction of the use of ozone-depleting substances, primarily chloro fluoro carbons (*CFCs*), halon, carbon tetrachloride, trichloro ethane and methyl bromide. The Montreal Protocol has been ratified by about 50 nations.

mother liquor 母液 *muye*

在某一化學過程，如過濾或相分離時分離產生的濃廢液。通常，母液的濃度比一般廢水高。

A concentrated aqueous waste stream which is separated during a chemical process, typically in a filtration or a phase separation. Mother liquors are generally more concentrated than general effluents.

moving bed adsorption column 移動床吸附塔 *yidongchuang xifuta*

由上流的液體使吸附劑(如活性炭)處於懸浮狀態的吸附塔。待淨化的液流用循環泵維持循環流動。

An adsorption column in which the adsorbent, e.g. activated carbon, is kept in suspension by the upwards flow of the liquid in the column. The flow of the liquid to be purified is maintained by circulation pumps.

moving grate incinerator 移動爐箅式焚燒爐 *yidong lubishi fenshaolu*

見：grate incinerator。

See *grate incinerator*.

MS 質譜分析 *zhipu fenxi*

見：mass spectrometry。

See *mass spectrometry*.

multi-purpose equipment 多功能設備 *duogongneng shebei*

可在若干不同場合和過程中使用的工藝設備。(參見：過程專用設備)

Processing equipment which can be used for several different applications and processes. (See also *process-dedicated equipment*)

multiple hearth incinerator 多爐床焚燒爐 *duoluchuang fenshaolu*

適用於用泵輸送的粘稠廢物如污泥和漿狀廢物的焚燒爐。它不能接受處理體積龐大的廢物。這種焚燒爐的主要部分為一圓柱形的垂直焚燒室，內襯耐火磚，並安設有許多平盤式的水平爐床。在焚燒室的中心，在垂直軸四周安裝了一個旋轉爐耙和刮板系統，以使廢物在爐床周圍移動。在焚燒過程中，廢物被從一層爐床刮到下一層爐床。爐灰在焚燒爐底部取出。焚燒空氣從爐底向上逆流循環。多爐床焚燒爐一般用於焙燒金屬礦石。在污染控制中，已應用它來處置廢水處理產生的剩餘污泥。其設計先進的爐型，見*複合技術焚燒爐*。

An incinerator for wastes which have a consistency suitable for pumping, e.g. sludges and pasty waste materials. Bulky wastes are not acceptable. The main part of the incinerator consists of a cylindrical, vertical, brick-lined incineration chamber with a number of flat, tray-like, horizontal hearths. In the center of this chamber, and constructed around a vertical axis is a system of rotating rakes or scrapers used to move the waste around the hearths. Waste is introduced at the top and falls onto the upper-most hearth. During the incineration process, it is scraped down from hearth to hearth to the lower-most hearth. Ash is taken out at the bottom. Air for incineration circulates countercurrent from the bottom up. Multiple hearth incinerators are generally used to roast metal ores. In pollution control, they have been applied to dispose of excess sludge from effluent treatment. For an advanced design see *hybrid technology incinerator*.

municipal effluent treatment plant 城市污水處理場 *chengshi wushui chulichang*

為處理一般城市污水而特別設計的污水處理場。城市污水具有流量大、污染負荷較小的特點，但是其流量和負荷的晝夜變化大。如用*活性污泥法*進行處理，曝氣時間需要2小時或更長時間。問題可能來自各小行業和小工廠的排水；因此必須小心避免毒物的衝擊負荷。在這些工廠中推行*清潔生產計劃*，將會對末端處理提供有力的支持。

An effluent treatment plant designed specifically for the treatment of general municipal sewage. Municipal waste water is characterized by a high flow volume and a relatively low pollution load. There are, however, strong diurnal variations of flow and load. With the treatment based on *activated sludge technology*, aeration times of 2 hours and more are required in the aeration basins. Problems may arise from different small trade and small industrial factories; care must be taken to avoid toxic shock loads. Instruction and communication programs for *clean production*, directed at these factories, often provide a valuable support to technical end-of-pipe measures.

municipal incinerator 城市(廢物)焚燒爐 *chengshi (feiwu) fenshaolu*

由市政當局擁有和運作的用於處置城市垃圾的焚燒爐。它大多是爐箅式焚燒爐，但也正在推廣應用其他類型的焚燒爐。

An incinerator owned and operated by a municipality to dispose of *municipal trash*. Municipal incinerators are in most cases *grate incinerators*; however, other types are being promoted.

municipal sewage 城市污水 *chengshi wushui*

一個城市(包括小工廠)排出的一般性的混合污水，其主要成分是人的糞便，因此易於生物降解。城市污水的濃度一般比工業廢水低，但問題可能來自小量的含有毒組分的工業廢水(如電鍍工業廢水)的影響。

The general mixed sewage of a city including small trade. A large fraction of municipal sewage consists of human excrement and is therefore easily biodegradable. The overall concentration is generally lower than that of *industrial effluents*. Problems may result from small trade effluents, if these contain toxic elements, as for instance from electroplating industries.

municipal trash 城市垃圾 *chengshi laji*

城市中收集的一般廢物的統稱術語。其數量可從每人每年大於600公斤至300公斤或以下，由社會的富裕程度、廢物分離收集設施、人們受教育的水平和廢物處置*費用*等因素所決定。城市垃圾應只包括一般家庭廢物和來自小型商業與工業的類似廢物，應避免含有任何可能會造成問題的廢物成分，即不應含*特殊廢物*。(參見：城市廢物管理)

A common term for general wastes collected in a city. The quantity varies from over 600 kg to 300 kg or less per person per year. This quantity is influenced by factors such as the affluence of a society, facilities for separate waste collection, instruction of the population and *fees* for waste disposal. These wastes should only contain general household wastes and similar wastes from small trade and industry.

Any waste components leading possibly to problems (*special wastes*) should be avoided. (See also *municipal waste management*)

municipal waste disposal fees 城市廢物處置費 *chengshi feiwu chuzhifei*

　　市政部門處置一般生活垃圾而分攤到各戶家庭需支付的費用，它根據所送去處置的廢物的數量來決定收費的多少，這符合*污染者付費原則*。

Fees which have to be paid by the private households for the disposal of common trash and garbage by the municipality. These fees are in line with the application of the *polluter-pays-principle* and they depend on the quantity of waste given for disposal.

municipal waste management 城市廢物管理 *chengshi feiwu guanli*

　　保證各類城市廢物按環境可接受的方式進行處置而採取的協調管理行動。其總體管理概念係面向一般居民和小型企業，包括採取以下行動：（1）教育居民在可能的情況下積極推行廢物的分離收集和廢物的再利用；（2）提供可再利用物質如馬口鐵罐、鋁罐、玻璃瓶、紙和塑料等分離收集的設備；（3）推行家庭小規模堆肥；（4）指導小型企業採用清潔工藝和減少廢物；（5）建立廢物收集和運輸系統，實行廢物的定期收集；（6）提供有效和足夠的處置技術，如受控填埋場和／或具有煙氣控制的城市焚燒爐；（7）如果政策上可行的話，按污染者付費原則繳納處置費。

Coordinated management activities which guarantee an environmentally acceptable disposal of all types of municipal wastes. An overall concept requires a number of activities, directed at the general population and at small enterprises: (a) instruction and information of the population to promote waste separation and waste reuse whenever possible; (b) the availability of facilities for separate collection of reusable materials, such as tin cans, aluminum cans, glass bottles, paper, plastic etc.; (c) promotion of small scale composting by households; (d) instruction of small trade enterprises on clean processes and waste reduction; (e) a waste collection and transport system for regular collection of wastes; (f) having available efficient and adequate disposal technology: either controlled landfills and/or a municipal incinerator with emission control; and (g) if politically feasible, the application of disposal fees based on the polluter-pays-principle.

municipal zoning laws 城市區劃法 *chengshi quhuafa*

　　市政級的立法，它規定了城市中具有不同使用功能的建築區：工業區、貿易和小廠區、居民區（可能有2-3種類型）和商業區。區劃法對環境保護具有重要作用，它們為居民提供並保護了最適宜的生活條件。

Legislation, generally on the municipal level, specifying building zones for different usage in a municipality. Zones may be: industrial use, trade and small factories, residential zones (possibly 2 to 3 types), business zones. Zoning laws are important instruments in the context of environmental protection. They serve to provide and safeguard optimal living conditions for the population.

N

nanofiltration 毫微米級過濾 *haoweimiji guolü*

　　基於*半滲透膜*的分離過程，其分離能力在*超濾*和*逆滲透*之間。它能截除分子量300–500的物質，因此氯化鈉之類的一價鹽可通過膜，而大多數有機化合物將被截留下來。毫微米級過濾主要應用於化學產物（如染料）的濃縮和脫鹽以及廢水的預處理。與其他的膜過程一樣，毫微米級過濾也是靠膜兩側的壓力差來驅動的。

　　A separation process based on *semi-permeable membranes*. The separation capabilities of nanofiltration are between *ultrafiltration* and *reversed osmosis*. The characteristic cut off is in the molecular weight range of 300 to 500. Therefore, monovalent salts, e.g. sodium chloride pass through the membrane while most organic molecules are held back. Important applications are to concentrate and de-salt chemical products (e.g. dyestuffs) and to pretreat waste water.

　　As in other membrane processes, nanofiltration is driven by a pressure difference across the membrane.

nanogram 毫微克 *haoweike*

　　10^{-9}克，環境痕量分析使用的單位。例如，把1克鹽溶於一個體積為1百萬立方米（200米×200米×25米深）的小湖泊中，其鹽濃度大約為每升1毫微克。

　　The 1000 millionth part of one gram; 10^{-9}g; a unit in environmental trace analysis. As an illustrative example: dissolving one gram of salt in a small lake of 1,000,000 m^3 (200m x 200 m x 25 m) will result in a concentration of salt of about one nanogram in each liter.

nanometer, nm 毫微米 *haoweimi*

　　長度單位，10^{-9}米，用以度量和確定諸如光波長的單位。

　　A unit of length, 10^{-9} meter; a unit used to measure and specify e.g. the wavelength of light.

natural treatment systems for effluents 廢水自然處理系統 *feishui ziran chuli xitong*

基於在自然環境中如淺塘、生長水生植物的沼地或土壤(包括植物的根部)中發生的生物和生物化學過程的廢水處理系統。它具有投資小、能耗少但佔地較大的特點,可用於處理含易生物降解污染物的廢水,也可對用傳統方法處理過的廢水作最終淨化處理。

Treatment systems for effluents which are based on biological and biochemical processes taking place in nature, for instance, in flat ponds, or a marsh with aquatic plants, or soil including roots of plants. The systems require lower investments and less energy but more land. This treatment can be applied to effluents with easily biodegradable, non-critical pollutants. It can also be used as a polishing treatment for effluents that have been treated before by conventional means.

negative ionizing electrode (electrostatic precipitator) 負離子化電極(靜電除塵器) *fulizihua dianji (jingdian chuchenqi)*

靜電分離器中的負電極,通常它是由金屬絲構成的電極。由於負電極和正電極(集塵電極)之間很大的電位差,使得塵粒荷電並被庫倫力吸引到 *沉積電極*。

The negative electrode in an electrostatic separator. It is generally designed as a wire electrode. As a result of the high difference of the electric potential between the negatively charged electrode and the positive collecting electrode, dust particles are electrically charged and are then attracted by Coulomb forces to the *precipitation electrode.*

neutralization equivalents 中和當量 *zhonghe dangliang*

用以表示廢水酸度的一項參數。酸度是用化學中和當量(毫克當量/升或毫價/升)來測定和表示的,它也可以表示為所需要的氧化鈣或氫氧化鈣的量。中和當量與廢水處理場進水中和時所需要的鹼量直接有關。

A parameter used to describe the acidity of an effluent. The acidity is measured and expressed in terms of chemical neutralization equivalents (mEq/liter or mVal/liter). It may also be expressed as the amount of calcium oxide or calcium hydroxide needed. Neutralization equivalents are directly related to the amount of alkali needed to neutralize the effluent input at an effluent plant.

NGO 非政府組織 *feizhengfu zuzhi*

    Non-governmental organization的縮寫。指聲稱代表公眾或某些*有利 害關係者(團體)的利益*的一些私人組織團體。非政府組織的目標是很廣 泛的，有些組織還設定了環境目標。許多非政府組織團體在國際範圍內 活動，其目標往往超越了現今的國家法規。

    Non-governmental organization. A term used for some privately organized groups claiming to represent the interests of the public or of specific *stakeholder groups*. The scope of the objectives of NGOs is very wide, some follow environmental goals. Many NGO groups are operating on an international scale. Very often their objectives go much beyond current national legislation.

nitrification 硝化(作用) *xiaohua (zuoyong)*

    利用特殊微生物，如*亞硝化桿菌和硝化桿菌*，將氨氮生物氧化為硝 酸鹽的過程。硝化作用可在廢水生物處理裝置的好氧段進行，但為了完 全去除氮，硝化段必須與反硝化段聯合使用。

    The biological oxidation of ammonia nitrogen to nitrate by the use of special microorganisms, e.g. *nitrosomonas* and *nitrobacter*. Nitrification may take place in the aerobic zone of a biological effluent plant. To remove nitrogen totally, the nitrification zone must be combined with a denitrification zone.

nitrification-denitrification process 硝化-反硝化過程 *xiaohua-fanxiaohua guocheng*

    聯合應用*硝化和反硝化*兩個階段，從廢水中去除含氮物質的過程。

    The process of removing nitrogeneous substances from an effluent through the combination of the two steps *nitrification* and *denitrification*.

nitrifying bacteria 硝化細菌 *xiaohua xijun*

    在土壤和水生生境中存在的一類好氧細菌。硝化細菌包括一些不同 的細菌，它們通過將氨或亞硝酸鹽氧化為硝酸鹽而獲得能量。在廢水脫

氮的硝化階段，硝化細菌起著重要作用。（參見：硝化作用、反硝化-硝化過程）

A family of aerobic bacteria which occur in soils and some aquatic habitats. The family includes several different bacteria. The organisms obtain energy by oxidizing ammonia or nitrite to nitrate. Nitrifying bacteria are instrumental in the nitrification step of effluent denitrification. (See also *nitrification*; *denitrification-nitrification process*)

nitrobacter 硝化桿菌 *xiaohua ganjun*

在土壤和某些水生生態系統中存在的一種*硝化細菌*。

One genus of *nitrifying bacteria*, occurring in soils and certain aquatic eco-systems.

nitrogen cycle 氮循環 *danxunhuan*

元素氮在生態圈中的閉路循環。簡單地說，氮循環的最重要環節是：某些微生物對空氣中分子氮的同化作用，氮以蛋白質等含氮化合物的形式在*食物鏈*中的轉移，以及微生物對死亡動物的蛋白質進行的最終代謝作用（*反硝化/硝化*）使之又回到分子氮狀態。實際上，這一過程是非常複雜的，還包括有許多其他環節。（參見：氧循環、碳循環）

The closed cycle of the element nitrogen in the ecosphere. In a much simplified representation, the most important links of the nitrogen cycle are: the assimilation of molecular nitrogen from air by certain microorganisms, the transfer of nitrogen through nitrogenous compounds, e.g. proteins, along the *food chain*, the final metabolism of proteins of dead animals, again by microorganisms (*denitrification/ nitrification*) back to molecular nitrogen. In reality, this picture is much more complex and contains many additional steps and links. (See also *oxygen cycle*; *carbon cycle*)

nitrosomonas bacteria 亞硝化桿菌 *yaxiaohua ganjun*

屬於硝化桿菌族中的一類硝化細菌，它可在好氧過程中將氨氧化為亞硝酸鹽，然後亞硝酸鹽被另一類細菌*硝化桿菌*氧化為硝酸鹽。在廢水*硝化-反硝化*處理中利用的就是這兩種細菌。（參見：硝化細菌）

A class of nitrifying bacteria belonging to the family of nitrobacteriacea. Nitrosomonas bacteria are able to oxidize ammonia to nitrite in an aerobic process. Nitrite is then oxidized to nitrate by another group of bacteria called *nitrobacter*. Both are used in the *nitrification-denitrification* treatment of effluents. (See *nitrifying bacteria*)

NOEC 無可見作用濃度 *wukejian zuoyong nongdu*
見：no observable effect concentration。
See *no observable effect concentration.*

NOEL 無可見作用水平 *wukejian zuoyong shuiping*
見：no observable effect level。
See *no observable effect level.*

no observable effect concentration (NOEC) 無可見作用濃度 *wukejian zuoyong nongdu*
對靶生物不會產生可察覺影響(急性或慢性效應)的某種物質。在毒理學研究和職業衛生研究中常常使用這一術語。

The concentration of a substance under investigation at which no effect (neither acute nor chronic) upon the *target organism* can be observed. The term is used in different types of toxicological studies and also in *occupational health* studies.

no observable effect level (NOEL) 無可見作用水平 *wukejian zuoyong shuiping*
與毒性研究如廢水毒性有關的一項術語。與無可見作用濃度不同，無可見作用水平是指在不產生可見作用的情況下廢水的稀釋倍數，它可以通過將廢水系列稀釋後對魚進行毒性試驗來確定。其值用稀釋比表示，如1份廢水：512份水。

A term used in connection with toxicity studies, for instance of effluents. In contrast to the NOEC value, the NOEL value refers to a dilution of an effluent, which produces no observable effect. The NOEL may be determined by testing a series of dilutions of the effluent for their toxicity, e.g. versus fish. NOEL is reported as a dilution ratio, e.g. as 1 part of effluent in 512 parts of water.

noise 噪聲 *zaosheng*

任何令人厭煩和影響健康的聲音。噪聲會損害或破壞人的聽力，造成身體或精神緊張。它所產生的不舒適感取決於聲音的強度（可以用物理方法測量）和個人的因素如人的個性、健康狀況和對聲音所持的態度。另外，也還與人的活動和噪聲的信息量有關。同一聲級和同一類型的聲音（如某些迪斯科音樂），既會使一些人感到愉快，又會使另一些人感到極度厭煩。

Any unwanted sound, affecting human comfort and health. Noise can impair or destroy human hearing and it can cause physical or mental stress. The discomfort depends on the intensity of sound (which can be measured physically) and on personal factors, such as personality, health and attitude towards sound. Other factors are related to the activity of the person and the information content of the noise. The same level and type of sound (e.g. some music in a DISCO) may be pleasant to one person, and a horror to another one.

noise, ambient 環境噪聲 *huanjing zaosheng*

在某一地點所能聽到的由各種聲源產生的總噪聲。

The total noise resulting from all different sources of sound which can be heard at a given point.

noise, ambient (immission) limits 環境噪聲（散發）限值 *huanjing zaosheng (sanfa) xianzhi*

以*分貝*表示的環境噪聲的限值。在*職業衛生*中利用此限值來保護職工聽力不受損害。另外，噪聲污染作為一個環境問題，在*城市區劃法*中對*噪聲防護區*也按其使用的類型規定有各自的限值。

Limiting values for ambient noise levels, given in *decibel(A)*. Such limits are used in *occupational health* to prevent damage to the hearing of employees. In relation to noise pollution as an environmental problem, limits may be defined in *noise protection zones* in *municipal zoning laws*. The limiting values depend on the type of usage of a zone.

noise, background 背景噪聲 *beijing zaosheng*

當聲源處在寂靜無聲狀況時某一地點的聲級。

The sound level at a given point and under given circumstances when the emission source under investigation is silent.

noise, disturbance limits 噪聲，干擾限值 *zaosheng, ganrao xianzhi*

以分貝表示的噪聲限值，超過該限值時通常會對人產生不良影響。如超過30-40分貝，會干擾睡眠；超過45-55分貝，會干擾集中注意力；超過50-60分貝，會干擾交談；而在更高的聲級下甚至會損害健康，這取決於暴露於噪聲下時間的長短。

Limits in dB(A) above which noise generally has a negative impact on well being: disturbance of sleep, above 30 to 40 dB(A); disturbance of concentration, above 45 to 55 dB(A); disturbance of communication, above 50 to 60 dB(A); and health impairment occurs at higher levels, depending on the duration of the exposure.

noise emission limits 噪聲散發限值 *zaosheng sanfa xianzhi*

見：sound emission limits。

See *sound emission limits*.

noise monitoring 噪聲監測 *zaosheng jiance*

對某一地點聲壓級的測量（以分貝表示）。永久性的噪聲監測係在機場附近以及在大工廠（如煉油廠）進行，而其他的噪聲監測可根據具體調查研究對象而定。在評價其測定結果時，還必須考慮噪聲的峰值及其出現的頻率。

The measurement of the sound pressure level in dB(A) at a given location. Permanent noise monitoring is carried out in or near airports or large factories, (e.g. refineries). For other investigations, campaigns of measurements are carried out for special investigations. For the evaluation of the results, peak values and their frequency also must be taken into consideration.

noise pollution 噪聲污染 *zaosheng wuran*

　　噪聲對人的干擾。由於對噪聲的感受隨個人的觀點、感覺和敏感程度有很大不同，所以很難對噪聲污染下確切的定義。但無論如何噪聲污染會使*生活質量*大大下降。噪聲的控制使用*噪聲防護措施*技術及防止聲音傳播的措施。在*城市區劃法*中對*噪聲防護區*規定了各自的噪聲限值。

　　The disturbance of the population through *noise*. As the perception of noise depends very much on personal views, feelings and sensitivity, a clear definition of noise pollution is difficult. Nevertheless, noise pollution may lead to a significant reduction of the *quality of life*. The control of noise pollution is based on technical *noise protection measures* and on measures to inhibit the *sound transmission*. Limiting values are defined for *noise protection zones* in *municipal zoning laws*.

noise protection measures 噪聲防護措施 *zaosheng fanghu cuoshi*

　　在發生源減少聲音發散的技術與管理措施。技術措施包括：在噪聲防護和隔離建築物中密閉強聲源，採用消聲器，改進設備結構，更換強聲設備，以及設置屏障以阻止*聲音的傳播*。管理措施包括：規定*聲音發散源*的限值，對城市的某些開發區(*噪聲防護區*)規定噪聲級別，以及規定在強聲源下如在飛機起飛或載重運輸車輛下工作的時限。

　　Technical and administrative measures to reduce the emission of sound at the source. Technical measures include the enclosure of loud sound sources, silencers, construction measures in machines, the placement of loud machines in protected and insulated buildings as well as the construction of barriers to *sound transmission*. Administrative measures include specifications and limits for the *sound emission* of sources, specification of noise levels for certain areas in urban development (*noise protection zone*) and time restrictions on the operation of loud sources, e.g. on the take-off of airplanes or on heavy traffic.

noise protection zones 噪聲防護區 *zaosheng fanghuqu*

　　城市規劃中規定不許超過某些*聲壓級*的分區。通常，一個城市可劃分為若干個具有不同聲壓級的區域，如居民區(35-45分貝)、混合區(40-50分貝)和工業區(上限70分貝)。另外，夜間的噪聲限值往往比白天低。對於脈衝噪聲和持久噪聲，也可規定較低的限值。

Zones in municipal planning in which certain *sound pressure levels* must not be exceeded. A city usually has several different zones with levels for pure residential areas (35 to 45 dB(A)), mixed zones (40 to 50 dB(A)) and industrial zones (up to 70 dB(A)). Limiting values at night are often lower than daytime limits. Lower limits are also applied to noise with impulses of sound, and noise with persistent tones.

non-biodegradable compounds 不可生物降解化合物 *bukeshengwu jianjie huahewu*

不能*生物降解*的物質。有些化合物僅在*好氧*條件下是不可生物降解的，但可通過*厭氧*過程進行分解。例如，氯代脂肪烴可在厭氧條件下脫氯，然後再在好氧系統中完全降解。許多化合物可在好氧和厭氧條件下降解。

Substances which are not amenable to *biodegradation*. Certain compounds are non-biodegradable under *aerobic* conditions only, but can be broken down by *anaerobic* processes. Chlorinated aliphatic hydrocarbons can be taken as an example. They are dechlorinated under anaerobic conditions; this is followed by complete degradation in an aerobic system. Many compounds can be degraded under aerobic as well as anaerobic conditions.

non-dispersive infrared analysis 非色散紅外分析 *feisesan hongwai fenxi*

見：infrared analysis, non-dispersive。

See *infrared analysis, non-dispersive*.

non-economic legal instruments 非經濟的法律手段 *feijingjide falü shouduan*

基於容許做什麼或禁止做什麼這一概念的傳統法律控制手段，它也稱為命令和控制手段(管制法)。與此相反的是*環境立法中的經濟手段*。

Traditional legal control instruments which are generally based on the concept of allowing or forbidding certain actions. These systems are also called command and control ('police law') instruments. In contrast to these, see *economic instruments in environmental legislation*.

non-governmental organization 非政府組織 *feizhengfu zuzhi*

見：NGO。

See *NGO.*

non-reactive wastes 非反應性廢物 *feifanyingxing feiwu*

見：wastes, non-reactive。

See *wastes, non-reactive.*

non-renewable resources 不可再生資源 *bukozaisheng ziyuan*

見：resources。

See *resources.*

non-secure landfill 非安全填埋 *fei'anquan tianmai*

　　沒有任何特別預防措施來保護地下水以及沒有特殊管理與控制的廢物堆放或填埋。（參見：受控填埋、廢物管理）

　　A waste dump or landfill without any special precautions to protect groundwater and without specific management controls. (See also *controlled landfill; waste management*)

non-sensitive catalysts 非敏感性催化劑 *feiminganxing cuihuaji*

　　對*催化劑毒物*不敏感的催化劑；或不會產生中毒性不可逆損害的催化劑。

　　Catalysts which are not sensitive to *catalyst poisons*; catalysts which are not damaged irreversibly by poisoning.

NO$_x$ 氮氧化物 *danyanghuawu*

　　具有不同氧化度的氮的氧化物，如一氧化二氮、一氧化氮和二氧化氮的總稱，它是一項*綜合指標*。測定廢氣中氮氧化物的自動分析儀賦予這三種氧化物以不同的權重。這些儀器的測定原理不同，因此測定的結果並不總是可以相比的。為了準確測定各種氮的氧化物的量，必須使用特定的方法。

　　A *group parameter* for the total of the oxides of nitrogen with different degrees

268

of oxidation: N_2O, NO and NO_2. Automatic instruments to measure NO_X in an off-gas attribute different weights to the three oxides. They are based on different principles and the results are therefore not always comparable. To determine accurate quantities of the individual oxides, specific methods must be used.

nutrients in effluent treatment 廢水處理的營養物 *feishui chulide yingyangwu*

微生物新陳代謝所需的營養物。在*活性污泥技術*中，微生物所需的營養物主要是氮和磷，此外，還需微量的其他物質。在廢水生物處理中，通常應維持碳：氮：磷等於100:5:1。

Microorganisms require certain nutrients for their metabolism. In *activated sludge technology*, these nutrients include mainly nitrogen and phosphorous. Additional substances are needed only in trace quantities. As a rule, a ratio of carbon to phosphorous to nitrogen of 100 to 5 to 1 should be maintained in biological effluent treatment.

O

object off-gas 設備廢氣 *shebei feiqi*

見：apparatus off-gas。

See *apparatus off-gas*.

occupational health 職業衛生 *zhiye weisheng*

有關工人在企業工作期間健康保護事項的總稱。健康損害可能是由於工人接觸有毒或有害化學品，或接觸有害工作條件而造成的。廣義地說，還包括職業安全在內。

The sum of the aspects which relate to the health maintenance of workers during their occupation in an enterprise. Health impairment may be the result of an exposure of workers to toxic or noxious chemicals or to health-damaging working conditions. In a wider sense, occupational safety is also included.

octanol-water coefficient 辛醇-水系數 *xinchun-shui xishu*

表徵某物質親脂性的參數，它反映某化合物在辛醇和水相之間的分配狀況，以該化合物在正辛醇和水中的平衡濃度之比的對數來計算。系數值大（大於3），表示*親脂性*高，則在*食物鏈*中*生物富集*的可能性大。

A parameter for characterizing the degree of lipophilicity of a substance. The octanol-water coefficient reflects the distribution of a compound between n-octanol and an aqueous phase. It is calculated as the logarithm of the ratio of the equilibrium concentrations of the compound in n-octanol and in water. A high coefficient (a value greater than 3) indicates high *lipophilicity* and a high *bioconcentration* potential in the *food chain*.

odor 氣味 *qiwei*

可用嗅覺來感知的物質的性質。不同人對氣味的敏感性很不相同，它的描述隨人區別很大，可按令人愉快到令人厭惡的分級範圍加以說明。氣味是不能用儀器來測定的。化工廠周圍的氣味往往是令人討厭的而且與毒性有關。（參見：臭味測定法）

The property of a substance which is noticed by the sense of smell. Different persons have highly different sensitivities. The description of smells depends very much on the person and odors are described on a scale ranging from pleasant to highly repulsive. Odor is not measurable by instrumental methods. Odors around chemical factories are often taken as repulsive and are associated with toxicity. (See also *olfactometry*)

odor index 臭味指數 *chouwei zhishu*

每一有機化合物特有的濃度值。臭味指數反映的是物質的臭閾濃度，這是一般人剛能感覺出某化合物的最低濃度。在已經發表的文獻資料中，列有許多揮發性有機化合物的臭味指數的數據。

A concentration value which is characteristic for each organic substance. The odor index reflects the threshold odor concentration of a substance. This is the lowest concentration at which a compound is still noticeable by the average person. Lists of the odor indices for a large number of volatile organic chemicals have been published.

odor recognition threshold limit 惡臭閾限值 *echou yuxianzhi*

見：odor index。

See *odor index*.

OECD 經濟合作與發展組織 *jingji hezuo yu fazhan zuzhi*

經濟合作與發展的國際性組織，其總部設在巴黎。它主要開展經濟領域的活動，但現在也在積極開展環境保護工作。經濟合作與發展組織頒佈了許多有關環境分析的標準分析方法，也還提出了諸如飲用水中污染物最大容許濃度的限值。

The international Organization for Economic Cooperation and Development has its headquarters in Paris. Initially, the OECD was active mainly in the economic field. Today, it is also active in environmental protection. It has issued a number of documents on standardized analytical methods for environmental analysis. OECD also proposed limits of maximum allowable concentrations of pollutants, e.g. for drinking water.

OECD-BOD Test 經濟合作與發展組織－生物需氧量測定 *jingji hezuo yu fazhan zuzhi-shengwu xuyangliang ceding*

由 *經濟合作與發展組織* 提出的測定物質長期可 *生物降解性* 的試驗方法。它是在2升的玻璃燒杯中於攪拌和連續曝氣的條件下，進行為期至少28天的試驗。為了得到可實際應用的結果，必須從處理該廢水的廢水處理場取來 *接種污泥*。然後測定過濾後的樣品的總有機碳，根據試驗過程中總有機碳的去除率，就能判別降解的程度。

A test to determine the long term *biodegradability* of a substance, as proposed by *OECD*. The test is carried out in a 2 liter glass beaker under stirring and with continuous aeration for at least 28 days. To obtain results which are applicable to the practical case, the necessary *inoculum* must be prepared from the effluent treatment plant which will be used to treat the waste water. The degree of degradation is controlled by measurement of TOC removal in a filtered sample.

off-gas 廢氣 *feiqi*

從某排放源排出的污染氣流或空氣流，它們可以區分為不同類型的廢氣。

A polluted gas stream or air stream from an emission source. Different off-gas classes can be distinguished.

off-gas analysis 廢氣分析 *feiqi fenxi*

空氣污染控制中對 *排放廢氣* 的測定。廢氣分析取決於排放源的特徵、工藝過程、污染物的種類，也還取決於測定的目的。

常規的空氣污染物為顆粒物、二氧化硫、氮氧化物、重金屬和有機污染物。可能的廢氣排放源包括工業排放源（如反應容器、空氣污染控制設備或整個生產廠房）、水泥窯、危險廢物焚燒爐、電廠、貯罐。廢氣取樣要比廢水取樣難，而且它大多是直接在廢氣流中（ *取樣口*）進行連續測定。測定時必須考慮生產運行的狀況。

所採用的分析方法取決於污染物的化學性質，但通常採用氣相色譜法。另外，對測定的結果進行正確的 *統計評估* 也是很重要的。（參見：空氣質量目標）

Measurements of *emissions* in connection with air pollution control. The analyses depend on the characteristics of the emission source, the process, the type of pollutant and on the objectives of the measurements.

Conventional air contaminants are: particulate matter, sulfur dioxide, nitrogen oxides, heavy metals and organic contaminants. Possible emission sources are industrial sources (e.g. reaction vessels, APC units, or whole production buildings), cement kilns, waste incinerators for different hazardous wastes, power plants, storage tanks. Sampling is more difficult than sampling of liquid discharges. Analysis is carried out mostly by continuous measurement, directly in the off-gas stream (*sampling ports*). The status of the manufacturing operation must be taken into consideration.

Analysis methods vary with the chemical nature of the contaminant, often, gas chromatography is used. Proper *statistical evaluation* of the results is important. (See also *air quality goals*)

off-gas classes 廢氣類別 *feqi leibie*

為了有效地應用空氣污染控制技術，可把廢氣分類為*室內空氣*、*工藝廢氣*和*裝置廢氣*。通常，這三類廢氣所需處理的氣量與污染物濃度是不相同的，因此需用不同的空氣污染控制方法進行處理。

For the purpose of applying efficient air pollution control technology it is advantageous to classify off-gas streams into: *room air*; *process gas*; *apparatus off-gas* or *object off-gas.* These three categories usually have different volumes and different concentrations of pollutants to be treated and therefore need different APC methods.

off-gas combustion 廢氣燃燒 *feiqi ranshao*

用焚燒方法消除廢氣中有機污染物的技術。將廢氣與輔助燃料和輔助空氣一起加入到*焚燒室*，使有機污染物(*揮發性有機化合物*)在其中氧化分解。現代化的高溫燃燒系統，都伴有蒸氣發生設備以節能和回收有機物的能量。如果廢氣中有機物(揮發性有機化合物)的濃度變化不定，則需要有完善的過程控制系統。(參見：鍋爐房中廢氣的燃燒，催化燃燒)

A technology used to destroy organic contaminants of an off-gas stream by incineration. The off-gas stream is fed into an *incineration chamber*, together with additional fuel and possibly additional air. Organic contaminants (*VOC*) are destroyed by oxidation. Modern high temperature systems are combined with the generation of process steam to economize energy and to recover the energy content of the organic contaminant. Off-gas streams with variable concentrations of VOC require elaborate process control systems. (See also *combustion of off-gases in boiler house*; *catalytic combustion*)

off-gas data sheet 廢氣數據卡 *feiqi shujuka*

見：*排放清單*。但對每一廢氣流編寫如廢物數據卡或廢水數據卡一樣的廢氣數據卡是不可行的。

See *emission inventory* . Editing off-gas data sheets for individual off-gas streams, similar to waste data sheets or effluent data sheets, is not practicable.

off-gas scrubbers 廢氣洗滌器 *feiqi xidiqi*

見：wet APC-systems。

See *wet APC-systems*.

off-gas separation concept 廢氣分別處理概念 *feiqi fenbie chuli gainian*

對生產現場產生的不同廢氣分別作為單獨的廢氣流進行處理的基本概念。它與把所有廢氣集中在一個大的綜合處理裝置中進行處理的觀念不同，使得能對不同的廢氣進行個別特殊的處理，從而更具有靈活性和經濟性。

The basic concept of treating different off-gas streams of a manufacturing site as separate streams of emissions. This contrasts to the idea of combining all streams and treating them together in one large multi-purpose unit. Off-gas separation allows for specific individual treatments; the concept gives more flexibility and is therefore more economical.

off-gas treatment 廢氣處理 *feiqi chuli*

對廢氣進行處理以回收或去除其中的污染物。（參見：空氣污染控制技術）

The treatment of off-gas streams to recover, modify or remove pollutants. (See also *APC technologies*)

off-gas velocity 廢氣流速 *feiqi liusu*

空氣污染控制中用於測定廢氣流量的因子。當利用流速數據計算流量時，必須考慮廢氣管道中的不規則流動和擾動。（參見：風速計）

Measurements used in APC for the determination of off-gas flows and volumes. Irregular flow and turbulence in an off-gas duct have to be taken into consideration when calculating volume flows from velocity data. (See also *anemometer*)

off-site treatment 廠外處理 *changwai chuli*

在廠外不屬於該工廠擁有的設施中進行的廢物處理。在這種情況下，管理控制應包括廠外處理。在處理*特殊廢物*時，應遵循"*從搖籃到墳墓*"*的原則*。在重要的情況下，應對廠外處理設施進行*外部審計*。

Treatment of waste streams in facilities outside the factory, often not owned by the factory. Management controls should in such cases also cover off-site treatment. In the treatment of *special wastes*, the *cradle-to-grave-principle* should be applied. In critical cases, off-site treatment units should be subject to an *external audit*.

off-site treatment in remediation 場外恢復補救處理 *changwai huifu bujiu chuli*

將污染的土壤或地下水送至場外處理設備進行處理的恢復補救措施。

Remediation or decontamination of contaminated ground or groundwater by transporting the contaminated material to an off-site treatment facility.

off-specification products 不合格產品 *buhege chanpin*

生產出的不合乎質量規格要求因而被認為是廢物的產品。這類廢物根據其性質的不同，應該以另一種方式加以再利用，或者必須作為*特殊廢物*進行處理和處置。

Products from production, which do not meet the quality specifications and which are therefore considered as waste material. Basically, this type of waste should be reused in another way, or it must be treated and disposed of as *special wastes*, depending on its properties.

olfactometry 惡臭測定法 *echou cedingfa*

臭味的評定。臭味是主觀性很強的感覺，不能用儀器來測定。為了定量地、再現地評定臭味（*臭味指數*），使5人一組的鑒定人員在標準條件下暴露於含有不同濃度的測試物質的大氣中，由他們確定剛能聞辨出的某種惡臭物質的最低濃度。

The assessment of odor. Odors are very subjective sensations and cannot be measured by instruments. To arrive at a reproducible quantification of odor (*odor index*) a panel of five test persons is exposed under standardized conditions to an atmosphere, containing the test substance in variable concentrations. The persons in the panel have to determine the minimum concentration of a bad smelling substance which can still be noticed.

oligotrophic 貧營養的 *pinyingyangde*

按字義指缺少食物或缺乏營養。這是湖泊生態系統缺少磷、氮營養物質的一種狀態。通常，未開發的山區湖泊多處於貧營養狀態。同樣，有機碳濃度很低的環境也屬於貧營養狀態。（參見：富營養的）

Literally lacking food or nourishment. The state of a lake ecosystem with minimal input of fertilizing substances, e.g. phosphorous and nitrogen. Often, undisturbed mountain lakes are in an oligotrophic state. Also, environments with a very low concentration of organic carbon are oligotrophic. (See also *eutrophic*)

on-column injector (in GC) 柱上注射器（氣相色譜法中）*zhushang zhusheqi (qixiang sepufazhong)*

毛細管氣相色譜法中使用的技術和部件。將某溶液的全樣品在低溫和無樣品損失的情況下，通過一特殊進樣口注射入毛細管柱中。在進樣時，載氣流會受到擾動。這種操作方法的優點是可以進行全樣品分析，而且敏感物質可幾乎不被分解。

A technique and module used in capillary gas chromatography. The total sample of a solution is injected through a special injection port into the capillary column at a low temperature and without any loss of sample. During the injection, the carrier gas stream is interrupted. The advantage of this mode of operation is that the total sample is used for the analysis and there is little or no decomposition of sensitive substances.

on-line analysis 在線分析 *zaixian fenxi*

見：analysis, on-line。

See *analysis, on-line.*

on-site treatment 現場處理 *xianchang chuli*

在生產現場屬於企業所擁有的設施進行的廢物處理。（參見：廠外處理）

Treatment of waste streams by units located on the manufacturing site and generally owned by the enterprise. (See also *off-site treatment*)

on-site treatment in remediation 現場恢復補救處理 *xianchang huifu bujiu chuli*

直接在污染場地的處理設施中對污染的土壤或地下水進行處理的恢復補救措施。（參見：現場恢復補救）

Remediation or decontamination of contaminated ground or groundwater in a facility directly at the contaminated site. (See also *in-situ remediation*)

open hearth incinerator 平爐焚燒爐 *pinglu fenshaolu*

見：multiple hearth incinerator。

See *multiple hearth incinerator.*

open mass flow system 開式物流系統 *kaishi wuliu xitong*

物料由一端進入從另一端流出的過程。它是一種無直接反饋的線性、開式的物流系統。從環境觀點來看，以*閉合物流系統*為佳。

A process in which material enters in one end and leaves at the other. It is a linear, open mass flow system without any direct feedback. For environmental reasons, *closed mass flow systems* are to be preferred.

operating costs 運行費用 *yunxing feiyong*

　　見：total product manufacturing costs。

　　See *total product manufacturing costs.*

operating procedure 操作規程 *caozuo quicheng*

　　生產工藝文件資料中有關生產技術操作的內容。操作規程詳述了各個工藝步驟，列出了生產過程中的關鍵問題和檢查控制點，它還必須包括與污染控制和安全生產有關的一切問題。

　　In technical manufacturing operations, a section of the *process documentation.* The operating procedure gives a description of the individual process steps. The operating procedure lists critical aspects of the process as well as checkpoints and controls to be carried out. It must also include a full description of all aspects which are relevant for pollution control and safety.

optical sensor 光傳感器 *guangchuanganqi*

　　光測量系統的信號檢測單元，它有許多不同的類型，從簡單的光電池到高靈敏的光電培增管。其原始信號可以用電子儀器放大。在分析方面，它可應用於光傳輸、折射或色散系統，也可應用於螢光測定。

　　The signal detection unit of an optical measurement system. Optical sensors include a variety of types ranging from simple photocells to highly sensitive photo-multipliers. The primary signal may be amplified electronically. The application is in analytical light transmission, reflection or dispersion systems and also in fluorescence measurements.

Organization for Economic Cooperation and Development (OECD) 經濟合作與發展組織 *jingji hezuo yu fazhan zuzhi*

　　見：OECD。

　　See *OECD.*

OSHA 美國政府職業安全與衛生署 *meiguo zhengfu zhiye anquan yu weisheng shu*

　　The Occupational Safety and Health Administration of the United States Government 的縮寫。

osmosis 滲透（作用）*shentou (zuoyong)*

如兩溶液被一半滲透膜隔開，則溶劑會自發的流向溶液，或稀溶液會自發的流向濃溶液，而其中所含的較大的分子將會被阻留下來。滲透作用對生物機體的細胞和細胞壁起著重要作用。在一閉合系統中，如果穿過膜的滲透壓達到一定程度（取決於溶液的性質和膜兩側的濃度差），則液體的滲流將停止。（參見：逆滲透）

The spontaneous flow of solvent into a solution, from a more dilute solution into a concentrated one, if the two solutions are separated by a semi-permeable membrane. Larger molecules are held back. Osmosis plays a fundamental role in living organisms at the level of the cells and cell walls. In a closed system, the osmotic flow of liquid is stopped if the *osmotic pressure* across the membrane reaches a certain level which depends on the nature of the solution and the concentration difference on both sides of the membrane. (See also *reversed osmosis*)

osmotic pressure 滲透壓 *shentouya*

當用半滲透膜進行分離時，為了防止溶劑穿過膜流向溶液，必須施於溶液上的額外壓力。它等於溶劑停止流過半透膜時達到的壓力差。滲透壓主要決定於半滲透膜兩側物質的濃度。

The excess pressure which must be applied to a solution to prevent the passage into it of solvent when separated by a semi-permeable membrane. This is the same as the pressure difference reached across a semi-permeable membrane at the point where the flow of solvent is stopped. The osmotic pressure depends mainly on the concentrations of matter on both sides of the semi-permeable membrane.

overhead costs 一般管理費用 *yiban guanli feiyong*

工廠中與某一特定產品或生產過程沒有直接關係的那些費用要素。例如，不同層次的管理費用，質量控制費用，環境服務（如溝通／交流）費用以及某些基礎設施費用。實際上，它還包括不能分攤到產品中的一些其他的費用要素。

Those cost elements within a factory which cannot be directly associated with a particular product or production process. Examples are: management at different levels, quality control, central environmental service costs (e.g. communication), certain infrastructure costs. In actual practice, overhead costs also include some other elements which cannot be allocated to a product.

oxidation ditch 氧化溝 *yanghuagou*

　　廢水處理場一種特殊設計的 *活性污泥池*。它是一個圓形（或橢圓形）池，截面呈淺平長方形。污泥 *混合液* 靠旋轉的葉輪在池內循環，葉輪還同時起著推進液體流動和充入空氣的作用。這個系統的輔助設備基本上與其他形式的活性污泥系統相同。

A special design of an *activated sludge basin* of an effluent treatment plant. The oxidation ditch is constructed as a circular (or oblong) basin with a relatively flat rectangular cross section. The *mixed liquor* is circulated through the basin with the help of rotating paddle-wheels, serving at the same time to propel the liquid and to introduce air. Auxiliary elements of the system are basically as in any other activated sludge plant.

oxidative off-gas treatment 尾氣氧化處理 *weiqi yanghua chuli*

　　尾氣的惡臭往往是由 *厭氧過程* 產生的，它可以利用濕式空氣污染控制洗滌系統進行氧化 *脫臭* 處理，即在洗滌段於鹼性洗滌液中投加氯或過氧化氫作為氧化劑來脫臭，但必須小心避免氯過量。

Often, odorous substances in off-gas streams are the result of *anaerobic processes*. There is a possibility to *deodorize* such off gas streams by oxidation processes, using wet APC scrubbing systems. Chlorine or hydrogen peroxide in an alkaline scrubbing liquor may be used as an oxidizing agent in a scrubbing stage. Care must be taken to avoid excess chlorine.

oxygen cycle 氧循環 *yangxunhuan*

　　元素氧在生態圈中的閉合循環。簡單地說，氧循環的主要環節是：通過好氧微生物的氧化作用吸收和利用大氣中的氧，人和一切動物呼吸利用氧並排出二氧化碳（含結合態氧），以及綠色植物生長過程同化二氧化碳同時釋放分子氧到生態系統。但實際上這種聯繫是非常複雜的。（參見：氮循環、碳循環）

The closed cycle of the element oxygen in the ecosphere. In a simplified picture, some main steps of the oxygen cycle are: the absorption and use of atmospheric oxygen by aerobic microbiological oxidation. Use of oxygen by man and all animals for breathing, the production of CO_2, containing bound oxygen, the assimilation of CO_2 in the growth process of green plants and the simultaneous release of molecular oxygen into the ecosystem. In reality the interconnections are much more complex. (See also *nitrogen cycle*; *carbon cycle*)

ozone 臭氧 *chouyang*

由三個氧原子組成的 O_3 分子，它是一種具有極強氧化能力的特殊形態的氧。臭氧常常被認為與低層大氣中出現的*夏季煙霧*以及高層平流層中發生的*臭氧耗竭*有關。它也常常用作為*臭氧氧化技術*的氧化劑。

The molecule of O_3, composed of three atoms of oxygen. Ozone is thus a special form of oxygen with an extremely high power to oxidize other compounds. Ozone is often referred to in connection with *summer smog* situations in the lower atmosphere, as an oxidant in *ozone oxidation technology* and in connection with *ozone depletion* in the upper stratosphere.

ozone, stratospheric 平流層臭氧 *pingliuceng chouyang*

在正常未受污染的大氣中，臭氧主要存在於低層平流層中（在高度15公里至30公里處）。它是在短波紫外輻射作用下由分子氧形成的。在平流層，臭氧吸收了波長240毫微米至320毫微米的有害的太陽紫外輻射，因此對地球表面具有保護作用。平流層臭氧的含量以*多勃遜單位*表示。（參見：臭氧耗竭）

In the normal unpolluted atmosphere, ozone is mainly present in the lower stratosphere (at an altitude of 15 to 30 km) where it is formed from molecular oxygen by short wave ultraviolet radiation. In the stratosphere, ozone absorbs harmful solar UV radiation of a wavelength between 240 and 320 nm. It therefore has a protective effect on the surface of the earth. The quantity of stratospheric ozone is reported as Dobson Units. (See also *ozone depletion*)

ozone depletion 臭氧耗竭（破壞） *chouyang haojie (pohuai)*

臭氧耗竭是由於在平流層發生的複雜的游離基反應和光化學過程而

產生的，它會大大減弱臭氧對到達地球上的紫外輻射的保護屏障作用。
而氯氟烴化合物作為引發劑，對這一反應配位起著重要作用。（參見：
氯氟烴、蒙特利爾議定書）

Ozone depletion is the result of complex radical and photochemical processes taking place in the stratosphere. This is resulting in a significant reduction of the protective shielding function of ozone against ultraviolet radiation on earth. Certain chloro fluoro carbon compounds play an important role as initiators in the reaction complex. (See also *chloro fluoro carbons*; *Montreal Protocol*)

ozone hole 臭氧空洞 *chouyang kongdong*

報刊雜誌上用於説明因*臭氧耗竭*造成平流層臭氧濃度下降的專門用語。

A journalistic description of the lowering of the stratospheric ozone concentration as a result of *ozone depletion*.

ozone in summer smog 夏季煙霧中的臭氧 *xiaji yanwuzhongde chouyang*

在存在*夏季煙霧*的情況下，通過光化學反應在低層大氣中所產生的臭氧。

Ozone produced in the lower atmosphere through photochemical reactions in situations of *summer smog*.

ozone oxidation technology 臭氧氧化技術 *chouyang yanghua jishu*

臭氧具有高氧化電位，因此可用於氧化過程。它被大規模地應用於飲用水和游泳池用水的消毒處理。在中試試驗研究中，也利用臭氧來全部或部分氧化廢水中的污染物，但目前還未得到廣泛應用。在各種應用場合下，臭氧都是就地用臭氧發生器由乾燥空氣或氧於高壓放電下產生的。

As a result of its high oxidation potential, ozone may be used in oxidation processes. Ozone is used on a large scale for sanitizing purposes, in the treatment of water for drinking and for swimming pools. Experimental systems in the piloting stage also use ozone for full or partial oxidation of pollutants in effluents; however, this technology so far is not widely applied. For all applications, ozone is produced at the place where it is used from dry air or oxygen in a high voltage electrical discharge in an ozone generator.

P

packaging waste 包裝廢物 *baozhuang feiwu*

　　貨物包裝產生的廢物。如果包裝材料未受污染，那麼它不會引起環境問題；如果包裝材料已被污染，則包裝廢物需按特殊廢物進行處理，處理方法依污染的性質和程度而不同。

　　Waste material resulting from packaging of goods. If not contaminated, this is a waste without problems. If contaminated, packaging waste needs to be treated in a similar way as special wastes, depending on the nature and the quantity of the contamination.

packed column 填料塔 *tianliaota*

　　在化工過程如蒸餾、精餾中使用的柱狀工藝設備，它也可應用於空氣污染控制中或作為生物反應器。填料塔的一部分（*填充段*）裝填有填料，為相間接觸提供巨大表面積。無規填料塔的填料有不同的類型，它們的特徵可用傳遞效率、壓降、負荷、大小尺寸、材質等主要參數來表示。這些數據通常係由供應商提供。填料也還可起到作為微生物的載體的作用。（參見：有規填料塔、無規填料塔）

　　A cylindrical process unit (column) for use in chemical processing, e.g. in distillation, rectification, air pollution control, or as a bioreactor. Part of the unit is filled in the *packing zone* with a packing material to provide a high surface area for close contact between the phases. Different types of filling for randomly packed columns are characterized by key parameters like mass transfer efficiency, pressure drop, loading rates, size, material data, etc. Data is normally available from the supplier. The packing may also function as a support material for micro-organisms. (See also *structurally packed column*; *randomly packed column*)

packed column scrubber 填料洗滌塔 *tianliao xidita*

　　在洗滌塔用*洗滌液*從廢氣中洗滌去除污染物的空氣污染控制系統。當廢氣經過*填充區*時，污染物通過質量傳遞由氣相進入到洗滌液液相中。系統可以採用*並流和逆流*方式操作，而系統的*壓降*是系統的一個重要參數。洗滌液係用泵打到填料塔頂部噴灑到填充區上，並且進行循環使用。根據情況的不同，可以向洗滌液投加對污染物特定有效的化學物質。在運作過程中需不斷地排掉一部分洗滌液，通常在把它排入水體之前應作為污染廢水進行處理。

An air pollution control system which removes pollutants from an off-gas stream by washing with a *scrubbing liquor* in a scrubbing tower. The off-gas is passed through a *packing zone* in which a mass transfer of the pollutants from the gas phase to the liquid phase of the scrubbing liquor takes place. Systems use *co-current* and *counter-current flow*. The *pressure drop* of the system is a critical parameter of a system. The scrubbing liquor is circulated by pumps and sprayed onto the packing zone at the top of the tower. Depending on the situation, different pollutant-specific chemicals may be added to the liquor. Part of the liquor is constantly removed. This generally needs treatment like a polluted effluent before discharge to a waterbody.

packing zone 填充段（區）*tianchongduan (qu)*

濕式洗滌塔（或者一般地說，單元操作塔）中廢氣與洗滌液密切接觸的部分。在填充段裝填有特殊形狀的填料，如*鞍形填料、鮑爾環、刺猬形填料*。

That section of a wet scrubber (or a processing column in general) in which the off-gas comes into intimate contact with the scrubbing liquid. The packing zone is filled with a special type of *column packing saddles, pall ring, hedge hog.*

PACT 粉末活性碳處理 *fenmo huoxingtan chuli*

見：powdered activated carbon treatment。

See *powdered activated carbon treatment.*

PAH 多環芳烴 *duohuanfangting*

Polyaromatic hydrocarbons 或 polycyclic aromatic hydrocarbons 的縮寫。這些化合物可由不完全燃燒過程產生，例如，存在於汽車排放尾氣的痕量殘餘物中。多環芳烴與某些癌症的發生有關。（參見：致癌多環芳烴）

Polyaromatic hydrocarbons, polycyclic aromatic hydrocarbons. These compounds may result from incomplete combustion processes, e.g. as trace residues in transport emissions. These compounds are relevant because some of them are connected to the appearance of certain forms of cancer. (See also *cPAH/carcinogenic PAH*)

pall ring column packing 鮑爾環填料 *bao'erhuan tianliao*

無規填料塔如洗滌塔所用的一種具有最佳傳質效果的填料。其每一個組件的直徑和長度是不同的，大約在幾厘米左右。鮑爾環由不同的材料如鋁、不銹鋼和各種類型的塑料製成，它們的形狀像截成很短的短管，內側具有切口或孔眼以及輔助結構（凸起、架橋結構）。市場上有不同尺寸和設計的鮑爾環供應，其特性由一些重要性能參數表示。（參見：*鞍形填料*，*刺猬型填料*）。

A type of packing for randomly *packed columns*, e.g. for scrubbers, with the function to optimize mass transfer. Diameter and length of each element vary around a few centimeters. Pall rings are manufactured of different materials, e.g. aluminium, stainless steel and different types of plastic. They are shaped like short cuts of tubing with cut-outs or perforations and additional elements (protrusions, bridges) on the inside. Different sizes and designs are available commercially and are characterized by certain key performance parameters. (See also *column packing*; *saddles*; *hedge hog filling*)

Paques System 帕克斯系統 *pakesi xitong*

厭氧廢水處理系統的荷蘭商業名稱。（參見：升流式厭氧污泥床系統）

A Dutch trade name for a commercially available anaerobic waste water treatment system. (See also *upflow anaerobic sludge blanket system*)

parameter 指標（參數）*zhibiao (canshu)*

表示某一事物性質的指標。在環境保護上，它們被用來描述廢水、廢物、廢氣、土壤等的特性。有兩類指標：*綜合指標和單一要素指標*。

An indicator to qualify the properties of an object. In environmental protection, parameters are used to describe the quality of effluents, wastes, off-gases, soil, etc. Two types are: *group parameters* and *single element parameters*.

parameter, single element 單一要素指標（參數）*danyi yaosu zhibiao (canshu)*

用以表徵廢物流特性的單一要素的指標，如汞離子濃度或六價鉻濃度。（參見：綜合指標）

A parameter for qualifying a waste stream, specifying only one single element, e.g. the concentration of mercuric ion, or of chromium-VI. (See also *group parameters*)

particulates 顆粒物 *keliwu*

大氣中或氣流中極細微的固體顆粒物。它是一項綜合指標，用以表示氣流中具有不同化學組成的細微固體顆粒物的濃度。

Microscopically small particles of solid matter in the atmosphere or in a gas stream. Group parameter for characterizing the concentration of such particles of different chemical composition in a gas stream.

particulates in off-gas, continuous measurement 廢氣中的顆粒物，連續測定 *feiqizhongde keliwu, lianxu ceding*

對廢氣中的顆粒物一般採用光學儀器進行連續測定。它將一光束照向氣流，由於顆粒物使光反射而產生一分散光束，測定該分散光的強度，就可以相應求得顆粒物的濃度。這種方法不能測得絕對濃度，除非進行仔細的校準。它適合於在實際條件下對焚燒爐煙氣中粉塵進行連續測定。

Instruments for the continuous measurement of particulates in off-gas are generally based on an optical system. A light beam is directed into a gas stream. A dispersed beam of reflected light results from the particulate matter. The intensity of the dispersed light is measured; it is related to the particulate concentration. The method does not provide an absolute concentration unless careful calibration is carried out. It is suitable for continuous measurements under practical conditions for comparison, e.g. for dust in the flue gas of an incinerator.

pay back time 回收期 *huishouqi*

評估投資的經濟風險或合理性所用的術語。它是回收原始投資資本所需要的時間，以年表示。回收期與投資收益率有關。有關環境投資評估的一些論點也可在這裏應用。(參見：投資收益率)

A term used in the evaluation of the economic risk or soundness of an investment. It is the time in years needed to pay back the capital of the original investment. The term is related to the return on investment. The arguments regarding the evaluation of environmental investments apply here too. (See also *return on investments*)

PCB 多氯聯苯 *duolülianben*

Polychlorinated biphenyl 的縮寫。多氯聯苯是具有良好介電性的一類化合物(以前用在變壓器和電容器中作為電介質)，易溶於脂肪，難溶於水中(辛醇系數小)，而且生物降解率極低，因此它們會在生態系統的疏水體中積累。由於它們具有毒性，所以會產生不良影響。另外，在燃燒時多氯聯苯還會生成四氯二苯對二惡英(TCDD)。目前，它們已為其他電介質所取代。(參見：親油性、辛醇系數)

Polychlorinated biphenyl ($C_{12}H_{10}$-nCl_n). A group of compounds with favorable properties as dielectrics (formerly used in electrical transformers and condensers). As a result of their good solubility in fats and their poor solubility in water (low *octanol coefficient*) and the extremely low rate of biodegradation, PCBs tend to accumulate in the hydrophobic compartments of the ecosystems. Due to their toxic properties, they cause adverse effects. In addition, PCBs may also form TCDDs in case of a fire. Nowadays, PCBs are replaced by other dielectrics. (See also *lipophilic*; *octanol coefficient*)

PCB-wastes 多氯聯苯廢物 *duolülianben feiwu*

電力設備中的廢多氯聯苯或被多氯聯苯污染的設備。這類廢物是危險廢物，需採用特殊的方法或技術進行處置。

Waste-PCB from electrical equipment, or equipment contaminated with PCB. Such wastes are hazardous wastes. Their disposal requires special methods or technologies.

per mille or per mil 千分率 *qianfenlü*

千分之一，類似"百分率"的術語。

One part in one thousand parts. A term in analogy to 'percent'.

percent 百分比 *baifenbi*

百分之一。

One part in one hundred parts.

performance goals or limits 工作目標或限值 *gongzuo mubiao huo xianzhi*

對某項工作或效率規定的限值，例如，對廢水處理要求 BOD_5 去除率達到95%。

Goals or limits specifying a certain performance or efficiency, e.g. for effluent treatment the requirement of 95% BOD_5 removal.

performance report 工作報告 *gongzuo baogao*

一種可能的控制體系。環境項目年度報告，與企業管理經濟事項的年度報告相類同。它由該上報工廠的有關人員編制完成，內容應包括實測或估算的"三廢"排放量的系統而全面的數據，以及諸如規章的制定、成功的經驗與失敗的教訓等的說明。工作報告可以用作為定型化的控制體系，以及作為環境規劃的基礎。

A control possibility. A system for annual reporting of environmental topics, similar to the annual reporting on economic aspects to management of an enterprise. The report is completed by the staff of the reporting site. It should contain systematic and comprehensive data on measured or estimated emission quantities (air, water and wastes) and descriptive elements, such as legal developments, successes and failures. Performance reporting may be used as a formalized control system and as a base for environmental planning.

performance standards 工作標準 *gongzuo biaozhun*

所期待的工作水平的標準，如水污染控制和空氣污染控制中的工作標準。它也可應用於其他管理領域中。

Standards for the level of an expected performance, e.g. in water pollution control or in air pollution control. Performance standards are also applicable in other management fields.

period costs 當期費用 *dangqi feiyong*

用作非直接費用的一項術語。

A term also used for *indirect costs*.

permanganate value 高錳酸鹽值 *gaomengsuanyanzhi*

在酸性條件下用高錳酸鉀測定污染水COD(*化學需氧量*)值的方法。高錳酸鹽法測得的COD值,一般比重鉻酸鉀法測得的COD值低,這兩種方法測得COD值的關係,主要取決於污染物的性質。另外,高錳酸鹽值也總是比TOC(*總有機碳*)值低。

有些國家規定必須採用這種方法進行廢水的測定。其標準測定方法見*美國材料試驗學會標準*。

A method to measure the quantity of the *COD* in contaminated water, using potassium permanganate under acidic conditions. The permanganate value generally gives lower values than the COD method using potassium dichromate. The relationship strongly depends on the nature of the pollutants. The permanganate value is always lower than the TOC value.

Several countries oblige the generators of effluents to use this method. Standardized methods are described in the *ASTM Standards*.

permeate 滲透液 *shentouye*

逆滲透系統中通過半透膜的水相(稀液)。滲透液中污染物的濃度比原溶液低。(參見:逆滲透過程、半滲透膜、濃縮液)

The liquid dilute, usually aqueous phase passing the semi-permeable membrane of a reversed osmosis system. (See also *reversed osmosis process*; *semi-permeable membrane*; *concentrate*) The permeate has a lower concentration of pollutants than the initial solution.

permeating water 滲透水 *shentoushui*

見:water, permeating。

See *water, permeating*.

permeation tube in gas chromatography 氣相色譜法使用的滲透管 *qixiang sepufa shiyongde shentouguan*

產生校準標準的設備。它將某恆量物質通過一多孔膜管擴散進入氣體中，得到含不同濃度揮發性化合物的氣流。可以用硅橡膠管作為多孔膜。此方法可用來定量測定工作場所中的危險物質（如硫酸二甲酯）。另外，當用*連續氣相色譜法*測定反應器中的氣態產物時，也可用它來提供參比信號。

A device to generate calibration standards. Different concentrations of volatile compounds in a gas stream are obtained by diffusing a constant amount of a substance through a porous membrane tube into the gas. A tube of silicon rubber may be used as a porous membrane. The method is applied for instance to measure quantitatively dangerous substances (like dimethyl sulfate) in the workplace. It may also be used to provide a reference signal when measuring process gases in a reaction vessel by *continuous gas chromatography*.

personal monitoring 個人監測 *geren jiance*

*職業衛生*的一項內容，即對職工於某一時期在化工工作場所環境中接觸化學品的情況所進行的監測。（參見：隨身監測）

An element of *occupational hygiene*. Monitoring of the exposure of an employee to compounds in the work atmosphere at his workplace during a certain time. (See also *man monitoring*)

personal objectives for individual employee 職工的人員目標 *zhigongde renyuan mubiao*

它是環境綜合管理的一個組成部分。人員目標具體確定各職工要達到的目標，例如，"年終前完成"三廢"排放數據庫的修改工作"，"在某時限前由某工作小組完成廢水處理場的計劃工作"。人員目標應與*人員激勵制度*相結合。

An element of integral environmental management. Personal objectives specify goals to be reached by individual employees. For example in the present context: 'the complete updating of an emission database by the end of the year', or 'management of a team to complete the planning aspects of an effluent plant by ...'. Personal objectives may be linked with *personnel incentive systems*.

personnel incentive systems 人員激勵制度 *renyuan jili zhidu*

採用獎勵辦法來激發人員做出成績的激勵制度。環境獎勵應與環境保護成績，如成功地實現了個人環境目標相關聯。

A motivating system through the application of incentives on the personnel level to achieve certain results. Environmental incentives should be related to environmental achievements, e.g. the successful reaching of personal environmental objectives.

personnel participation schemes 人員參與制度 *renyuan canyu zhidu*

人事管理，尤其是綜合環境管理的一個組成部分。其目的在發現問題和解決問題過程中考慮職工的意見和調動職工的積極性。

An element of personnel management, in particular of integral environmental management. Any system which actively aims at the integration of the ideas and contributions of individual employees in the identification of problems and in the solving process.

pH

表示水溶液酸性和鹼性的一項指標。pH標度與溶液中氫離子或氫氧根離子濃度的對數值有關。pH7為中性，pH小於7為酸性，pH大於7為鹼性。

A parameter for the acidity or alkalinity of an aqueous solution. The pH scale is related in a logarithmic way to the concentration of hydrogen ion or hydroxyl ion in the solution. pH 7 is neutral, pH below 7 acidic, pH above 7 is alkaline.

pH meter pH計 *pHji*

測定水溶液pH值的儀器。通常，用電位計測定氫離子敏感玻璃電極和參比電極的電位差。

An instrument to measure the pH-value of an aqueous solution. Generally, the potential difference of a hydrogen ion sensitive glass electrode and a reference electrode is measured in an electrical potentiometer.

phase boundary 相界面 *xiangjiemian*

兩互不相溶的液相之間的分界層。當用有機溶劑從水相中萃取某物質時,其相界面就是水相與有機溶劑相之間的界面。(參見:邊界層)

The boundary layer between two liquid phases which are immiscible. In an extraction of a substance from an aqueous phase with an organic solvent, it is the boundary between the aqueous phase and the organic solvent layer. (See also *boundary layer*)

phase separation 相分離 *xiangfenli*

兩互不相溶的液相的分離過程,例如,萃取過程。一個完善的分離過程對於減輕水相中的有機負荷和最大限度地減少產品損失具有重要作用。(參見:邊界層、聚結器)

The process of separating two immiscible liquid phases, e.g. in an extraction process. A perfect separation process is important to reduce the organic load in the aqueous phase and to minimize the loss of product. (See also *boundary layer*; *coalescer unit*)

phosphate removal 磷酸鹽去除 *linsuanyan quchu*

磷酸鹽是水體*富營養化*的一個重要因素。磷酸鹽的去除係基於兩種原理。首先,重要的是要防止廢水或逕流水中存在過量的磷酸鹽。如在農業、畜牧業中小心使用糞尿和磷肥,在家庭中使用低磷洗滌劑,都能達到這一目的。其次,在廢水處理場中也可以包括除磷過程。典型的辦法是在進水口投加鐵鹽或鋁鹽進行沉降處理,或者投加鐵鹽或鋁鹽到活性污泥池在進行生物處理的同時進行沉降處理。如果磷酸鹽的去除效率要求很高,那麼需另外進行絮凝和過濾處理。(參見:廢水三級處理)

Phosphates are an important factor in the *eutrophication* of a waterbody. Their removal is based on two principles. Of first importance are solutions to avoid excess phosphates in the effluents or in run-off water. Measures in agriculture, animal husbandry, (careful use of liquid manure and phosphate fertilizers) and in the households (use of detergents with low phosphate content) serve this objective. Secondly, process steps may be integrated in an effluent treatment plant. Typical solutions are precipitation with iron or aluminum salts in the in-flow, or simultaneous precipitation by addition of iron or aluminum to the activated sludge basin. If high performance phosphate removal is required, then additional flocculation and filtration process steps are required. (See also *tertiary treatment of effluents*)

photochemical smog 光化學煙霧 *guanghuaxue yanwu*

在美國洛杉磯地區夏季熱天出現的一種類型的煙霧。光化學煙霧或夏季煙霧是汽車尾氣中所含的污染物（主要是氮氧化物、揮發性有機碳）和其他來源的化合物通過在低層大氣中發生的一系列複雜的化學反應而生成的。其一些反應步驟是由紫外輻射引發和促進的。夏季煙霧的主要組成是臭氧、氮氧化物和光氧化劑。

A smog type which was recognized as such in hot summer days in the Los Angeles area. Photochemical or summer smog results from pollutants (mainly NO_X and VOC) contained in exhaust gases of motor vehicles and other sources through a number of complex chemical reactions in the lower atmosphere. Some reaction steps are initiated and/or promoted by UV radiation. Critical components of summer smog are *ozone*, *NO_X* and *photooxidants*.

photometer 光度計 *guangduji*

在標準條件下測定溶液吸收紫外光或可見光的儀器。濾光光度計或比色計測定的是濾光器所透過的某一光譜帶的吸收能力，而分光光度計測定的是整個可見光或紫外光的吸收光譜。

An instrument to measure the absorption of UV or visible light of a solution under standard conditions. Filter photometers or colorimeters measure the absorption over a spectral band transmitted by the light filter. Spectrophotometers on the other hand measure an absorption spectrum over the whole visible or UV range of light.

photooxidants 光氧化劑 *guangyanghuaji*

對夏季煙霧或光化學煙霧的形成起主要作用的物質。光氧化劑是在大氣中紫外光的照射下形成的。前體物質（各種有機化合物、溶劑、烴類、揮發性有機化合物）在一系列光化學游離基反應中被氧化，包括生成過氧游離基和氧化氮。一種叫做過氧乙醯硝酸酯的化合物，是大氣中光化學反應的重要中間體。

Substances playing a role in the formation of *summer smog* or photochemical smog. Photooxidants are formed under irradiation by UV light in the atmosphere. Precursor substances (various organic compounds, solvents, hydrocarbons, VOCs) are oxidized in a chain of photochemical free radical reactions, including peroxyl radicals and nitrogen oxides. A class of compounds called 'peroxyacyl nitrates' are important intermediates in the reactions in the atmosphere.

photovoltaic solar energy 光電太陽能 *guangdian taiyangneng*

用太陽能電池直接由太陽輻射生成的電能。

Electrical energy which is produced directly from solar radiation in solar cells.

phytocenose 植物群落 *zhiwu qunluo*

某群落生境中各種不同的植物。（參見：生物群落）

The different plants (flora) of a given biotope. (See also *biocenose*)

phytoplankton 浮游植物 *fuyou zhiwu*

在水生生態系統如海洋、湖泊或池塘中能自由漂浮的微小植物。它們是*自養生物*，能同化二氧化碳生成細胞物質，為*浮游動物*提供食物。因此，浮游植物是水生*食物鏈*中最低的一環，它可以看作是維持海洋一切生命的原動力。水中的一些污染物可被浮游植物攝取而不發生轉化，並傳遞給食物鏈中較高級的一環。在此過程中，由於*生物放大*作用會使污染物的濃度增高。

Small plants which are free floating and drifting in an aqueous ecosystem, the oceans, a lake or a pond. Phytoplankton organisms are *autotrophic* and are able to assimilate CO_2 to build cell mass to provide food for *zooplankton*. Phytoplankton is thus the lowest level of the aquatic *food chain*. It can be looked at as the motor sustaining all life in the oceans. Some contaminants in the water are taken up by phytoplanktons without conversion and are passed on to higher levels of the food chain. In this process, their concentration may increase as a result of *biomagnification*.

phytoremediation 利用植物恢復補救 *liyong zhiwu huifu bujiu*

建議利用高等植物進行恢復補救的一項技術，這是基於某些特殊的植物具有從污染土壤中將金屬提取出來並在植物體內積累的這一性質。這種方法的缺點是：由於植物中含有金屬，因此它們必須在特殊的條件下進行處置。目前這項技術尚處於試驗階段。

A proposed remediation technique using higher plants. Certain special plants have the property to extract metals from contaminated soils and accumulate them in the plant material. A disadvantage of this method is that the metal is then contained in the plants, which then have to be disposed of under special conditions. The technique is still in the testing phase.

phytotoxicity 植物毒性 *zhiwu duxing*

物質對植物的毒性。

The toxicity of a substance towards plants.

picogram 微微克 *weiweike*

10^{-12} 克。

The 10^{-12} part of a gram.

piezometer 地下水位計 *dixiashuiweiji*

在地下水調查研究中使用的取樣和監測井。井的直徑較小，約為2.5厘米，常常配備有儀器來測量地下水位的高度。對於直徑10厘米左右較寬的監測井，則配有泵用於地下水取樣。

A sampling and measuring well used in groundwater investigations. The diameter of the well is relatively small, about 2.5 cm. The piezometer is often equipped with instrumentation to measure the height of the groundwater table. Monitoring wells with a wider diameter of ca. 10 cm often are equipped with a pump for sampling groundwater.

plankton 浮游生物 *fuyou shengwu*

在水生生態系統如海洋或湖泊中能自由漂浮的微小生物。浮游生物包括*浮游植物*和*浮游動物*在內的各種各樣的生物,在*食物鏈*中它們是高等動物的主要食物源。水中的污染物首先會被浮游植物吸收而不發生轉化,並通過*生物放大*過程在浮游動物和*捕食魚類*中積累而增高其濃度。

Small organisms which are free floating or drifting in a water ecosystem, e.g. in the oceans or in a lake. Plankton includes a diversity of organisms, including *phytoplankton* and *zooplankton*. Plankton serves as the primary source of food for higher animals in the *food chain*. Contaminants in the water may be absorbed unchanged first by phytoplankton and, through the process of *biomagnification*, may be accumulated in zooplankton and in *predator* fish in increasing concentrations.

plasma incineration technique 等離子體焚燒技術 *dengliziti fenshao jishu*

焚燒某些會產生嚴重問題的有害廢物的高新技術。實際焚燒是在電弧的等離子體區中進行的,其溫度約為1600–2000℃。等離子體技術用電作為主要能源,其優點是產生的煙氣數量較少。(參見:焚燒)

A high-tech method for incineration of certain highly problematic hazardous wastes. The actual incineration takes place in the plasma region of an electrical arc. The temperature is around 1600–2000℃. Plasma technology uses electricity as a primary source of energy. Its advantage lies in the fact that the flue gas volume is relatively low. (See also *incineration*)

plastic waste codes 塑料廢物標記 *suliao feiwu biaoji*

見:codes for plastic wastes。

See *codes for plastic wastes*.

plastic wastes 塑料廢物 *suliao feiwu*

主要由廢合成有機聚合物材料組成的廢物。其產生途徑有:(1)直接作為生產過程的副產物而產生;(2)塑料製品(如塑料袋)使用後產生;(3)作為廢物分離過程的產物而產生,如來自廢鉛蓄電池加工中的聚丙

烯。為了便於進一步處理和循環利用，必須將塑料廢物分離成不同類型的聚合物材料。(參見：塑料廢物編碼)

Wastes consisting mainly of waste synthetic organic polymeric material. Such wastes are generated: directly as a byproduct of a manufacturing process; after use of plastic material (e.g. plastic bags); or as a product of a waste separation process, e.g. polypropylene from processing of used lead accumulators. To facilitate further treatment and recycling, plastic wastes must be separated into different categories of polymeric material. (See also *plastic waste codes*)

plate columns 板式塔 *banshita*

内部具有板式和盤式結構的垂直加工塔，它廣泛用於蒸餾和精餾工藝。其分離效率取決於所謂"理論板"數及其結構。(參見：理論板)

A vertical processing column with plate- and tray-like constructions inside. Distillation and rectification technology makes extensive use of plate columns. The separation efficiency depends on the number of so-called 'theoretical plates' and on their construction. (See also *theoretical plate*)

plate deposition test 平板沉積測試 *pingban chenji ceshi*

在工作場所中測定塵粒沉積的簡單方法。測試時，將25×25厘米的玻璃平板或鋁制平板水平地放置在工作場所中的重要地點，暴露於環境空氣中若干天。然後，可對平板上的塵粒直接進行計量或者作進一步分析。它是對工作場所中粉塵狀況進行監測的一種簡單方法。

A simple test to measure the deposition of dust particles in a site. In this test, glass or aluminum plates of 25 by 25 cm are exposed horizontally and in strategic locations of a site to the ambient air for a number of days. The dust on the plate can be quantized directly or used for further analysis. The plate deposition test is a simple method for monitoring the dust situation in a site.

point emission source 點排放源 *dianpai fangyuan*

從某一"點"，如從一個煙筒或煙囪排放的廢氣源。通常，一個工廠有許多不同的點排放源。(參見：擴散排放源)

A source of off-gas emissions from which emissions originate from a single 'point', e.g. a stack or a chimney. A factory usually has a number of different 'point sources'. (See also *diffuse emission source*)

poisoning of catalysts 催化劑中毒 *cuihuaji zhongdu*

由於某些元素或簡單化合物使催化劑失活或使催化劑活性受到不可逆破壞。（參見：催化過程、催化劑毒物）。

Rendering a catalyst inactive or destroying the catalytic activity irreversibly by certain elements or simple compounds. (See also *catalytic processes*; *catalyst poisons*)

police filter 控制過濾器 *kongzhi guolüqi*

在要求嚴格的情況下設置在廢氣過濾器後的二級過濾器。它通常具有較細的孔隙度和較大的阻留能力。（參見：過濾器穿透）

A second filter, which, in critical cases, is added after an off-gas filter. It generally has a finer porosity and higher retention. (See also *filter breakthrough*)

police laws 管制法 *guanzhifa*

指規定容許做什麼和不容許做什麼的法律，有時也稱為"命令和控制"法律。許多傳統法律都基於這一概念。在環境保護方面，管制法典型地規定了廢氣排放限值、廢水排放限值和必須遵循的管理制度，不符合規定要求就要給與懲罰。另一類根本不同的法律是基於*經濟刺激*的概念。

A term used for a type of laws which defines actions which are allowed and actions which are not allowed. The term 'command and control' legislation is sometimes used. Many traditional laws are based on this concept.

In environmental protection, police laws typically specify emission limits, discharge limits as well as administrative procedures to be followed. Penalties for non-compliance range from fines to personal imprisonment, depending on the penal law of the country. A fundamentally different type of legislation is based on *economic incentives*.

policy statement on environmental protection 環境保護政策報告 *huanjing baohu zhengce baogao*

見：environmental policy statement。

See *environmental policy statement.*

pollutant 污染物 *wuranwu*

在介質中非天然存在的物質或化合物。環境污染物可能是人為產生的（如農藥），或者也可能是自然產生的（如叢林火災）。環境保護涉及的是人為的污染物。

A substance or compound which is not naturally present in a medium. Pollutants of the environment may be of *anthropogenic* origin (e.g. pesticides), or of natural origin (e.g. bush fires). Environmental protection deals with anthropogenic (man-made) pollutants.

polluter 污染者 *wuranze*

見：generator of pollution。

See *generator of pollution.*

polluter-pays-principle (PPP) 污染者付費原則 *wuranzhe fufei yuanze*

與環境污染有關的費用支付的基本原則。這一原則規定：污染者（造成污染的責任方）必須支付污染預防或控制的費用，以及因污染造成的損害費用。但鑒別污染者常常是不容易的。（參見：內部的污染者付費原則、社會費用原則、污染者之鏈）

A basic principle for the payment of costs related to environmental pollution. The polluter-pays-principle stipulates that the 'polluter' i.e. the party responsible for pollution, must pay the costs for the prevention or control of the pollution and/or the damage costs resulting from this pollution. The identification of the polluter is often not straightforward. (See also *internal polluter-pays-principle*; *social cost principle*; *chain of polluters*)

pollution control technologies 污染控制技術 *wuran kongzhi jishu*

防止或減輕生態系統污染的技術，包括：在源頭預防污染的 *完整的 工藝過程開發*、*水污染控制*、*空氣污染控制*、*廢物管理與處置*。

Technologies serving to prevent or reduce the pollution of ecosystems. The main elements are: *integral process development* to prevent pollution at the source, *water pollution control*, *air pollution control*, and *waste management and disposal*.

pollution load 污染負荷 *wuran fuhe*

在一定期間排放的污染物的總量，如噸COD/天或噸VOC（揮發性有機化合物）/年。

The total quantity of pollutant emitted during a given time span, e.g. tons of COD per day or tons of VOC per year.

pollution prevention 污染預防 *wuran yufang*

污染控制概念中一個重要而基本的內容。（參見：源頭污染預防、完整的工藝過程開發）

An important and fundamental element of an overall pollution control concept. (See also *PPS*; *integral process development*)

pollution prevention at the source 源頭污染預防 *yuantou wuran yufang*

見：PPS。

See *PPS*.

polyamide membranes 聚醯胺膜 *juxian'anmo*

由聚醯胺材料製成的膜。它在化工技術上有著許多方面的應用，如應用於逆滲透、毫微米級過濾、超濾、氣體分離、全蒸發和蒸汽滲透。

Membranes made of polyamide material. Uses in chemical process technology are manifold: reversed osmosis, nanofiltration, ultrafiltration, gas separation, pervaporation and vapor permeation.

polyelectrolyte 聚電解質 *judianjiezhi*

一種含離子基的有機聚合物 *絮凝劑*。市場上有各種不同類型的聚電解質供應。具有不同分子量的陰離子和非離子型聚電解質，常常用作澄清劑。陽離子型聚電解質可用於污泥脫水。

A *flocculating agent* based on an organic polymer, containing ionic groups. Different types are available commercially. Anionic and non-ionic polyelectrolytes with different molecular weights are often used for clarification. Cationic substances may be used for sludge dewatering.

population, ecological 生態種群 *shengtai zhongqun*

一種生物物種的個體群。這一術語原來只用於指人類群體。現在，在生態學中它也用於指其它物種。

A group of individuals of one biological species. The term was originally used only to describe human populations. In ecology, it is now used also for other species.

population of test data 試驗數據總體 *shiyang shuju zongti*

在同樣試驗條件下對某一狀況進行重複測量所得到的一組個別數據。試驗數據總體是*統計評估*的基礎數據。為了進行統計評估，必須知道試驗數據的確切來源。必須注意測試數據總體應真正代表了真實值的總體。

A group of individual values resulting from repeated measurements of a situation under the same test conditions. The population of test data is used as the basic input to a *statistical evaluation*. For this evaluation, the exact source of the test data must be known. Care must be taken that the population of test data is a true representation of the population of the real values.

post-combustion chamber 後燃燒室 *houranshaoshi*

廢物焚燒爐的一個組成部分。它是一個大的*燃燒室*，直接設置在一次燃燒室之後，其作用是使離開一次燃燒室的煙氣中的有機物完全氧化。為此，必須使煙氣在後燃燒室於溫度800-1200°C下停留幾秒鐘。通常，需要有額外的*輔助燃料燒咀*才能達到上述溫度。

An element of a waste incinerator. The post-combustion chamber is constructed as a large *combustion chamber* immediately after the primary combustion chamber.

It serves to complete the oxidation of organic material in the flue gases as they leave primary incineration. To achieve this, the flue gases must remain in the post combustion zone at a temperature of 800–1200°C for a few seconds. Generally, this temperature requires additional *support fuel burners*.

potential risk 潛在風險 *qianzai fengxian*

在萬一發生事故時會顯現的對人、財產和環境的風險。潛在風險可用對環境的潛在影響的程度和事故概率來表徵。在此基本範疇內，可能有各種不同的組合。

The risk to persons, property and to the environment which may manifest itself in the event of an accident. A potential risk is characterized by the magnitude of the potential impact on the environment, and the probability of an accidental event. Within this matrix, different combinations are possible.

powdered activated carbon treatment (PACT) 粉末活性碳處理 *fenmo huoxingtan chuli*

將物理化學處理與生物處理相結合的廢水處理。它通過將粉末活性碳投加到曝氣池中，來提高傳統活性污泥廢水處理裝置的效率。同時，粉末活性碳可作為供微生物生長的載體，以及作為有毒和難降解物質的吸附劑。但粉末活性碳不能回收，因此會大大增加處理成本。

A combined physical-chemical and biological treatment for aqueous effluents. The efficiency of a conventional activated sludge effluent plant is increased by the addition of powdered activated carbon to the biology basin. The activated carbon powder may act at the same time as a support for the growth of microorganisms and as an adsorbent for toxic and refractory substances. The carbon cannot be recovered and may add considerably to the treatment costs.

POX 可氣提的有機鹵素化合物 *keqitide youji lusu huahewu*

Purgeable organic halogen compounds的縮寫。它是水污染控制中使用的一項綜合指標。它所測定的是水樣中能用氮氣氣提出的那部分揮發性污染物。測定的方法是將上述氣提出的氣體連續燃燒，然後像測定可

吸附的有機鹵素化合物那樣，進行鹵化物的分析。（參見：可吸附的有機鹵素化合物、可提取的有機鹵素化合物）

Purgeable organic halogen compounds. A group parameter used in water pollution control. POX is determined as that part of the pollutants which is volatilized from the aqueous sample by stripping (purging) with nitrogen and continuous combustion of the stripping gas. The analysis of halides is then carried out essentially as with AOX. (See also *AOX*; *EOX*)

ppb 十億分之（幾分）*shiyifenzhi (jifen)*

十億分之一，即 10^{-9}。舉例說，5人大約相當於全世界人口的十億分之一（注：在美國，1 billion 等於十億；而在歐洲，1 billion 通常表示1萬億）。

One part per (American) billion, i.e. one part in 10^9 parts, or one part in one thousand millions. As a descriptive example, one can say that 5 persons represent about 1 ppb of the world population. Note that in the American use one billion is equal to one thousand million, while in Europe, it is generally one million millions.

ppm 百萬分之（幾分）*biawanfenzhi (jifen)*

百萬分之一，即 10^{-6}。例如，克/噸；12人與上海市人口之比。

One part per million, i.e. one part in 10^6 parts. Example: one gram per ton, or about 12 persons in relation to the population of the city of Shanghai.

ppm by volume 百萬分之幾（體積比）*baiwanfenzhiji (tijibi)*

體積比百萬分之幾。通常，氣體測定的數據是用這一單位表示的。1 ppm（百萬分之一，體積比）等於1立方米體積中含1毫升氣體X。

The unit 'parts-per-million' applied to volume ratios. Generally, data on gas measurements are given in vol.-ppm. One ppm by volume equals 1 ml of the gas X in one cubic meter of gas.

PPP 污染者付費原則 *wuranze fufei yuanze*

見：polluter-pays-principle。

See *polluter-pays-principle*.

PPS 源頭污染預防 *yuantou wuran yufang*

Pollution prevention at the source的縮寫。(參見：清潔工藝、完整的工藝過程開發)

Pollution prevention at the source. (See also *clean technologies*; *integral process development*)

ppt 億萬分之(幾分) *yiwanfenzhi (jifen)*

億萬分之一，即 10^{-12}。例如，5克方糖溶於長500米、寬500米、深200米的湖中，其濃度為1 ppt。

One part per trillion, i.e. one part in 10^{12} parts, or one part in one million millions. As a descriptive example, one can say that one 5 gram cube of sugar dissolved in a lake of 500 m (length) x 500 m (width) x 20 m (depth) represents a concentration of about 1 ppt.

preacquisition audit 購買獲得前審計 *goumai huodeqian shenji*

買主在購買獲得某工廠之前在該廠進行的*審計*。這種審計是代表買主進行的，其目的在於驗核潛在的*債務*。在環境保護方面，確定工廠的污染狀況(包括生產場地、地下水和生態系統的污染)是審計的重要內容。

An *audit* carried out at a factory before the acquisition of said factory by a buying enterprise. The preacquisition audit is carried out on behalf of the potential buyer. It serves to identify potential *liabilities*. In the context of environmental protection, contaminations (of the site, of groundwater, of an ecosystem) at the site are important points and must be identified.

precipitating agent 沉澱劑 *chendianji*

能與可溶性污染物形成不可溶沉澱物的化學物質。其相互作用可能是化學鍵合或者物理化學作用。通過除去沉澱物而使污染物得到分離。(參見：凝聚劑、絮凝劑)

Chemical substances which form an insoluble precipitate with another dissolved pollutant. The interaction may be a chemical bonding or be of a physico-

305

chemical nature. By removing the precipitate, the pollutant is separated. (See also *coagulants*, *flocculants*)

precipitation electrode 沉積電極 *chenji dianji*

　　靜電除塵器的正電極。由於受正沉澱電極的吸引，可使所處理廢氣中荷負電的粒子分離。正電極通常為一金屬板，並可通過震打除去所收集的粉塵。(參見：科特雷爾技術)

The positive electrode of an electrostatic precipitator. Negatively charged particles from the off-gas to be treated are attracted by the positive precipitation electrode and are thereby separated. The positive electrode is generally built in form of metal plates which can be rapped (shaken) in order to dislodge the collected dust. (See also *Cottrell Technology*)

precipitation of particulates 微粒沉降 *weili chenjiang*

　　用物理分離方法從廢氣中分離塵粒。為了提高分離程度，可能需要進行預處理。(參見：靜電分離器、沉降室、旋風分離器)

The separation of particles (dusts) from an off-gas stream by physical separation. Certain pretreatment steps may be necessary to increase the degree of separation. (See also *electrostatic separators*; *drop out boxes*; *cyclones*)

precipitation technology for effluents 廢水沉澱技術 *feishui chendian jishu*

　　投加沉澱劑形成不溶性沉澱物，並採用過濾、沉降、離心或浮選方法去除沉澱物的廢水處理技術。

A treatment technology for effluents which is based on the formation and removal of insoluble precipitates formed by adding a *precipitating agent*. The removal can be effected by *filtration*, *sedimentation*, *centrifugation* or *flotation*.

precipitation water 降水水 *jiangshui shui*

　　來自雨或融雪的水，它含有相當量的大氣污染物(見：酸雨)，並且會進一步被地面上的各種污染物所污染。(參見：逕流水)

Water from rain or melting snow. Precipitation water may contain considerable quantities of pollutants from the atmosphere (see *acid rain*) and be further con-taminated with various pollutants taken up on the ground. (See *runoff water*)

precision gas standards 精密氣體標準 *jingmi qiti biaozhun*

市場上供應的裝於小不銹鋼瓶中的氣體或標準混合氣體，它們在廢氣分析中作標準用。

Gases or standardized gas mixtures which are available commercially in small stainless steel cylinders. Precision gas standards are used for calibration purposes in off-gas analysis.

precision of a test method 試驗方法的精密度 *shiyang fangfede jingmidu*

精密度是指分析方法的重複性，但是高精密度並不相當於高準確性。某一方法可能是精密的，也就是說它可始終得到同樣的結果（包括相同的誤差），但是它卻可能是不準確的。（參見：準確度）

Precision refers to the repeatability of an analytical method. A high precision is, however, not equivalent to high accuracy. A method may be precise, i.e. it may always give the same result including the same error, but it is not accurate. (See also *accuracy*)

preclarifier 前澄清池 *qianchengqingchi*

廢水處理場的第一沉降池，它通過簡單沉降以分離不溶物及中和、絮凝後產生的污泥。前澄清池具有不同的形式和設計結構，但一般都設有一個在不受擾動下使物質沉降的池，以及使分離後的污泥慢慢移除的機械（大多為刮泥機）。

The first sedimentation basin of an effluent treatment plant. It serves to separate insoluble matter and sludge after neutralization and flocculation by simple sedimentation. Preclarifier basins may be constructed in different forms and designs. All have in common a basin allowing undisturbed sedimentation of material and a mechanism for slowly removing the separated sludge, in many cases with a scraper mechanism.

preconcentration 預濃縮 *yunongsuo*

母液*焚燒*或*濕式空氣氧化*處理前的預處理步驟。由於經濟上的原因及為了節省能源，進行這種預處理可能是可取的。可以採用*逆滲透過程*進行預濃縮，但也可以採用如蒸發這類更為傳統的方法。

A pretreatment step prior to *incineration* or *wet air oxidation* of aqueous mother liquors. For economic reasons and to avoid high energy consumption, such a pretreatment may be advisable. Preconcentration can be effected with the *reversed osmosis process*, but also with more conventional processes like evaporation.

preconcentration processes in waste incineration 廢物焚燒預濃縮過程 *feiwu fenshao yunongsuo guocheng*

在廢物焚燒前採用過濾、離心、蒸發或逆滲透等方法分離水相的過程。例如：為了節能，剩餘*生物污泥*在焚燒前需進行預濃縮。在大多數情況下，對分離出的稀水相也需進行處理。

Processes, like filtration, centrifugation, evaporation or reverse osmosis which are suitable to separate an aqueous phase prior to incineration of a waste. Example: for reasons of energy economy, excess *biosludge* has to be preconcentrated before incineration. In most cases the separated dilute aqueous phase also needs treatment.

preconditioning of off-gases 廢氣預處理（調節）*feiqi yuchuli (tiaojie)*

廢氣淨化中用以提高污染物去除效率的過程，包括增加顆粒的*凝聚*或用*文丘里洗氣法*使顆粒預濕並吸收在小液滴上。選擇有效的預處理方法，有助於減少系統的總*壓降*，從而降低能耗和減少費用。

In off-gas purification, one or several process steps may serve to enhance the removal efficiency for pollutants. Preconditioning may include steps to increase particle *agglomeration* or *venturi type treatment* to pre-wet particles and absorb them in liquid droplets. The selection of effective preconditioning helps to reduce the overall *pressure drop* of the system and as a consequence the energy consumption and the costs.

predator 捕食者 *bushizhe*

主要以其它物種(其被捕食者)為食物的某種動物。捕食者群體與被捕食者群體之間存在著一種*捕食者與被捕食者的關係*。

An animal species which feeds primarily on other species, its prey. The populations of predators and prey are related to each other by the *predator-prey relationship*.

predator-prey relationship 捕食者與被捕食者的關係 *bushizhe yu beibushizhe de guanxi*

某*捕食者*群體與其生活在同一群落環境中*被捕食者*群體之間的關係。簡單地說,在存在一大群被捕食者的條件下,其捕食者群體將迅速增長,因此被捕食者群體的數量將減少,而這又將導致捕食者群體的減少。因此,這兩類*群體*處於*動態平衡*之中。它們將在不同時間達到其最大和最小的數量。

The relationship between a population of a *predator* and a population of its *prey* occupying the same biotope. In a simplified way, one can say that with a large population of prey, the population of its predator will grow rapidly, thus reducing the magnitude of the population of prey, which will diminish. This in turn will again lead to a reduction of the predator population. Both *populations* are in a *dynamic equilibrium*. The maxima and minima of both populations will be reached at different times.

prepaid disposal fee for consumer goods 消費品的預付處置費 *xiaofeipinde yufu chuzhifei*

購買某些消費品時所付的特殊費用。它是根據*污染者付費原則*,作為消費品在使用後以環境可接受的方式進行處置所需要的費用而徵收的。所徵收的錢將作為一項特別專款供以後用來支付處置費用。如果不能保證對消費品進行正確處置的話,則要對它徵收預付處置費。現在,在許多國家利用這項費用來處置廢乾電池、舊汽車、冰箱和計算機等。(參見:費用、收費、污染者付費原則)

A special fee, paid at the moment of purchase of certain consumer goods. The fee is collected for covering the payment of the cost of environmentally acceptable disposal of the goods after their use, according to the *polluter-pays-principle*. The money thus collected flows into a special fund from which the disposal costs will be paid later on. Such fees are collected for consumer goods if the correct disposal is otherwise not guaranteed. At present such fees are used in a number of countries to pay for the disposal of used dry cells, automobiles, refrigerators, computers, etc. (See also *fee*; *levy*; *polluter-pays-principle*)

pressure difference in APC units 空氣污染控制裝置的壓差 *kongqi wuran kongzhi zhuangzi de yacha*

　　空氣污染控制系統中兩點之間的壓力差，如廢氣進口側與出口側之間的壓力差。（參見：壓降）

The difference in pressure between two points. In air pollution control systems, the pressure difference between the input side of the off-gas and the output side. (See also *pressure drop*)

pressure drop 壓降 *yajiang*

　　空氣污染控制處理系統進口側與出口側之間的壓力差。風機的耗電量主要取決於壓差，而廢氣必須克服壓差才能被傳送通過系統。因此，系統的壓降是空氣污染控制裝置耗能與運轉費用的一項指標。低壓降與高效率是現代空氣污染控制技術開發的目標。

The pressure difference (delta-p) between the input side and the output side of a treatment system in air pollution control. Off-gas has to be transported through the system against the pressure difference. The consumption of electric power by fans depends strongly on the pressure difference. The pressure drop across the system is therefore also an indication of the energy requirements and the running costs of an APC unit. Low delta-p and high efficiency are the objectives of the development of modern APC-technology.

pretreatment of effluents 廢水預處理 *feishui yuchuli*

　　在廢水處理場進行傳統的生物降解處理之前，採用物理、化學或生物方法聯合進行的廢水處理。

　　在工藝過程末端進行的預處理步驟，包括針對該過程特定進行的化學預處理。根據情況的不同，可採用許多不同的方法，如沉澱、凝聚、絮凝、沉降和過濾，以及能促進兩相系統相分離的聚結、蒸餾分離、萃取或化學預處理（如部分氧化）。在廢水處理場進行的末端處理，其預處理步驟，包括中和、一次沉降、均衡和脫色。

The combination of physical, chemical or biological process steps in effluent treatment prior to the conventional biodegradation in a water treatment plant.

Pretreatment steps at the end of the process include process-specific chemical pretreatments. A range of processes are applicable, depending on the individual case: precipitation, coagulation, flocculation, sedimentation and filtration, coalescers to improve phase separation of two phase systems, distillative separations, extractions, or chemical pretreatments like partial oxidation. At the level of the end-of-pipe effluent treatment plant, common process steps include neutralization, primary clarification, buffering or equalization, and possibly decolorization.

pretreatment of wastes 廢物預處理 *feiwu yuchuli*

能改變廢物性質以改進其處置狀況或減少其數量的技術或過程。預處理方案的選擇取決於廢物的類別、組成和性質，以及擬採用的最終處置方法。主要的預處理方法包括：*分離、去毒、穩定化和預濃縮*。

A group of technologies or process steps suitable for modification of wastes in order to improve their disposal properties. Another aspect is to reduce their quantity. The selection of pretreatment options depends on the type, composition and properties of the wastes and on the planned method of final disposal. The main groups of processes include *separation processes*, *detoxification processes*, *stabilization processes*, and *preconcentration processes*.

preventive maintenance 預防性維修 *yufangxing weixiu*

以保持技術設備的經濟價值和正常功能為目標的管理方法。系統的預防性維修要求在設備出現故障前對設備進行定期檢查，以防止發生故障。在預防性維修計劃中，也應包括污染控制設備。（參見：預防性維修制度）

A management approach aiming at maintaining the economic value and proper functioning of technical equipment. In systematic preventive maintenance, equipment is checked regularly before any failures manifest themselves. Such failures are prevented. Preventive maintenance plans must also include pollution control equipment. (See also *preventive maintenance system*)

preventive maintenance system 預防性維修制度 *yufangxing weixiu zhidu*

技術設備維修的管理和控制制度，其主要內容包括：列出技術設備中需進行定期巡視和維修的檢查點和部件；預先確定的維護檢查時間表；明確賦予的職責；明確規定的維修維護的步驟和責任 (如果需要的話) 。

A formalized system to manage and control maintenance of technical equipment. A preventive maintenance system incorporates the elements: a list of checkpoints and elements of technical equipment which need to be looked at periodically and maintained; a predetermined time schedule to carry out the maintenance checks; clearly assigned responsibilities; and clearly defined procedures and responsibilities for repairs and maintenance, if necessary.

preventive measures 預防措施 *yufang cuoshi*

為避免某些不良影響而採取的防範措施。在許多領域都可以應用這種預防性措施。(參見：源頭污染預防、損害預防、預防性維修)

Measures taken as a precaution to avoid certain negative effects. Preventive measures can be applied in different fields. (See also *PPS*; *damage prevention*; *preventive maintenance*)

preventive safety measures 預防性安全措施 *yufangxing anquan cuoshi*

安全管理總體系統的一個部分。它是一組綜合措施，與*緊急響應管理*一起，以避免事故和緊急事件的發生，以及減少它們的潛在影響。在生產工藝方面，這包括進行*風險評價*，把安全問題納入工藝過程開發 (*完整的工藝過程開發*) 中，以及採取適當的技術措施。

當發生事故時，應採取*緊急事故遏制措施*，以限制事故排放的影響。

One aspect of an overall system of safety management. Together with *emergency response management*, it is the combination of measures which are taken to avoid accidents and emergencies and to reduce their potential impact. On the level of the process, this includes a *risk analysis*, the integration of safety aspects in process development (*integral process development*) and the adaptation of suitable technical measures.

For the case of an accident, emergency *containment measures* serve to limit the impact of the emissions.

prey 被捕食者 *beibushizhe*

供另一物種(其捕食者)作為食物的某種動物。一種物種可以既是捕食者，同時又是被捕食者。它們之間存在著一種*捕食者與被捕食者的關係*。

An animal species which is the food of another species, its *predator*. One species can at the same time be predator as well as prey. The relation of predator and prey is described by a *predator-prey relationship*.

primary clarifier 一次澄清池 *yici chengqingchi*

廢水處理場中位於生物處理前的沉降池。(參見：前澄清池)

The sedimentation basin of an effluent treatment plant which is placed prior to the actual biological treatment. (See also *preclarifier*)

primary source energy 一次能源 *yici nengyuan*

見：energy, primary source。

See *energy, primary source.*

problem wastes 成問題的廢物 *chengwentide feiwu*

見：special wastes。

See *special wastes.*

procaryotic cells 原核細胞 *yuanhe xibao*

比*真核細胞*簡單得多的一種細胞，它是一切細菌和藍綠藻的構成單元。

A cell which is much simpler than the *eucaryotic* type. Procaryotic cells are structural units in all bacteria and in blue green algae.

process development 過程開發 *guocheng kaifa*

見：integral process development。

See *integral process development*.

process documentation 生產工藝文件資料 *shengchan gongyi wenjian ziliao*

化工生產中有關某一工藝過程所有最新的正式文件資料。其主要內容包括：工藝過程說明、物料數據、各步操作說明、輔助設備如污染控制設備的操作說明、以及安全、環境(廢物及其排放)與職業衛生等方面的資料。

A formal documentation including all relevant and updated information to carry out a process in chemical production. The main chapters include: a description of the process itself, data on the materials used, a step-by-step operating description, operating instructions for auxiliary equipment, e.g. for pollution control, safety data, environmental data (wastes, emissions), and occupational health data.

process effluents 工藝廢水 *gongyi feishui*

在生產過程末端或某一段工藝過程末端直接產生的廢水和母液。在該處，所產生的工藝廢水彼此未相混合，濃度往往較高。工藝廢水可用特定的末端工藝過程設備進行處理。在許多情況下，可將廢水中的物質加以回收、淨化和再利用。

Effluents and mother liquors generated directly at the end of a manufacturing process or a process step. At this point, process effluents are not mixed with each other and the concentrations are often relatively high. Process effluents can be treated by specific end-of-process equipment. In many cases, materials can be recovered, purified and reused.

process gases 工藝廢氣 *gongyi feiqi*

一種特殊類型的廢氣。在工藝廢氣中，污染物濃度很高，空氣含量很低。工藝廢氣含有由化學反應產生的具有化學計量數量的污染物，而且它只在某些生產步驟中排放。通常，其流量是無規律的。例如，重氮化反應產生的二氧化氮，氯化反應產生的氯化氫等等。工藝廢氣的氣量一般較小，在1-100米³/小時範圍內。（參見：廢氣類別）

A specific type or class of off-gas. In process gas streams, the pollutant concentration is high and the fraction of air in the gas stream is low. Process gases contain pollutants from the chemical reactions in stoichiometric quantities. Emissions arise only during certain production steps. Often, the flow is irregular. Examples are: NO_2 from a diazotization reaction, HCl from a chlorination reaction, etc. Generally the flow volumes of process gases are relatively low and range from 1 to 100 m³/h. (See also *off-gas types*)

process optimization 過程優化 *guocheng youhua*

見：integral process development。

See *integral process development*.

process-dedicated equipment 過程專用設備 *guocheng zhuanyong shebei*

專用於某一特定生產過程的工藝設備，這與多功能設備完全不同。

Equipment and processing units which are used specifically for one particular manufacturing process. This is in contrast to *multi-purpose equipment*.

product assessment 產品評價 *chanpin pingjia*

對某一產品在其正常使用期間的環境影響和可接受性的評價。在產品研究開發階段，要根據產品的使用和潛在影響，進行產品的生態毒性及其在生態系統中性狀變化的一系列特殊試驗，依據這些試驗結果可對該產品進行評價。

The assessment and evaluation of the environmental impact and acceptability of a product during its normal use. The assessment is based on a range of specific tests carried out during the product research and development phase. The tests cover ecotoxicity and the behavior of the product in the ecosystem. They depend on the use and the potential impact of a product.

product carry-over 產物攜帶 *chanwu xiedai*

在由若干段過程組合的廢氣處理單元中，產物或污染物從上一段向下一段的傳輸。在一個組合的空氣污染控制系統中，產物攜帶會導致去除效率降低，因此應在各段之間進行適當分離以減少產物的攜帶。（參見：液滴分離器、組合空氣污染控制系統）

The transport of products or pollutants from one stage to the next one of an off-gas treatment unit containing several process steps. In a combined APC-system, product carry-over will lead to lower removal efficiencies. It should be reduced by appropriate separation units between the stages. (See also *droplet separator*; *APC-systems, combined*)

product ecobalances 產品生態平衡 *chanpin shengtai pingheng*

見：ecobalance。

See *ecobalance*.

product life cycle analysis 產品生命周期分析 *chanpin shengming zhouqi fenxi*

見：ecobalance。

See *ecobalance*.

product stewardship 產品管理 *chanpin guanli*

從產品的初始概念、研究開發、生產、銷售、直到最終處置為止，沿著產品整個*生命周期*不斷跟蹤產品所進行的全部管理活動。通過產品管理能儘早在開發階段發現問題從而消除其有害影響。產品管理考慮了與產品有關的一切方面，包括原料使用、經濟問題、顧客的要求和質量、產品使用時對環境的影響、廢物處理與處置問題。（參見：產品評價、生命周期分析）

The total of the management activities needed to follow a product from the initial idea through research and development (R&D), production, marketing to the moment of the final disposal, along its whole *life-cycle*. Product stewardship serves to identify possible problems early in the development phase and to eliminate unwanted effects. It integrates all aspects relating to the product: raw material use, economic aspects, customer satisfaction and quality, the impact of the product on the environment during its use, including questions of waste handling and disposal. (See also *product assessment*; *life cycle analysis*)

product-line analysis 產品線生命周期分析 *chanpinxian shengming zhouqi fenxi*

見：life cycle of a product; ecobalance。

See *life cycle of a product*; *ecobalance*.

project assessment, environmental protection 工程項目評價，環境保護 *gongcheng xiangmu pingjia, huanjing baohu*

見：company internal impact assessment。

See *company internal impact assessment*.

project management remediation 恢復補救，項目管理 *huifu bujiu, xiangmu guanli*

見：remediation, project management。

See *remediation, project management*.

protists 原生物 *yuanshengwu*

生物學中生物分類上使用的術語。原生物構成了所謂五個生物王國之一，是一切生物的主要組成部分。它主要包括單細胞的*真核生物*，如原生動物、真菌和一些藻類。它們在水生生態系統的食物鏈中和在土壤中具有重要的作用。

A term used in the classification of organisms in biology. Protists constitute one of five so-called kingdoms, the main divisions of all organisms. Protists include mainly single cell *eucaryotic organisms*, such as *protozoa*, *fungi* and some *algae*. They have an important function in the *food chain* of aquatic *ecosystems* as well as in soils.

protozoa 原生動物 *yuansheng dongwu*

由一個單細胞組成的*原生物*王國中的原生微生物，它包括約3萬個不同的種類，大多生活在水中，但也能生活在土壤中。其中許多種在結構上具有使它們可在水中運動的能力。原生動物可分為5大類：根足蟲類、鞭毛蟲類、*纖毛蟲類*、吸管蟲類和孢子蟲類。纖毛蟲作為*活性污泥*中的生物在污染控制方面是很重要的。此外，它也還是河流中的*指示生物*。

Primitive microorganisms of the kingdom of *protists*, consisting of one single cell. Protozoa include about 30,000 different species living mostly in water, but also in soil. Many of them are equipped with structures to allow them to move in the water. Protozoa are subdivided into five principal groups: rhizopoda, flagellata, ciliata, suctoria and sporozoa. The *ciliata* are of interest in the context of pollution control as organisms in *activated sludge*, and also as *bioindicator organisms* in rivers.

public interest groups 公共利益團體 *gonggong liyi tuanti*

代表一般公眾利益並且獨立於政府部門和工業界的團體，它們可以代表著範圍廣泛的利益和目的，其運作方式也是很不相同的。（參見：非政府組織）

Stakeholder groups which represent certain interests of the general public and which are independent of administration and of industry. Public interest groups may represent a wide range of interests and objectives. Their mode of operation varies very much. (See also *NGO*)

pure oxygen biological effluent treatment 純氧廢水生物處理 *chunyang feishui shengwu chuli*

將純氧引入活性污泥曝氣池中的一種特殊的廢水生物處理系統。純氧曝氣池通常被分成幾個小間，並加蓋密閉。其曝氣效率比敞口式傳統活性污泥系統高。應用這一系統的條件是要具備有氧氣。（參見：純氧曝氣系統、優諾克斯系統、林多克斯系統）

A special application of biological effluent treatment in which technically pure oxygen is introduced into the activated sludge basin. These basins are generally subdivided into compartments and are closed. The aeration efficiency is higher than in open conventional activated sludge systems. A condition for the application of such a system is the availability of technical oxygen. (See also *aeration systems with oxygen*; *UNOX*; *LINDOX*)

318

purgeable organic halogen compounds 可氣提的有機鹵素化合物 *keqitide youji lusu huahewu*

見：POX。

See *POX*.

pyrolysis 熱解 *rejie*

在無氧條件下加熱使有機物部分分解。熱解溫度比焚燒溫度低，在400-700°C範圍內。(參見：廢物熱解)

The partial decomposition of organic matter by heat and in the absence of oxygen. Pyrolysis temperatures are below incineration temperatures and range in the order of 400-700°C. (See also *pyrolysis of wastes*)

pyrolysis of wastes 廢物熱解 *feiwu rejie*

在有限供氧或完全無氧條件下廢物的熱分解。其熱解溫度為500-800°C左右，比焚燒溫度低。有機廢物的熱解產物含有可分離、回收和再利用的物質。然而，廢物熱解技術目前尚處於中試階段，成功運行的工業化裝置還很少。

A thermal destruction or modification of wastes under a limited supply of oxygen, or without oxygen at all. The pyrolysis temperature is in the order of about 500-800°C and is lower than the temperature of incineration. The products of pyrolysis of organic wastes may contain substances which can be separated, recovered and reused. The technology is as yet in a piloting stage, few full size facilities are in successful operation.

Q

quality criteria of analysis 分析的質量標準 *fenxide zhiliang biaozhun*

　　不同分析步驟和不同實驗室所得到的分析數據，其質量是不同的，它可用結果的離散性、分析的 *準確度* 和分析的 *可靠性* 來表示。分析方法、分析程序和實驗室的質量，可以由 *實驗室之間的相互測試* 來檢驗。

Different analytical procedures and different laboratories provide analytical data of different quality. The quality of the results can be described by the dispersion of the results, the *accuracy* of an analysis and the *reliability* of the analysis. The quality of methods and procedures and of laboratories can be verified also by *inter-laboratory tests*.

quality goals 質量目標 *zhiliang mubiao*

　　環境空氣、河流、水體、飲用水等質量控制的定量目標。質量目標由一些重要指標(參數)的數值表示；作為長遠目標它們對環境保護工作起到指導作用。(參見：環境空氣、排放)

Quantitative goals for the quality of ambient air, rivers, waterbodies, drinking water, etc. Quality goals are defined by numerical values for critical parameters; and these goals serve as long range objectives for directing environmental protection activities. (See also *ambient air*; *emission*; *immission*)

quality of life 生活質量 *shenghuo zhiliang*

　　討論環境保護目標時使用的一般性術語，其定義受文化傳統、個人愛好和個人 *價值* 影響很大。生活質量包括 *無形價值*、可獲得的物質上的東西和服務，而每一項具有不同的權重。

A general term used in the discussion of environmental protection objectives. The definition of the term is strongly influenced by cultural traditions, personal preferences and personal *values*. Quality of life includes *immaterial values* as well as the availability of material goods and services, each with different weights.

quench 驟冷 *zhouleng*

　焚燒爐煙氣處理的第一步過程。在驟冷單元中，熱煙氣由大約1000°C冷卻到較低的溫度(取決於後續步驟的類型與作用)。驟冷室的建造材料必須能耐高的熱應力和氣體腐蝕，在有些情況下需用石墨元件。

　Usually the first process step in the treatment of flue gases from an incinerator. In the quenching unit the hot flue gas is cooled from a temperature around 1000°C down to a lower level. This lower level depends on the type and function of the subsequent process step. The material of the quench has to withstand high thermal stress and an attack by corrosive gases. In some cases, graphite elements are used.

R

rain, acid 酸雨 *suanyu*

見：acid rain。

See *acid rain.*

rainwater overflow 雨水溢流 *yushui yiliu*

城市廢水處理系統的一個組成部分。雨水溢流可以是下水道系統的一部分，或者可以位於廢水處理場的進口側。較老的傳統管道系統通常包括雨水在內，因此污水量會有很大波動。雨水溢流則起到分流作用，它把雨水直接排入受納水體，以減少進入廢水處理場的廢水量，使之不超過其處理能力。在新系統中，污染嚴重的初期雨水被保存在集水池中，然後流入下水道至廢水處理場。(參見：分流式下水道系統、集水池)

An element of a municipal effluent treatment system. The rainwater overflow can be part of the sewer system or be at the input side of the effluent treatment plant. Older traditional canalization systems generally include precipitation water. The sewage volumes therefore may fluctuate considerably. The rainwater overflow serves to divert that part of the volume of the incoming effluent which exceeds the hydraulic capacity of the treatment plant. Diversion takes place directly to a receiving waterbody. In newer systems, the highly polluted first part of the precipitation water is retained in a catch basin and fed back to the sewer and the treatment plant later. (See also *separation sewer systems; catch basin*)

random error in analysis 分析中的偶然誤差 *fenxizhongde ouran wucha*

重複測量或測定時產生的實驗誤差或波動。偶然誤差似乎是不可控制的，它的產生也似乎沒有明確的原因。但是如果進行周密的調查研究，往往會發現造成這種誤差的某些新的因素，清除這些因素就能獲得良好的測定結果。(參見：系統誤差)

Experimental errors and fluctuations which occur when measurements or determinations are repeated. Random errors appear to be uncontrollable and without transparent cause. Careful investigation of the fluctuations will often reveal new causes which may then be eliminated to optimize a measurement. (See also *systematic errors*)

randomly packed column 無規填料塔 *wugui tianliaota*

　　裝填無規則形狀材料的填料塔（如生物反應器）。填料可用塑料（如聚氯乙烯、高密度聚乙烯）、膨脹陶瓷材料或任何其他適合的材料製成。在所謂的*生物濾池*中，可以用泥炭沼、椰子纖維或類似材料作為填料。活性碳也可以用作為無規載體。

　　A column, e.g. a bioreactor, which is packed with a random shaped material. The packing may be made of plastic (e.g. PVC, HDPE), of expanded ceramic material or any other material which is suitable for the desired application. In a so-called *biofilter*, twigs of peat moss, coconut fibers or similar material may be used. Activated carbon may also serve as a random support.

ranking criteria 排序判據 *paixu panju*

　　在評價技術方案的適用性時所採用的客觀判據，如：成本、效率、可靠性、零備件的易獲得性、維修維護要求、操作的難易程度等。它常在*決策技術*和*排序方法*中使用。有兩類判據：強制性的判據（必須滿足的）和"最好具有的"判據（它是重要的但為非強制的）。排序判據需根據具體情況加以選擇，並給與不同的權重。

　　Objective criteria which are used in evaluating of the suitability of a technical option. Examples are: costs, efficiency, reliability, availability of spare parts, maintenance requirements, ease of operation, etc. Ranking criteria are used in *decision techniques* or *ranking methods*. Two groups are: compulsory criteria (which must be fulfilled) and 'nice-to-have' criteria, which are important but optional. Ranking criteria need to be selected for the specific case and be given different weights.

ranking methods 排序方法 *paixu fangfa*

　　在進行*決策技術*和應用某些*排序判據*時所採用的方法。對要排序的方案根據所選擇的判據一一進行評價與考核，然後將各個方案滿足每一判據的程度通過各種數學計算進行換算，構成一個總的排序。

Methods used in *decision techniques*, making use of certain *ranking criteria*. Options to be placed in a ranking order are each evaluated and qualified against the selected criteria. The degree of fulfillment of each criterion by each option is then transformed by different mathematical calculations and transformations into an overall ranking order.

RBC 生物轉盤 *shengwu zhuanpan*

見：rotating biological contactor。

See *rotating biological contactor*.

reaction-gases 反應氣 *fanyingqi*

一類廢氣。（參見：工藝廢氣）

A type of off-gas. (See also *process gas*)

real-time data 實時數據 *shishi shuju*

由監測器、檢測器和任何其他在線測定儀無滯後提供的數據。實時數據對那些不容許有時間滯後的控制活動，如直接過程控制或嚴重排放事故的直接監測是必須的。

Data, provided by monitors, detectors and any other on-line measuring instrument without any delay. Real-time data is needed for control activities where a time delay is not acceptable, e.g. direct process control or direct monitoring of critical discharges.

recalcitrant compounds 難降解化合物 *nanjiangjie huahewu*

見：non-biodegradable compounds。

See *non-biodegradable compounds*.

receiving waterbody 受納水體 *shouna shuiti*

處理後或未經處理的廢水被排入的水體，例如河流、運河、湖泊、港灣等。

A waterbody into which treated or untreated effluent is discharged. Examples: a river, canal, lake, estuary, etc.

recombination of molecular fragments 分子碎片的重組 *fenzi suipiande chongzu*

　　廢物焚燒時在不完全氧化條件下發生的化學過程。分子碎片的重組會導致生成新的分子。一般認為在有些老焚燒爐中 2, 3, 7, 8－四氯二苯對二惡英的形成就是由於這一過程引起的。

　　A chemical process, which may take place under conditions of incomplete oxidation in the incineration of wastes. The recombination of molecular fragments may lead to the formation of new molecules. It is thought that the formation of highly controversial TCDD in some older incinerators is due to this process.

recombination processes 重組過程 *chongzu guocheng*

　　在某些高溫焚燒情況下，一些有機物的分子最初被分裂為較小的分子，並部分被氧化為分子斷片。根據焚燒條件的不同，這些斷片可在煙氣中重新組成新的化合物。廢物焚燒爐煙氣中的 2, 3, 7, 8－四氯二苯對二惡英，被認為是重組過程的產物。

　　Under certain conditions of high temperature *incineration*, some organic molecules are initially split into smaller, partially oxidized molecular fragments. Depending on the incineration conditions, such fragments may recombine in the flue gas to form new compounds. The problem of TCDD in flue gases from waste incinerators is thought to be the result of such processes.

recorder, analog 模擬記錄器 *moni jiluqi*

　　將分析儀器的模擬信號繪製出來的記錄器。它在記錄紙上畫出的是一根曲線，如紫外光譜。模擬數據的加工處理和評估，需要人工説明。

　　A recorder for plotting the analog signals from an analytical instrument. The result is a curve on a sheet of paper, e.g. UV spectrum. Further processing and evaluation of analog data requires manual interpretation.

recovery rate of a method 方法的回收率 *fangfade huisoulü*

　　表示分析質量的一項參數。它是為了控制目的而外加某痕量污染物於樣品中所回收的該污染物的百分比。

　　A parameter for the quality of an analysis. The recovery rate is that percentage of a pollutant X which is recovered when an additional trace quantity of this pollutant X is added to a sample for control purposes.

recycling 循環利用 *xunhuan liyong*

　　廣義地說，指重複利用廢物的一切活動，而不是採用處置方法對廢物進行處置。循環利用包括的內容很廣，如舊玻璃瓶的再利用，廢塑料再利用於生產新的塑料制品，金屬（如鋁）的回收與再利用，以及廢物焚燒回收熱能等。

All activities serving to reutilize wastes (in the broadest sense of the word) instead of disposing of them by waste disposal methods. The term recycling covers topics as diverse as the reuse of an old glass bottle, the reuse of plastic waste materials in the manufacture of new plastic goods, the recovery and reuse of metals, e.g. aluminium, to the recycling of energy by recovering the energy content of wastes in a waste incinerator.

recycling technologies 循環利用技術 *xunhuan liyong jishu*

　　用於廢物回收和再利用的技術，它主要包括：（1）對物質進行分離、回收和淨化並用來替代新物料（真正的循環利用）的技術，如鋁、鐵、鉛和玻璃的循環利用。（2）在不作大的變化下利用廢物作為要求較低的新產品的技術，這有時也叫做"下游循環利用"技術，如再利用廢塑料來生產廉價的塑料袋，用廢紙來生產質量較低的紙張。（3）能量循環利用技術，即在廢物焚燒爐中以熱回收方式利用廢物來回收能量。

　　物質的循環利用受經濟、能源和質量等因素制約。在有些情況下，由於質量原因對於所回收的物質很難找到使用的用戶。

Technologies which serve to recover and reutilize waste materials. The main types include firstly, technologies which serve to separate, recover and purify materials which can then be used as an identical replacement of a new material (true recycling); and examples are recycling of aluminum, of iron, of lead, and also of glass. Secondly, technologies which aim to use waste materials without major change in a new and less demanding application. This is occasionally called 'downcycling'. An example of this group is the reuse of waste plastic material to manufacture cheap plastic bags, or the use of waste paper to produce lower quality paper. Thirdly, energy recycling which refers to using waste materials to recover their energy content in waste incinerators with heat recovery.

Recycling of materials is limited by economic, energy and quality considerations. In some cases, it is difficult to find a user for recovered materials for quality reasons.

refractory ionic compounds 難降解的離子化合物 *nanjiangjiede lizi huahewu*

不可生物降解或難*生物降解*的有機離子化合物，如萘、磺酸或某些表面活性劑。

Ionic organic compounds, e.g. naphthalene sulfonic acids, or certain surfactants, which are not, or only poorly, *biodegradable*.

refractory lining 耐火襯 *naihuochen*

某些焚燒爐如*回轉窯焚燒爐*燃燒室用耐火磚砌成的保護襯層。耐火磚含有相當量的氧化鋁和二氧化矽，能耐高溫和含鹼金屬的液態熔渣。

The protective lining of the combustion chamber of certain incinerators, e.g. *rotary kiln incinerators*, with refractory ceramic bricks. Refractory ceramic bricks contain a higher fraction of alumina and silica and are more resistant to high temperature and liquid slag containing alkali metals.

refractory pollutants 難降解污染物 *nanjiangjie wuranwu*

廢水中或生態系統中不能生物降解和礦質化的污染物。對廢物流中這類污染物，需通過在源頭改進工藝過程，即通過*完整的工藝過程開發*來避免產生，以及通過*分離技術*、*再生技術*和化學或熱分解等其他方法來去除。

Pollutants in effluents or in ecosystems in general, which are not amenable to biological degradation and mineralization. Their removal from waste streams requires other methods, e.g. avoidance through process modifications at the source *integral process development*, *separation technologies*, *regenerative technologies* and or chemical or thermal destruction.

regeneration of activated carbon 活性碳再生 *huoxingtan zaisheng*

為了去除（解吸）活性碳所吸附的化合物和使活性表面獲得再生而進行的 *粒狀活性碳再生* 的過程。再生需要特殊的設備。在熱再生系統中，把水蒸汽噴入熱反應器，通過與水發生化學反應，將活性碳表面的碳除去，恢復其新的活性表面。所吸附的無機物只能通過洗滌加以除去。對於小用戶，通常通過簽定商業合同由供貨廠商進行再生。在再生過程中活性碳大約會損失 5-10%。

The process of regenerating granular activated carbon (*GACT*) with the objective to remove (desorb) adsorbed compounds and regenerate an active surface. The regeneration technology requires special equipment. Some thermal regeneration systems inject steam into the hot reactor. Through a chemical reaction with water, some carbon from the surface of the carbon is removed chemically, leaving a new active surface. Inorganic adsorbed material can only be removed by washing. For small users, regeneration is usually carried out by the supplier on a commercial contract basis. In the regeneration process, about 5-10% of activated carbon is lost.

regenerative technologies 再生技術 *zaisheng jishu*

用於回收和再生污染物或廢物以供再利用的技術，這與破壞或礦化廢物的污染控制技術不同。作為一種特殊情況，可以把具有熱回收的廢物焚燒裝置認為是一種再生技術；因為它回收了廢物中所含的能量。再生技術在廢物 *循環利用* 和 *源頭污染預防* 方面是很重要的。

Technologies which serve to recover and regenerate pollutants or wastes for further use. Regenerative technologies contrast with pollution control technologies which destroy or mineralize waste matter. As a special case, incineration of waste in a facility with heat recovery may be considered; in this case the energy contained in the waste is recovered. Regenerative technologies are important with *recycling* and *PPS* (*pollution prevention at the source*).

rehabilitation 恢復 *huifu*

見；remediation。

See *remediation*.

rehabilitation measures 恢復措施 *huifu cuoshi*

見：remediation measures。

See *remediation measures*.

reliability 可靠性 *kekaoxing*

如果某種分析方法在不同實驗室使用時或在將來使用時都始終會得到*準確*的結果，則稱這種方法是可靠的。

An analytical method is called reliable if it will always produce *accurate* results when used in different laboratories, also in future use.

remediation 恢復/補救 *huifu/bujiu*

清除和淨化由過去操作造成的污染以及恢復環境損害的活動。在全世界許多地方，由於過去採取不正確或不當方法處置廢物，在不同程度上造成了土壤、工廠場地、地下水體、河流和湖泊的污染。目前，補救活動集中於土壤和地下水的恢復上。（參見：超級基金計劃、污染場地）

Activities to remove and/or clean-up pollutions from past operations and restore environmental damages. In many places over the world, past practices of incorrect or careless handling or disposal of waste streams caused pollution of ground, factory sites, groundwater bodies, and rivers and lakes to a variable degree. Remediation today focusses on ground and groundwater remediation. (See also *SUPERFUND*; *contaminated sites*)

remediation, project management 恢復補救，項目管理 *huifu bujiu, xiangmu guanli*

對已污染的場地或地下水的恢復補救行動所進行的系統、靈活的管理過程。由於需要進行恢復補救的工作往往超過了技術和資金方面的可能能力，因此必須確定重點。初期工作應集中對已污染的場地進行歷史調查。然後，根據以下情況：（1）相對於未污染的背景值，場地或地下水體污染的類型和程度；（2）土壤的類型和性質，如吸附性；（3）擬在未來使用的情況；（4）污染的趨勢，來進行風險評價並確定恢復補救工作的重點。

在確定重點時，需要有關地下水或土壤污染程度和污染類型的數據資料。在已公佈的*荷蘭限值表*或類似的資料中列出的*干預值*，可作為決策的參考值。

A systematic and yet flexible management process to deal with the remediation of contaminated sites or groundwater. The setting of priorities is necessary in view of the need for remediation, which often exceeds technical and financial capacities. Initial activities must focus on a historical review of contaminated sites. Priorities then depend on a risk assessment, based on: the kind and the level of contamination of a site or a groundwater body in relation to unpolluted background levels, the type of soil and its properties, e.g. adsorption properties, the expected future use, and trends of contamination.

To arrive at priorities, data are needed on the degree and type of pollution of groundwater or soil. Reference values to assist in decision making are available, e.g. in *intervention values* as published e.g. in the *Holland List*, or similar documents.

remediation goals 恢復/補救目標 *huifu/bujiu mubiao*

對已污染的地下水和場地採取恢復補救行動所確定的目標。因為清除/淨化費用（*損害恢復費用*）與這些目標有很大的關係，所以確定一個現實的目標是很重要的。但是，這些目標必須包括長期*可持續發展*的概念。確定恢復/補救目標時必須考慮的因子有：（1）未被污染的介質（土壤或地下水）的背景值；（2）對擬使用的地下水或場地所進行的風險評價；（3）*費用–效益評價*。在已公佈的*荷蘭限值表*中，列有確定目標時所需使用的參考濃度值。

Goals defined for remediation activities of polluted groundwater and sites. As the cleanup costs (*damage repair costs*) depend very strongly on the level of these goals, a realistic goal-setting is important. Nevertheless, the goals must also include the concept of long term *sustainable development*.

Factors which have to be considered when determining goals are: the background level of unpolluted material (soil or groundwater), a risk assessment regarding the expected use of groundwater or the site, the feasibility of technical solutions, and a *cost-benefit* assessment. Reference concentration values for use in goal setting have been published in the *Holland List*.

remote sensing 遙感 *yaogan*

一些用於定性或定量地評估遠距離處情況的技術。真正的遙感可不必對所分析研究的介質進行取樣。例如，對地面現象的調查研究，可從飛機或衛星上用不同的光譜帶進行攝影，以及採用靈敏度和圖象技術進行解析。這在林業學、森林破壞、水污染和天氣預報上已有一些應用實例。在空氣污染方面，也已應用了*激光探測和測距系統*。

比較常用的遙感方法還包括介質（如廢氣）的遠程取樣（送中心分析）技術，或利用安設在管網不同點處的遙感器（如甲烷遙感器）進行監測並對信號進行集中處理的技術。

A number of techniques for the qualitative or quantitative assessment of situations at a distance. True remote sensing makes possible analyses without having to collect a sample of the medium under investigation. For the investigation of surface phenomena, photography from airplanes or from satellites, using different spectral bands, sensitivities and image techniques are used. Radar systems are also applied. Examples have been published in forestry, damages to forests, water pollution and also weather forecasting. In the air pollution field, *LIDAR* systems have been used.

More conventional remote sensing methods also include such techniques as remote sampling of a medium (e.g. an off-gas stream) for central analysis, or the use of remote sensors (e.g. for methane) at different points of a piping network and the central processing of the signals.

renewable energy 可再生能源 *kezaisheng nengyuan*

見：energy, renewable。

See *energy, renewable.*

renewable resources 可再生資源 *kezaisheng ziyuan*

見：resources。

See *resources.*

repair and maintenance 維修與維護 *weixiu yu weihu*

它是工廠中的一項重要工作，目的在於使所有設施保持良好與安全

的操作狀態。維修與維護計劃必須包括各類設備，如生產設備、基礎設施和污染控制設備。(參見：預防性維修制度)

Important activities in a factory. The objective is to maintain all facilities in good and safe operating conditions. Repair and maintenance programs must include all types of equipment: production, infrastructure and pollution control. (See also *preventive maintenance system*)

repeatability of results 結果的重複性 *jieguode chongfuxing*

如果同一樣品進行多次測定都始終得到同樣的結果，則稱該分析結果是重複的。然而，重複性好並不一定意味著*準確性*或*可靠性*高。

Results of an analysis are repeatable, if a number of measurements of the same sample consistently yield the same result. However, repeatability does not necessarily mean *accuracy* or *reliability*.

reporting channels in accidents 事故報告渠道 *shigu baogao qudo*

預先確定的報告重大事故和災害的信息渠道，以把泄漏或事故的情況報告政府主管部門、新聞媒介和可能直接受影響的居民。它是重大事件*警報計劃*的一個組成部分。它的效能應通過由各有關方面、工業部門、行政部門、新聞媒介參加的警報演習來加以檢驗。

Predetermined information channels which are used to report major accidents and disasters. The channels serve to inform the authorities, the news media as well as the population which may be directly affected by a spill or an accident. The channels form part of the *alarm plan* for the case of a major event. Their functioning should have been tested in alarm exercises with the participation of all parties concerned, such as industry, public administration and news media.

residence time in biological effluent treatment 廢水生物處理的停留時間 *feishui shengwu chulide tingliu shijian*

見：aeration, residence time。

See *aeration, residence time.*

residence time in chemical processing 化工過程的停留時間 *huagong guochengde tingliu shijian*

　　對於連續的加工過程，液體中某一假想體積元在反應器中所停留的平均時間。在化工過程的計算中，通過它使設備尺寸與反應動力學參數相關連。（參見：曝氣停留時間）

　　The average time which a hypothetical volume element of a liquid spends in a reactor for continuous processing. The term is used in calculations of chemical processing to relate the size of equipment to parameters of reaction kinetics. (See also *aeration, residence time*)

residual impact 殘存影響 *cancun yingxiang*

　　對於某種活動在採取了一切合理的、技術與經濟可行的措施來減少其影響之後仍然存在的對人和生態系統的影響。在分析某種活動的殘存影響時，必須把短期和長期的影響考慮在內。這種影響應盡可能小，而使管理又能承擔其全部責任。零污染的概念意味著殘存影響為零，然而在現實生活中這是不可能實現的。

　　The impact of an activity on man and the ecosystems which remains after all reasonable, technically possible and economically viable activities to reduce this impact have been taken. In the analysis of the residual impact of an activity, short term as well as long term considerations must be included. The impact should be so low that management of this activity can assume full responsibility for it. The concept of *zero pollution* implies a zero residual impact, which, however, is not possible in real life.

resources 資源 *ziyuan*

　　資源是用來生產物品的原料。窄義地說，資源包括諸如原油、天然碳氫化合物、礦物、礦石、煤、植物和動物產物、木材等原料。廣義地說，資源還包括用於生產過程並且會排出廢氣和廢水的空氣和水。而從更廣泛的意義上說，它還包括未受到干擾破壞的自然與景觀。

　　有些所謂的不可再生資源，雖然看起來似乎很豐富，但地球上具有的量是有限的，如煤、特殊金屬礦和原油。這些資源在它所在之處已儲藏有幾百萬年之久，而成為現在的礦藏；可以說這是它們在生態圈中的

最終去所。另一些所謂的可再生資源，是能夠不斷地再生的。它們是投入到地球生態系統的連續能流。水電、風能、太陽能、植物資源等都是可再生資源。而可持續發展則強調要以持續方式來管理一切資源。

Resources are raw materials which are used to manufacture goods. In the narrow sense, the term includes raw materials like crude oil, natural hydrocarbons, minerals, ores, coal, products of plants and animals and wood. In a wider sense, resources also include the air used to discharge off-gases and water to dispose of effluents.

In a still more general sense such values as undisturbed nature and landscapes are also included.

Some resources, called non-renewable, are available on earth in limited, although seemingly large quantities. Examples are coal, ores of specific metals, crude oil. These resources have been deposited millions of years ago in the present locations which are now mineral deposits; these might be called their *final sink* in the ecosphere. Other resources, called renewable, are constantly being produced anew. They are the result of the continuous energy input into the global ecosystem. Hydroelectricity, wind energy, solar energy, resources from plants and similar are renewable. *Sustainable development* puts emphasis on management of all resources in a sustainable manner.

respirometry 呼吸測定法 *huxi cedingfa*

測定*好氧微生物過程*（呼吸）所攝取的氧量的分析方法。目前市場上供應的呼吸計，其工作原理是不同的，有的基於補償測試期間氣壓的變化，有的基於平衡呼吸過程所釋放出的二氧化碳或其他氣體。呼吸測定法可應用於*廢水毒性*的研究、*活性污泥*活性的測定、*生物降解性*和植物生長的研究。

An analytical method to measure the uptake of oxygen by *aerobic* microbiological processes (respiration). Modern commercial respirometers are based on different principles. They also compensate for changes of barometric pressure during the test and for any carbon dioxide or other gases, which are released in the process. Respirometry can be applied in the study of *effluent toxicity*, the measurement of the activity of *activated sludge*, for *biodegradation* studies, as well as in research on the growth of plants.

Responsible Care Program 責任關懷計劃 *zeren guanhuai jihua*

1984年由加拿大化學工業提出的一項政策性計劃。現在，許多國家的工業組織都贊同了這一計劃。責任關懷計劃是在考慮了地方特殊情況的基礎上提出的一個全國性計劃。因此，它是由全國性的工業組織發起的。

其目的是保證化工企業以對其職工、環境、鄰近居民和其他公司負責任的態度進行運作。責任關懷計劃涉及到一個化工公司的全部產品和過程以及*產品生命周期*的各個階段。它的目標與現代的環境目標和*可持續發展*是相一致的。為了實際應用，已編寫了許多指南、準則、核對清單和工作指標供企業參考。

責任關懷計劃是工業組織和企業自願採納的。採納這一計劃的公司要公開做出遵守其原則的承諾。

A policy program developed in 1984 by the chemical industry of Canada. Today, industrial organizations in many countries subscribe to Responsible Care Programs which are always developed on a national basis, taking into account specific local aspects. The programs are thus the result of an initiative by a national industrial organization.

The Responsible Care Program aims to make certain that a chemical enterprise operates in a responsible manner with respect to its employees, its environment, its neighbors and other companies. The program thus has implications on all products and processes of a chemical company, throughout all phases of the *life cycle of the products*. These goals are in line with modern environmental objectives and with *sustainable development*. For practical use, a number of guidelines, codes, check-lists and performance indicators are available to the enterprises.

The Responsible Care Policy can be adopted on a voluntary basis by industrial organizations and enterprises. Companies that adopt the program are expected to pledge in public that they will follow the principles.

retentate in reversed osmosis 逆滲透過程的持留液 *nishentou guochengde chiliuye*

逆滲透過程為半滲透膜所阻留的濃液相，因此持留液中含有可進行回收的物質；而在應用於污染控制的場合，持留液則是含污染物的濃廢液。

The concentrated aqueous fraction which is held back by the semi-permeable membrane in the *reversed osmosis process*. The retentate thus contains the material to be recovered, or, in pollution control applications, the solution containing the pollutants.

retention basins 事故貯水池 *shigu chushuichi*

用以收集事故情況下所排放的廢水的大貯水池。這種污染嚴重的廢水可來自非正常生產的排水，或放有危險物品的儲存區或倉庫著火時的消防水。為此，可以單獨設立幾個貯水池，或在特殊情況下利用已有建築物的地下室，來收集事故廢水。對收集在事故貯水池中的廢水應進行分析，並在以後進行處理。

Large volume basins for collecting critical discharges of liquid effluents from an accident. Polluted effluents may result from abnormal events in production or from fighting a fire in a storage area or a warehouse which contains critical products. Separate basins, or, in special cases also the basement of an existing building, may be adapted to this purpose. The collected polluted water may be analyzed and treated later on.

retention time in gas chromatograph (GC) 氣相色譜法保留時間 *qixiang sepufa baoliu shijian*

某個別化合物通過氣相色譜柱的時間，它是從樣品注入瞬間至檢測出信號所測定的時間。保留時間是某一化合物的特徵值，是鑒別該化合物的重要參數。如果兩個化合物(一個是未知化合物，一個是參比化合物)具有相同的保留時間，即使改變色譜條件，它們也是相同的。

The time it takes for a particular compound to pass through the column of a gas chromatograph. The time is measured from the moment of injection of the sample to the detection of a signal. The retention time is a characteristic value of a compound and is a key element for its identification. Two compounds (an unknown and a reference compound) are identical if they show the same retention time even if the conditions of chromatography are changed.

return on investment (ROI) 投資收益率 *touzi shouyilü*

投資得到的盈利或收益。在一般費用管理中，投資收益率是以作為年利潤回收的投資資金的百分比表示的，可用它來評估經濟的合理性與投資的可行性。如單純以短期一年一年的經濟運作方式來看待問題的話，則污染控制設施的投資是沒有收益的，似乎是不經濟的。環境投資的收益與效益是長期的，因此用投資收益率來進行簡單的評價是不充分的，甚至可能是錯誤的。這應該與如果不進行污染控制所造成的投資收益率的損失相比較。（參見：時間範疇上的難題）

The profit, or the return, which results from an investment. In the general cost management situation, the return on investment (ROI) is expressed as a percentage of the invested capital which is returned as annual profit. The ROI serves to evaluate the economic soundness and viability of an investment. Looking at this in a purely short term year-by-year economic way, investments into pollution control facilities have no ROI and seem to be uneconomic. Gains and benefits from environmental investments are long term and a simple evaluation of the ROI situation is therefore not adequate and may even be wrong. This should be also compared to the loss of ROI if pollution control is not in place. (See also *time horizon dilemma*)

reuse 再利用 *zailiyong*

見：recycling。

See *recycling*.

reversed osmosis process 逆滲透過程 *nishentou guoceng*

自然*滲透*過程的相反過程，即向液體（如水）施以壓力使它抵抗*滲透壓*而通過*半透膜*的過程。在逆滲透過程中，液體通過了半透膜，但分子較大的污染物將被阻留下來。

A reversal of the natural process of *osmosis*. In this case, the liquid (e.g. water) is pressed through the *semi-permeable membrane* against the *osmotic pressure*. While the liquid passes through, larger molecular contaminants are held back.

reversed osmosis technology 逆滲透技術 *nishentou jishu*

逆滲透過程在技術上的應用。其主要組成部分為：泵系統，用以產生所需要的壓力將處理的液體傳輸通過半透膜；耐壓容器中的半透膜組件；必要的貯水槽和管道；以及過程控制單元。

逆滲透技術將溶液分離為稀相滲透液和濃相持留液。在實際應用中，半透膜必須能耐溶液腐蝕和抗壓力差，並且應根據具體情況選用不同的膜材料。

逆滲透主要用於海水脫鹽。在水污染控制中，它可用於濕式空氣氧化前對廢水進行預濃縮。其他方面的應用，還包括化工生產中的濃縮和淨化過程。(參見：微過濾和超濾)

The technical application of the *reversed osmosis process*. The main elements of this technology are: a system of pumps to generate the necessary pressure to transport the liquid to be treated across *semi-permeable membranes*, units of *semi-permeable membranes* in a containment which will withstand the necessary pressure, the necessary tanks and piping, and process control units.

Technical reversed osmosis separates a solution into a dilute *permeate* and a concentrated *retentate*. In the technical application, the membrane must be resistant to the solution and withstand the pressure difference. Depending on the application, different membrane materials are used.

Reversed osmosis is used mainly for desalination of seawater. In water pollution control, reversed osmosis is applied e.g. for the *preconcentration* of effluents prior to *wet air oxidation*. Other applications include concentration and purification processes in chemical manufacturing. (See also *microfiltration* and *ultrafiltration*)

reversed phase HPLC 逆相高壓液相色譜法 *nixiang gaoya yexiang sepufa*

見：gradient elution。

See *gradient elution*.

ring-jet unit 環形噴射器 *huanxing pensheqi*

空氣污染控制系統的一種處理設備。它是特殊設計的*文丘里型洗滌器*，其主體呈軸對稱狀，而文丘里氣體管道處在內外圓筒之間。它一個單元設備的長度約為70厘米，直徑約為30厘米，每一個單元可處理大約1500~2500米³／小時的廢氣。廢氣可被加速到100米／秒的速度。在氣速最高點通過一特殊噴咀引入洗滌液。環形噴射器可用於去除廢氣中的煙塵或氣溶膠。在技術應用上，可以把若干個環形噴射器安裝在洗滌塔的單一平面上進行平行處理。它是由汽巴-嘉基公司開發和市場化的。

A processing unit for *APC-systems*. The ring-jet is a special design of a *venturi type scrubber*. The main body has a rotational symmetry. A venturi type gas canal is shaped between the inner and the outer cylindrical elements. The length of a single unit is about 70 cm, the diameter ca. 30 cm. Each unit is capable of processing about 1500 to 2500 m^3/hr of off-gas. The off-gas is accelerated to velocities as high as 100 m/s. The washing liquid is introduced through a special nozzle at the point of highest gas velocity. The ring-jet unit serves to remove dust or aerosols from off-gas streams. In technical applications, several units can be mounted in a single plane of a scrubbing tower for parallel processing. The units are developed and marketed by CIBA-GEIGY.

ring-tests 環(實驗室)測試 *huan (shiyanshi) ceshi*

在若干實驗室中對分析步驟和方法進行的系統和協調的測試，目的是保證實驗室工作和分析方法的質量，使得到的結果可以相互比較。(參見：實驗室之間的測試)

Systematic and coordinated tests of analytical procedures and methods carried out in several laboratories. The objective is to ascertain the quality of the laboratory and the procedure for comparison of results. (See also *inter-laboratory test*)

RIO Conference 1992 1992年里約會議 *yijiujiu'ernian liyue huiyi*

1992年在里約熱內盧舉行的聯合國環境與發展會議，它是在聯合國主持下於1992年組織召開的一次國際會議，討論1972年*斯德哥爾摩會議*後二十年來的情況，重點是討論環境保護和與之密切相關的人類發展和經濟的一般問題。按照*可持續發展*的總概念和目標，會議通過了作為21世紀行動計劃的"地球首腦21世紀日程"和"環境與發展的里約宣言"。

UN Conference on Environment and Development 1992 in Rio de Janeiro. An international conference organized in 1992 under the auspices of the United Nations. The conference dealt with the situation twenty years after the *Stockholm 1972 Conference*. It focused mainly on environmental protection and the closely linked general topics of human development and economy.

Based on the overall concept and objectives of *sustainable development*, the conference adopted the 'Earth Summit Agenda 21', as a program of actions for the 21st century and the 'Rio Declaration on Environment and Development'.

risk 風險 *fengxian*

意外事件造成的潛在損害。風險是由兩個因子決定的，即潛在的損害程度和事件出現的概率。它可用這兩個因子的乘積來計算，但這種簡單的計算不適用於危害很大的情況。當危害很大時，對潛在的損害程度這一因子應賦予較大的權。

Potential damages caused by an unexpected event. Risk is determined by two factors: the potential magnitude of damage, and the probability of the occurrence of an event. The risk can be calculated as the product of the magnitude of the potential hazard and the probability that the hazard will occur. This simple calculation is not applicable to very large hazards. In these cases, the factor 'magnitude' has to receive more weight.

risk analysis 風險分析 *fengxian fenxi*

對某工藝過程或裝置進行系統評價以分析其潛在的*風險*。風險分析包括工藝過程本身固有的風險，以及技術故障、差錯或人為疏忽的潛在影響。

A systematic assessment of a process or an installation with the objective of analyzing potential *risks* of the process. The analysis includes risks inherent to the process itself, as well as the potential influence of technical failures, errors, human omissions or negligence.

river basin authority 河流流域管理部門 *heliu liuyu guanli bumen*

對河流以及某河流流域中的水污染控制具有管轄權的政府管理部門。河流流域可能延伸跨過其他國家的邊界(或省界、縣界、州界等)。

An official authority having jurisdiction over rivers and aspects of water pollution control in a river basin. A river basin may extend across other national boundaries or provinces, departments, cantons, states, etc.

river quality 河流質量 *heliu zhiliang*

見 : bioindicators in river ecosystems。

See *bioindicators in river ecosystems.*

river quality goal 河流質量目標 *heliu zhiliang mubiao*

見 : quality goals。

See *quality goals.*

ROI 投資收益率 *touzi shouyilü*

見 : return on investment。

See *return on investment.*

room air 室內空氣 *shineikongqi*

一種類型的廢氣。雖然其污染物濃度低,但必須根據工廠及其環境的具體情況,對密閉生產廠房內的空氣加以淨化。室內空氣的流量較穩定,在5,000-200,000米³/小時之間。

A type or class of off-gas. Although pollutant concentrations are low, the atmosphere in a closed production building has to be purified, depending on the particular situation of the factory and its environment. The air flow volumes of room air can range anywhere from 5,000 to 200,000 m³/h. The flow is fairly constant.

rotary kiln 回轉窯 *huizhuanyao*

見 : kiln, rotary。

See *kiln, rotary.*

rotary kiln incinerator 回轉窯焚燒爐 *huizhuanyao fenshaolu*

用於處理各種不同類型特殊廢物的一種焚燒爐，其主要部件是回轉窯。它採用連續運行方式。處理時，廢物被投入到緩慢旋轉的窯的上端，爐渣從窯的下端連續排出並在水槽中驟冷。載焰可保證達到所需要的燃燒溫度。除了窯體外，焚燒爐系統還包括：廢物加料設備、後燃燒室、熱回收設備、爐渣處理和煙氣處理。（參見：回轉窯、燃燒）

An incinerator for various types of special wastes. The main element is the rotating kiln. A rotary kiln incinerator is operated as a continuous reactor. Wastes are fed into the upper end of the very slowly rotating kiln. Slag is continuously taken off at the lower end and quenched in a water basin. Support flames serve to guarantee the necessary combustion temperature. In addition to the actual kiln, the unit includes a waste feeding mechanism, a post combustion chamber, heat economizing equipment, slag handling, and adequate flue gas treatment. (See also *kiln, rotary*; *combustion*)

rotating biological contactor (RBC) 生物轉盤 *shengwu zhuanpan*

廢水處理的一種生物反應器，其特點是用一旋轉的大寬盤作為微生物附著的載體。當盤子轉動時，其扇面交互浸入廢水中並與空氣接觸，從而進行氧的傳遞。剩餘污泥因紊流可從盤面上除去。

A bioreactor for treatment of effluents/waste water, which is characterized by a rotating support for the biologically active microorganisms. The support may have the shape of a large, broad disc. While rotating, the sectors of the support alternate between immersion in the waste water and contact with the air, thus facilitating oxygen transfer. The excess sludge is removed by turbulence.

rotating disc reactor 轉盤式反應器 *zhuanpanshi fanyingqi*

見：rotating biological contactor。

See *rotating biological contactor*.

Rotterdam Charter of the ICC 國際商會鹿特丹憲章 *guoji shanghui lutedan xianzhang*

國際商會於1991年通過的國際16點憲章，其目的是推行可持續發展。

An international 16-point charter adopted by the ICC (International Chamber of Commerce) in 1991. The objectives of the charter are in line with sustainable development.

run-off water 逕流水 *jingliushui*

工廠或城市污水系統所收集的雨水（包括融化的雪水），它們常常為許多微量物質如土壤、塵粒、植物、糞便、汽油、垃圾、殘餘的泄漏物、汽車磨蝕物等所污染，根據當地情況和污染程度，可能需要在污水處理場進行處理。在下大雨時，逕流水量可能會超過處理能力，這部分過量的逕流水通常將直接經由暴雨水分流排管排放。在現代設計的廢水處理系統中設有一個大的儲存池，用來收集和調節過量的逕流水，以便在以後進行處理。

Precipitation water collected in a factory or city sewage system, originating from rain or from melting snow. Runoff water is contaminated with traces of many different substances: soil, dust, plants, feces, motor oil, trash, remainders of spilled materials, abraded materials from vehicles, etc. Depending on the local situation and the amount of contamination, runoff water may require treatment in an effluent plant.

At times of heavy rain, the volume of runoff water may exceed the capacity of treatment. This excess is often discharged directly via a *stormwater overflow*. Modern designs of effluent treatment systems incorporate a large holding basin to collect and buffer the flow of this excess runoff. The collected contaminated runoff can then be treated later.

running costs 運行成本（費用）*yunxing chengben (feiyong)*

見：costs。

See *costs*.

S

saddle column packing 鞍形填料 *anxing tianliao*

填料塔中無規填充的填料。它是等大的，大致為球狀或似立方體，兩面的外形像兩個馬鞍，底面相連。其直徑為幾厘米。與其他類型的填料一樣，它由不同的材料（如陶瓷材料）製成。市場上有不同用途的鞍形填料供應；它們具有不同的性能參數。

Elements for random filling of *packed columns*. Each element is isometric, nearly spherical or like a cube, with two surfaces shaped like two saddles, combined base to base. The diameter is a few centimeters. As with other types, different materials, including also ceramic materials are used. Saddle filling for different applications is available commercially; and different types are characterized by key performance parameters.

safety audit 安全審計 *anquan shenji*

採用審計管理方法對工廠的安全狀況進行評價。（參見：審計）

The assessment and evaluation of the safety situation at a factory by application of the auditing management technique. (See also *audit*)

safety tests 安全測試 *anquan ceshi*

進行一系列的測試和分析以獲得有關某工藝過程的風險性以及在什麼樣的條件下該工藝過程不再是安全的信息資料。安全測試包括化學分析、爆炸試驗、熱力學參數評估、材料試驗、各種物理參數如導電率、蒸發熱等的測定。

A range of tests and analyses which serve to give information on the risks of a process and on the conditions, under which the process will not be safe any more. The tests include chemical analyses, explosion tests, the evaluation of thermo-dynamic parameters, material testing and combinations of different physical parameters e.g. electrical conductivity, heat of evaporation and others.

salinity 含鹽量 *hanyanliang*

廢水或水體中各種鹽(離子態物質)的含量。它是一項綜合指標,並不區分鹽的種類。鹽的總含量可通過*電導率*測定或*離子分析*來估算。鹽含量高的水不適合作為某些水生生物的生境,也不適合用作灌溉水和飲用水。無機鹽會導致水中鹽含量增高,它們不能用生物處理的一般方法從廢水中去除,因此必須在產生源採取措施減少其排放量。(參見:完整的工藝過程開發)

The content of different salts (ionic substances) in an effluent or a waterbody. The group parameter 'salinity' does not differentiate between different salts. The total salinity is assessed through *conductivity* measurement or *ion analysis*. A high salinity makes water unfit for many uses, e.g. as a habitat for certain aquatic organisms, for irrigation and as drinking water. Inorganic salts, leading to a high salinity, are not removed from effluents by standard methods, such as biological treatment. Their discharge must be reduced by measures at the source. (See also *integral process development*)

sample enrichment 樣品富集 *yangpin fuji*

見:enrichment step in analysis。

See *enrichment step in analysis*.

sampling and measuring ports 取樣和測量口 *quyang he celiangkou*

在建造空氣污染控制設備時考慮設置的開口、人孔或法蘭。在設計上應使測量設備的探頭能很容易地插入和固定。為了有效地進行測量,取樣口應設在易接近而且使廢氣氣流不致受到干擾之處。

Openings and ports, manholes or flanges, which are included from the beginning in the construction of air pollution control equipment. The design should be such, that the probes of measuring equipment can easily be introduced and fitted. To carry out measurements of efficiency, sampling ports should be located at a place which is easily accessible and where the flow of off-gas is not disturbed.

sampling of effluents 廢水取樣 *feishui quyang*

獲得分析所需的正確和有代表性的廢水水樣,是如同分析本身一樣重要的。取樣時，樣品必須真實代表取樣點的廢水，而且必須考慮到污染物濃度和廢水流量都在不斷變化。有不同的取樣方法（*瞬時樣、時間平均樣、流量平均樣*）。在分析之前必須正確地保存樣品，最好將它們保存在冷卻和避光條件下，以防止水樣變化。取樣規程應說明取樣的一切情況。（參見：取樣程序）

Drawing a correct and representative sample of an effluent for further analysis is as important as the analysis itself. The sample must truly represent the effluent at the point of sampling. Sampling must take into consideration variable concentrations of pollutants and variable flow rates. Different sampling techniques (*grab sample*; *time average sample*; *flow average sample*) are available. The sample must be stored correctly before analysis, in order to prevent deterioration. Cooling and protection from light is advisable. A sampling protocol includes all facts about sampling. (See also *sampling procedure*)

sampling procedure 取樣程序 *quyang chengxu*

取樣是分析中的一個重要問題。為了得到有意義的結果，樣品必須代表被分析介質的真實情況。取樣應遵循的一步一步的程序，一般在取樣規程中作出了規定。嚴格的取樣規程對於評估和解釋分析結果是很重要的。

取樣時，應該用取樣記錄本詳盡記錄實際的取樣過程，通常包括樣品來源、地點、取樣日期與時間、取樣的方法或類別、取樣者和分析者等資料，還應記錄取樣時的特殊情況和事件。

Sampling is an important aspect in analysis. To obtain meaningful results, the sample has to be a true representation of the medium to be analyzed. A step by step procedure, to be followed for taking a sample, is generally specified in a sampling protocol. A precise sampling protocol is important in the evaluation and interpretation of analytical results.

A sampling document is then used to record and document the actual details of the sampling process. The sampling document usually includes information on the source of the sample, location, sampling date and time, method or type of sample, the person having taken the sample and the analyses carried out. Special conditions or events during sampling should also be recorded.

sandfilter 砂濾池 *shalüchi*

　　用砂子作為過濾介質的過濾設備。為了在反沖洗周期中將砂子洗淨並將過濾分離的物質除去，水流應呈逆流流動。砂濾池可用於從河水制取工藝用水以及在*廢水三級處理*中去除懸浮物。（參見：雙介質過濾）

　　A filter unit, in which sand is used as the filter medium. For cleaning the filter and removal of the separated matter in the backflushing cycle, the flow of water is reversed. Sandfilters are used for instance for the removal of suspended solids in the preparation of process water from river water and also in *tertiary treatment of effluents*. For a special design see also *filtration, dual media*.

SCP 社會費用原則 *shehui feiyong yuanze*

　　見：social cost principle。

　　See *social cost principle*.

SCR process 選擇催化還原法 *xuanze cuihua huanyanfa*

　　表示選擇催化去除氮氧化物的術語。見*催化還原去除氮氧化物*。對於非催化法，見*選擇性非催化還原法*。

　　A term used for selective catalytic removal of nitrogen oxides. See *catalytic reductive removal of NO$_X$*. For non-catalytic processes, see *selective non-catalytic reduction process*.

scrubber 洗滌器 *xidiqi*

　　用於*濕式空氣污染控制系統*的一般術語。

　　A general term used for *wet APC-systems*.

scrubbing liquid 洗滌液 *xidiye*

　　在濕式空氣污染控制裝置中循環使用以吸收廢氣中污染物的液體，它通常為水溶液，也可能含有與廢氣中某些污染物發生反應的化學物質。

　　The liquid circulating in a wet APC process unit and which absorbs the pollutants contained in the off-gas stream. The scrubbing liquid is generally an aqueous solution. It may contain chemicals to react with certain pollutants in the off-gas stream.

SD 可持續發展 *kechixu fazhan*

見：sustainable development。

See *sustainable development*.

secondary clarifier 二次澄清池 *erci chengqingchi*

廢水處理中設在活性污泥池後面的澄清池，其作用是將*活性污泥*從*污泥混合液*中分離出來。分離出的大部分污泥將回流到曝氣池中。(參見：澄清池、最終澄清池、污泥回流)

In effluent treatment, the clarifier after the activated sludge basin. Its function is to separate *activated sludge* from the *mixed liquor*. The main part of the separated sludge is returned to the aeration basin. (See also *clarifier*; *final clarifier*; *sludge recycling*)

secure landfill 安全填埋 *anquan tianmai*

在採取了一切必要的措施防止土壤和地下水污染的填埋場中進行的廢物的堆埋。(參見：受控填埋場、廢物管理)

A deposit for wastes in a landfill which includes all necessary measures to prevent any contamination of soil and groundwater. (See also *controlled landfill*; *waste management*)

secure storage area 安全貯存區 *anquan chucunqu*

具有事故(如泄漏或火災)預防措施的倉庫或貯存區，它是工廠環境和安全綜合管理的一個組成部分。

現代預防措施包括：建立有效的管理體系(包括警報計劃和人員培訓)；制定有序的貯存計劃(包括指定貯存區)；將具有不同性質和不同潛在危害的物品分開存放在各自的貯存區；設立一貯存管理系統，以給出所貯存物品(包括貯存地點)的最新資料；發生重大事故(火災、嚴重泄漏)時的遏制措施，如火檢測報警器、灑水系統、消防水貯池和應急材料；配備一套工具和容器，以備發生少量泄漏情況下使用。這些措施可根據所貯存化合物的潛在問題而不同。

A storehouse or a storage area which includes precautionary measures for the case of an accident, such as a spill or a fire. Secure storage areas are one element of an integral environmental and safety concept of a factory.

Up-to-date measures include: an efficient management system including an alarm plan and instruction of personnel; an orderly storage plan with assigned storage areas; separate storage areas or compartments for goods of different nature and different hazard potential; a storage management system which gives up-to-date information on the goods stored, including information on the storage locations; containment measures for the case of a major accident (fire, major spill), such as fire detectors, sprinkler systems, retention basins for fire fighting water, and emergency material; as well as a set of tools and containers to take care of minor spills. These measures may vary according to the problem potential of the stored compounds.

sedimentation 沉降 *chenjiang*

通過重力使不溶物或沉澱物從懸浮液中分離的過程。*沉降速度取決於液相和懸浮顆粒之間的密度差以及懸浮顆粒的大小*。沉降是廢水處理中的一個重要過程。進行沉降處理的*澄清池*，其處理效果取決於沉降參數和澄清池中的垂直升流速度。

The separation of insoluble matter or precipitates from a suspension by gravity forces. The *sedimentation velocity* depends on the difference in density between the liquid phase and the suspended particles as well as on their particle size. Sedimentation is an important process in effluent treatment. The efficiency of *clarifiers* depends on sedimentation parameters and on the upwards vertical flow velocities in the unit.

sedimentation velocity 沉降速度 *chenjiang sudu*

懸浮液中懸浮物沉降的速度或速率。當基於*沉降*過程進行*澄清池*的設計時，沉降速度必須大於液體的升流速度。

The velocity or rate of sedimentation of suspended matter in a suspension. In the design of *clarifiers* based on a *sedimentation* process, the settling velocity must be greater than the velocity of the upward flow of the liquid.

seepage water 滲流水 *shenliushui*

滲流通過填埋場中堆埋廢物的雨水或融雪，也叫做"浸出液"。根據情況的不同，滲流水常常為廢物中可溶性組分所污染，然後它又會導致地下水污染。（參見：浸出液、廢物，浸出性質）

Rain water or melting snow which seeps through the deposited wastes of a landfill. The term 'leachate' is also used. Depending on the situation, such seepage water is often contaminated by soluble components from wastes. This may then lead to a contamination of groundwater. (See also *leachate*; *wastes, leaching properties*)

selective catalytic removal (SCR) process 選擇催化還原法 *xuanze cuihua huanyanfa*

見：SCR process。

See *SCR process*.

selective non-catalytic reduction (SNCR) process 選擇性非催化還原法 *xuanzexing feicuihua huanyuanfa*

用氨或尿素於750-900°C下在氣相中將一氧化氮和二氧化氮進行還原的過程。由於沒有催化劑存在，因此還原速度較低，而必須在高溫下進行。

選擇性非催化還原過程的效率通常在50-90%之間，而選擇催化還原過程的效率較高，在80-95%範圍內。（參見：催化還原去除氮氧化物、選擇催化還原法）。

A process to reduce nitrogen monoxide and nitrogen dioxide with ammonia or urea in the gas phase at 750-900°C. No catalyst is used, the rate of reduction therefore is lower and for this reason the process has to be run at elevated temperature.

The efficiency of the SNCR process is typically between 50% and 90%, while the efficiency of the SCR process is somewhat higher between 80 and 95%. (See also *catalytic reductive removal of NO_x*, *SCR process*)

self-purification of a waterbody 水體自淨 *shuiti zijing*

見：auto-regeneration。

See *auto-regeneration*.

semi-continuous gas chromatography 半連續氣相色譜法 *banlianxu qixiang sepufa*

見：gas chromatography, semi-continuous。

See *gas chromatography, semi-continuous*.

semi-permeable membrane 半透膜 *bantoumo*

能使某些分子通過並截留住一些其他分子的膜材料。因此，可以把半透膜看作是一種用於過濾分子大小的粒子的"過濾器"。半透膜在自然界、在生物體細胞內進行的*滲透過程*中起著重要的作用。自然系統中的膜材料具有複雜結構，含有改性碳水化合物、蛋白質和反應基的結構單元。

應用於工藝過程的半透膜是由合成高分子材料如纖維衍生物、聚醯胺、聚碸和硅樹脂等製成的。它們可應用於各種不同的分離過程，如*微過濾、超濾、毫微米級過濾、逆滲透、氣體分離、全蒸發、蒸汽滲透、膜蒸餾和電滲析*中。在選用膜材料時，必須考慮它們的化學和機械穩定性以及區分性問題。

A membrane which allows certain molecules to pass, while others are rejected. In this way, a semi-permeable membrane may be looked at as a 'filter' for molecular size particles. Semi-permeable membranes play important roles in nature, in *osmotic processes* in cells of living organisms. Membrane materials in natural systems have a complex structure including building blocks of modified carbohydrates, proteins and also reactive groups.

Semi-permeable membranes for processing technology are made of synthetic polymeric materials, such as cellulose derivatives, polyamides, polysulfones, silicones and others. They are used in a wide range of separation processes: *microfiltration, ultrafiltration, nanofiltration, reversed osmosis*, gas separation, pervaporation, vapor permeation, membrane distillation and electrodialysis. In the selection of such membranes, questions of chemical and mechanical stability and discrimination properties have to be taken into account.

separation 分離 *fenli*

　　將混合物分離成為其各組分的各種物理和化學單元過程。分離的機理隨被分離的混合物的類型而不同，簡單的如用篩子將砂子與礫石分開，複雜的如分離原油各餾分的煉製過程。

　　在污染控制技術中，採用分離方法進行處理往往比進行破壞處理為好，因為廢水或廢氣中分離得到的組分可以進行再利用，這意味著具有經濟效益。分離技術包括*蒸餾、過濾、逆滲透、沉澱、沉降、吸著過程、靜電分離、汽提、色譜法分離、液化/冷卻、萃取、浮選、破乳、離子交換*等。分離設備也有各種各樣，應根據所要解決的具體問題來加以選擇和採用。

A broad range of physical and chemical unit processes for separating mixtures into their components. The mechanisms of separation depend on the type of mixture to be separated. Separation can be as simple as separating sand from gravel by sieves, and as elaborate as a refining process for separating different fractions of crude oil.

In pollution control technology, the application of separation processes is often preferred to destruction. Separated components from effluents or off-gas streams may be reused, and they represent an economic gain. Individual techniques include *distillation, filtration, reversed osmosis, precipitation, sedimentation, sorption processes, electrostatic separation, stripping, chromatography, liquefaction/ cooling, extraction, flotation, emulsion splitting, ion exchange.*

There is a wide variety of equipment available. The selection and adoption depend on the specific problem which has to be solved.

separation of emulsions 乳液分離 *ruye fenli*

　　見：emulsion splitting。

　　See *emulsion splitting.*

separation processes in waste treatment 廢物處理的分離方法 *feiwu chulide fenli fangfa*

為了有利於廢物的最終處置，所採用的一類廢物處理方法，例如：污泥過濾或離心脫水，破乳分離有機相和水相。而對分離得到的每一相，必須進一步進行特殊處理。

A group of processes applied as a treatment of wastes in order to facilitate their final disposal. Examples are: de-watering of sludge by filtration or centrifugation and the breaking of emulsions in order to separate organic and aqueous phases. Each of the two or more separate phases must subsequently undergo special treatment.

separation sewer systems 分流式下水道系統 *fenliushi xiashuidao xitong*

由若干分流的管網組成的下水道系統，每個管網用以接納一類特定的廢水。在許多城市(或城市的各區)，通過分流式下水道系統將雨水與污水分開。在化工廠，污水分流系統包括如下各自的下水管道或排放溝渠：(1)對未污染的冷卻水，直接排到 *受納水體*；(2)對辦公樓、食堂、浴室和廁所排出的生活污水，直接排到 *活性污泥處理場*；(3)對污染的工藝廢水，在排到活性污泥處理場之前要進行特定的 *預處理*。根據工廠的具體情況，可能還要對工藝廢水作進一步分離。實驗室排出的廢水可以作為生活污水或工藝廢水對待，這要依具體情況而定。

A sewer system, which consists of several separate networks of piping or canals, each one for a specific type of effluents. In many cities (or parts of cities) rain water is separated from sewage by a separate sewer system. In chemical factories the concept includes individual piping, canals or ducts for uncontaminated water from cooling, for direct discharge to a *receiving waterbody*; sewage from office buildings, kitchens, showers and toilets, for direct discharge to an *activated sludge basin*; a system for contaminated process effluents, which are given specific *pretreatment* before they are discharged to the activated sludge plant. These discharges may again be separated further, depending on special considerations of a factory. Effluents from laboratories may be treated as sewage or as process effluent, depending on the circumstances.

separation step in analysis 分析中的分離段 *fenxizhongde fenliduan*

　　環境分析中的一個重要步驟，是從母體中如從廢水、工業廢物、土壤或地下水中將所關注的物質分離出來。分離技術包括氣提、揮發（*液上氣體分析*）、吸附劑的熱解析或固相萃取。在分離之後再用特定方法分析各個化合物或各類物質。

　　在所有的色譜技術（高壓液相色譜法、氣相色譜法和薄層色譜法）中，都是將混合物通過柱中的固相吸附劑並用氣體或溶劑作為洗脫劑來進行分離的。通過物質與固相的相互作用，可使混合物分離成各個組分。

In environmental analysis one of the essential steps is to separate the substances of interest from the matrix, e.g. from waste water, industrial waste, ground, or groundwater. Techniques for separation include stripping, volatilization (*headspace analysis*), thermo-desorption from an adsorbent, or solid phase extraction. After the separation step the individual compounds or classes of substances are analyzed by a specific method.

In all chromatographic techniques (HPLC, GC and TLC) separation is effected by passing the mixture of substances through a solid phase adsorbent in a column, with gas or solvents as the eluants. In each case the substance/solid phase interaction is responsible for the separation into individual components.

settling box dust eliminator 沉降除塵箱 *chenjiang chuchenxiang*

　　見：gravity settling box。

See *gravity settling box*.

sewage 污水 *wushui*

　　通常指由人類活動產生自家庭的廢水。（參見：廢水）

An aqueous waste stream generally being produced by human activities, having households as their origin. (See also *effluent*)

sewage flow volumes (design values) 污水流量（設計流量）*wushui liuliang (sheji liuliang)*

　　規劃廢水處理場需用的污水水量的不同值，這些值反映了在不同天氣條件下進入處理場的污水水量變化的情況，即(1)乾燥天氣季節的污水流量：廢水處理場的標準設計能力通常是乾燥天氣季節污水流量的兩倍；(2)一年最大雨量：每年中出現一次的最大雨量；(3)50年最大雨量：50年中出現一次的最大雨量。廢水處理系統的設計能力取決於以上這些水量值，此外也還與處理概念和下水道系統(大小、體積容量和分流情況)有關。(參見：暴雨水分流排管)

For planning of effluent plants, different values for sewage volumes are used. The values reflect the variation of input volumes of sewage in different types of weather. Dry weather flow volume: the standard design capacity of a treatment plant usually is two times the dry weather flow, i.e. the flow volume of sewage during dry weather. One-year maximum rain volume: the maximum rainweather flow occurring once every year. 50-year maximum rain volume: the maximum rain-weather flow occurring once in every 50 years. The design capacity of an effluent treatment system depends on these volumes as well as on the treatment concept and the sewer system (dimensions, volume capacity and branching). (See also *stormwater overflow*)

sewage management districts 污水管理區 *wushui guanliqu*

大城市的一個區，該區所排(或將要排)的所有污水都去到同一個城市污水處理場進行處理。在這一管理區中，對污水處理的管理必須進行全面協調。(參見：城市污水處理)

An area or district of a large municipality from which all sewage discharges are drained (or will be drained) to the same municipal effluent treatment station. All aspects of the management of sewage treatment in such a district are coordinated. (See also *municipal effluent treatment*)

sewage treatment works 污水處理廠 *wushui chulichang*

見：municipal effluent treatment plant。

See *municipal effluent treatment plant*.

sewer systems 下水道系統 *xiaoshuidao xitong*

收集居民區(城市)生活污水去進行處理的管道和管渠系統或網絡。

在城市中，傳統的單管道下水道系統所收集的污水，是污染程度不同的各種各樣排水的混合污水，也還包括地面逕流水。比較先進的下水道系統，應能分流不同類型的污水。（參見：分流式下水道系統、逕流水）

A system or network of pipes, ducts and canals to collect the sewage of a populated area (a city) for treatment. In a city, the collected sewage in the traditional single-pipe system is a mixture of all types of aqueous discharges, including different degrees of pollution and also including runoff water. More advanced sewer concepts allow for the separation of different types of discharges. (See also *separation sewer systems*; *runoff water*)

short term exposure limits (STEL) 短期接觸限值 *duanqi jiechu xianzhi*

見：STEL。

See *STEL*.

signal, analog 模擬信號 *moni xinhao*

見：analog signal。

See *analog signal*.

signal-to-noise ratio (in analysis) 分析的信噪比 *fenxide xinzaobi*

每一分析檢測系統即使在沒有物質存在時也會產生殘餘信號脈動，這在技術上稱為"背景噪聲"。每一種分析檢測方法都有它自己的背景噪聲，其值應愈小愈好。當加入少量測試物質時，就會產生一個信號。此信號與背景噪聲之比就叫做信噪比。（參見：檢測極限）

Each analytical detection system is characterized by a residual signal fluctuation even if there is no substance present. The technical term for this fluctuation is 'background noise'. Every method has its own background noise which has to be made as small as possible. On adding a small quantity of a test substance, a signal will result. The ratio of this signal to the background noise is called the signal-to-noise ratio. (See also *limit of detection*)

significance (percentage level) 顯著性(百分率水平) *xianzhuxing (baifenlü shuiping)*

顯著水平是對數據以及由該數據得到的結論的可靠性所作的一項判據。對於一組數據可以進行不同的統計檢定，來檢驗某一假設或結論在某一概率水平(如95%)下是否真實或有效。例如，根據4次測定的結果可以得到結論認為廢水的排放符合法律規定的要求，這在顯著水平為90%時可回答"對"，而在顯著水平為95%時可回答"不對"。這種分析使得它能從真實有效的結果中消除純偶然因素的結果。

The level of significance is a criterion which gives information about the reliability of data and of a conclusion drawn from the data. Different statistical tests can be carried out with a set of data, to check whether a certain hypothesis or conclusion is true or valid at a probability level of, for instance, 95%. As an example, a conclusion may be reached on the basis of 4 measurements that an effluent discharge is in compliance with the law. The answer may be 'yes, at a significance level of 90 percent' and at the same time 'no, at a significance level of 95%.' This type of analysis makes it possible to eliminate results which are based on pure chance from results which are true and valid.

single element parameter 單一要素指標(參數) *danyi yaosu zhibiao (canshu)*

見：parameter, single element。

See *parameter, single element.*

sink 去所 *qusuo*

見：final sink in the ecosystem。

See *final sink in the ecosystem.*

slag 爐渣 *luzha*

物料焚燒後殘存的無機殘渣。與灰不同，爐渣在焚燒溫度下是液體。其量與焚燒物料的組成有關。在焚燒城市廢物時，產生的渣量約為廢物初始重量的三分之一。

特殊廢物焚燒產生的爐渣含有原廢物中存在的各種無機物，應把這種爐渣看作是特殊廢物，並且應按特殊廢物處置的要求對它進行測試和選擇處置方法。

The inorganic residue remaining after incineration of material. In contrast to ashes, slag is liquid at the incineration temperature. The quantity depends on the composition of the incinerated material. In the incineration of municipal wastes, it amounts to about one third of the original weight of the wastes.

Slag from the incineration of special wastes contains all inorganic materials which were contained in the wastes. This slag itself has to be regarded as special wastes and must undergo the tests and selection procedures required for the disposal of special wastes.

sleeve filter 袖式過濾器 *xiushiguolüqi*

見：bag filter; wet sleeve filter。

See *bag filter*; *wet sleeve filter*.

sludge 污泥 *wuni*

一般指介於液體和固體之間的濃稠物質。污泥可以用泵輸送，但它很難通過沉降進行固液相的分離。在活性污泥法中，具有生物活性的污泥含有活的和死的微生物。

In general, material with a consistency between liquid and solid. Sludges can be pumped. In a sludge, there is little separation of phases through sedimentation. In the activated sludge technology, the biologically active sludge consists of living and dyed-off microorganisms.

sludge age 污泥泥齡 *wuni niling*

活性污泥在廢水生物處理場的曝氣池中的平均停留時間，以天表示。在實際的處理系統中，污泥泥齡可能達到15天之久。污泥泥齡是將曝氣池中污泥的總量（公斤）除以每天從該系統中除去的淨剩餘污泥量（公斤/天）而計算得到的。在計算時，還必須考慮系統中污泥的生成和污泥的內回流情況。

The mean residence time of activated sludge in the aeration basin of a biological effluent treatment plant, expressed in days. In practical systems, sludge ages of up to 15 days are possible.

The sludge age is calculated from the total amount of sludge in the aeration basin (kg) divided by the net amount of excess sludge removed from the system per day (kg/day). In this calculation, the generation of sludge in the system and the internal return of sludge also have to be taken into account.

sludge centrifugation 污泥離心分離 *wuni lixin fenli*

污泥處理中利用離心機有效濃縮剩餘活性污泥的固體物。離心機是具有水平旋轉軸的無孔轉鼓型離心設備，它是基於液相與污泥不溶物的密度不同而進行分離的。分離得到的濃縮污泥，可用焚燒法進行後續處理。適用的焚燒爐類型有*複合式焚燒爐、流化床焚燒爐、多爐床焚燒爐或回轉窯焚燒爐*。

分離得到的高污染的液相，通常需將它返回到處理場的最前部去進行補充處理。可以利用焚燒爐的廢熱將污泥預熱到60-70°C，以提高分離過程的效率。

In *sludge treatment*, the use of a centrifuge for efficient concentration of solid matter of excess activated sludge. The centrifuge, a solid bowl type, is a design with a horizontal axis of rotation. The separation is based on the different densities of the liquid phase and the insoluble part of the sludge. The concentrated sludge is separated for subsequent disposal of by incineration. Suitable incinerator types are: *hybrid technology incinerator, fluidized bed incinerator, multiple hearth incinerator*, or *rotary kiln incinerator*.

A highly polluted liquid phase is separated which needs to be given additional treatment. It is returned to the head of the treatment plant. The efficiency of the separation process can be enhanced by preheating the sludge stream to about 60-70°C with waste heat from the incinerator.

sludge conditioning 污泥調節 *wuni tiaojie*

污泥處理中改善固相和液相分離過程的方法。對於污泥過濾，最好用無機藥劑如硫酸鐵或石灰進行污泥的調節；對於污泥離心脫水，則用有機聚合物進行調節。

Any treatment which enhances the separation process of the solid and liquid phases in sludge processing. For filtration, conditioning with inorganic reagents, like iron sulfate and lime, is preferred; for dewatering by centrifugation, organic polymers are used.

sludge dewatering 污泥脫水 *wuni tuoshui*

見：sludge treatment。

See *sludge treatment*.

sludge digestion, aerobic 污泥好氧消化 *wuni haoyang xiaohua*

在廢水生物處理場，剩餘*活性污泥*的處置是整個廢水處理過程的一個重要部分。剩餘活性污泥必須通過好氧消化使它穩定化、失去活性和減小體積。

好氧消化有兩種可行的方式：（1）在廢水處理場活性污泥曝氣池中進行延時曝氣，曝氣時間可以延長到30小時；（2）將剩餘活性污泥在另一分開的池子中進行長期曝氣（30天左右）。以上兩種方法雖然可使污泥處置的問題得到減輕，但為此需消耗較多的能源，使得曝氣費用增高。（參見：污泥厭氧消化）

In a biological effluent plant, the disposal of excess *activated sludge* is an important part of the overall process. Excess activated sludge may be stabilized, rendered inactive and reduced in volume by aerobic digestion.

Two possibilities are feasible: (1) extended aeration in the activated sludge basin of the plant. Aeration may extend up to 30 hours. (2) long term (about 30 days) aeration of the excess sludge in a separate basin. In both cases, the problems of sludge disposal are reduced. However, to achieve this, more energy is required, leading to higher aeration costs. (See also *sludge digestion, anaerobic*)

sludge digestion, anaerobic 污泥厭氧消化 *wuni yanyang xiaohua*

用厭氧微生物法對剩餘活性污泥進行消化處理，處理後的殘留污泥不再具有活性。污泥厭氧消化會得到含65-80%的甲烷和25-30%二氧化碳的沼氣副產物。

Digestion of excess activated sludge by anaerobic microbiological processes.

Anaerobic treatment of sludge produces a residual sludge which is inactive. As a byproduct, *biogas* is also obtained, containing from 65-80% of methane in addition to 25-30% of carbon dioxide.

sludge drying bed 污泥乾化床 *wuni ganhuachuang*

利用太陽光使污泥乾化的方法處置廢水生物處理產生的剩餘活性污泥的簡單技術。但是污泥乾化床會產生惡臭問題。另外，對其排液必須加以收集並送回處理場進行處理。

A simple technology used to dispose of excess activated sludge from biological effluent treatment by drying the sludge in the sun. However, sludge drying beds may lead to odor problems. The drained liquid must be collected and returned to the treatment plant.

sludge fermentation 污泥發酵 *wuni fajiao*

用厭氧發酵法處理剩餘活性污泥。（參見：污泥厭氧消化、發酵）

The treatment of excess activated sludge by an anaerobic *fermentation* processes. (See also *sludge digestion, anaerobic*; *fermentation*)

sludge filtration 污泥過濾 *wuni guolü*

用過濾法處理剩餘活性污泥以除去污泥水份和提高污泥中有機固體物的含量。對於廢水處理的剩餘污泥，在過濾之前將它加熱到60-70°C，可以提高其過濾效率。在現代化的廢水處理場中，可利用焚燒爐的廢熱來加熱污泥。

某些污泥處置方法，要求首先必須對具有活性的剩餘生物污泥採用延時曝氣或投加氫氧化鈣進行穩定化處理。

Treatment of excess activated sludge by filtration in order to remove liquid and to increase the content of organic solid matter. The filtration efficiency for excess sludge from effluent treatment can be increased by heating the sludge to 60-70°C before filtration. In modern effluent treatment plants this is done by using waste heat from the incinerator.

For some methods of sludge disposal, excess biosludge, which is still active, may first have to be stabilized, either by extended aeration or by the addition of calcium hydroxide.

sludge fouling 污泥惡臭 *wuni echou*
　　見：*污泥發酵*。
　　See *sludge fermentation*.

sludge hygienization 污泥衛生處理 *wuni weisheng chuli*
　　見：hygienization, thermal, of activated sludge。
　　See *hygienization, thermal, of activated sludge*.

sludge recycling 污泥回流 *wuni huiliu*
　　廢水生物處理時，必須在*曝氣池*中維持一定濃度的活性微生物（*污泥揮發份濃度*）。因此，一般需將活性污泥從*二次澄清池*返回到曝氣系統中，只有一小部分多餘的污泥才被排到污泥處理系統去進行處理和處置。

　　In biological effluent treatment, a certain concentration of active micro-organisms (*MLVSS*) has to be maintained in the *aeration basin*. For this reason, the activated sludge is generally fed back from the *secondary clarifier* to the aeration system. A small fraction of surplus sludge is channeled to the sludge treatment system for further treatment and disposal.

sludge thickener 污泥濃縮池 *wuni nongsuchi*
　　剩餘活性污泥在過濾或離心之前進行預濃縮的沉降池。為了控制惡臭，污泥濃縮池應是密閉和氣密性的，污穢空氣在排放前應進行*脫臭*。

　　A sedimentation basin used for preconcentration of excess activated sludge before sludge filtration or centrifugation. For reasons of odor control, sludge thickeners should be closed and gas tight, and the foul air should be *deodorized* before discharge.

sludge treatment 污泥處理 *wuni chuli*
　　對廢水處理產生的剩餘活性污泥進行處理和最終處置的一系列過

程。對於問題嚴重的污泥，處理過程一般包括污泥濃縮、過濾或離心、焚燒和受控填埋場中焚燒殘灰的處置。對於問題不大的污泥，處理過程包括延時曝氣、過濾和填埋處置。

擬用作農業肥料的污泥，必須不含會在土壤和植物中積累的重金屬和其他殘留物，而且在施用前還必須進行滅菌處理。（參見：污泥滅菌）

A number of process steps for treatment and final disposal of excess activated sludge from effluent treatment. A standard sequence of steps for more critical sludges consists of sludge thickening, filtration or centrifugation, incineration and disposal of the incineration residues in a controlled landfill. For less critical sludges, sludge treatment may consist of extended aeration, filtration and disposal in a landfill.

Sludges which will be used for fertilization in agriculture must be free of heavy metals and other residues which might accumulate in the soil or in plants. Such sludges should also be sterilized before their discharge on the fields. (See *sludge sterilization*)

sludge volume index (SVI) 污泥體積指數 *wuni tiji zhishu*

測量*活性污泥*相對體積的一項參數。它是由測定污泥沉降30分鐘後的體積，將其值除以*污泥濃度（MLSS）*而求算得到的。因此，污泥體積指數表示的是單位乾活性污泥的相對體積。

A parameter to measure the relative volume of *activated sludge*. The SVI value is given by determining the sludge volume after sedimentation for 30 minutes and dividing this volume by the *MLSS* value. SVI therefore gives the relative volume of active sludge per unit of dry active sludge.

slurry 泥漿 *nijiang*

濃稠的固體懸浮液。通常，泥漿只能很慢地分離成液相和固相（沉積相），但它可用泵輸送。

A thick suspension of solid matter in a liquid. A slurry will normally separate slowly into a liquid and a solid, sedimented phase. Slurries can be pumped.

smog 煙霧 *yanwu*

　　大氣嚴重污染的一種特殊狀況，感覺像煙和濃霧狀。有兩類組成不同的煙霧：*夏季煙霧和冬季煙霧*。

　　A special condition of a heavy contamination of the atmosphere, feeling like smoke and heavy fog. The word SMOG is a combination of the two roots smoke and fog. There are two kinds of smog which differ in their composition: *summer-smog* and *winter-smog*.

smoke 煙霧 *yanwu*

　　由煙火或更一般地說由燃燒過程散發的排放物，它含有由燃燒過程產生的許多氣體組分，此外還含有細小而分散的顆粒物狀的各種不同組分。煙霧的可見度取決於顆粒物的濃度、水蒸汽的濃度、煙霧的溫度以及大氣的溫度和濕度。來自某一*點源*的煙霧，其技術名詞稱為*煙道氣*，兩者的組成是相似的。

　　The emissions from a fire, or, more general, from a combustion process. Smoke contains a number of gaseous components from the combustion process and, in addition, different components in the form of finely dispersed particulate matter.

　　The visibility of smoke depends on the concentration of this particulate matter, the concentration of water vapor, the temperature of the smoke and the temperature and humidity of the atmosphere.

　　The technical term for smoke from a *point source* is *flue gas*. The composition of both is similar.

SNCR process 選擇性非催化還原法 *xuanzexing feicuihua huanyuanfa*

　　見：selective non-catalytic reduction process。

　　See *selective non-catalytic reduction process*.

social cost principle 社會費用原則 *shehui feiyong yuanze*

　　環境費用支付的一種費用分攤原則。其含義是：環境費用(污染損害和污染預防的費用)總的來說應由國家或社會承擔。現在，社會費用原則一般已被*污染者付費原則*所代替。實際上，環境費用需把以上兩種原則結合起來進行分攤，但應強調後者。

A cost allocation principle for the payment of environmental costs. It implies that environmental costs (pollution damages and pollution prevention) should be paid by the state or by society in general. The social cost principle today is generally replaced by the *polluter-pays-principle (PPP)*. In reality, environmental costs are covered by a mixture of both mechanisms with strong emphasis on the PPP.

social costs from environmental damages 環境損害的社會費用 *huanjing sunhaide shehui feiyong*

由於不能確定直接污染者而不能歸於任何一方的 *損害費用*。因此，它是由整個社會來承擔的。同樣，與這類損害相關的 *損害恢復費用* 也是由整個社會來負擔的。(參見：污染者付費原則、社會費用原則)

Damage costs which are not attributable to any defined party because no direct polluter can be identified. As a result, these *damage costs* are born by society as a whole. Any *damage repair costs* for such damages also have to be paid for by society as a whole. (See also *polluter-pays-principle; social cost principle*)

soil 土壤 *turang*

支撐植物和維持它們不斷生長所需要的各種無機和有機組分的混合體。土壤的重要作用和特性是供給植物營養，它具有適當的結構、多孔性和持留水的能力，可提供作為微生物生存的場地。

有些污染物如重金屬會被植物攝取而進入 *食物鏈*。因此，*可持續發展的一個長遠重要目標，是保持土壤肥沃和不受污染*。

The mixture of different inorganic and organic materials needed to support plants and to sustain their growth. Important factors are the availability of plant nutrients, a suitable structure, porosity and binding capacity for moisture, and factors which allow for the existence of microorganisms.

Certain contaminants, e.g. some metals, are taken up in plants and may therefore enter the *food chain*. An important long term objective related to *sustainable development* therefore is to keep soils fertile and clean.

soil ecosystem 土壤生態系統 *turang shengtai xitong*

土壤的群落生境與生物群落的組合。各種不同的礦物質組分、水分、有機物質（主要來自死亡植物）、泥炭、土和其他組分，構成了土壤生態系統中生物群落生物體的群落生境。不同的微生物和微小生物，對於死亡有機物的降解，為植物提供某些營養物，以及對於土壤的結構和從空氣中固定氮，都具有重要作用。這些生物主要存在於土壤上層，而在土壤底層要少得多。

The combination of biotope and biocenose of soil. A variety of mineral components, moisture, organic matter, mainly material from dead plants, peat, mould and others form the biotope for the organisms of the biocenose in a soil ecosystem. Different microorganisms and small organisms are important for the degradation of dead organic matter, for providing some nutrients for plants, for the structure of soil and also for the binding of nitrogen from air. These organisms are found mainly in the top layer of soil, much less in the subsoil.

solid phase extraction 固相萃取 *guxiang cuiqu*

高壓液相色譜分析中富集微量化合物的技術。首先將溶解的樣品通過一個短的吸附柱，使被分析的化合物全部截留在柱子中。然後用溶劑將所有的持留物洗脫出來，再將此洗脫溶液注射到高壓液相色譜中作進一步分析。

A technique for the enrichment of trace compounds in *HPLC* analysis. The dissolved sample is first passed through a short adsorption column in which all compounds to be analyzed are retained. The total retained material is then eluted with a solvent and this solution is injected for further analysis into a HPLC apparatus.

solid special waste 固體特殊廢物 *guti teshu feiwu*

是固態的特殊廢物，即含有有害組分和具有潛在危害而需要進行特殊處理的固體廢物，如有毒的過濾殘渣、廢活性碳、污染的包裝材料、污染的電器設備（含多氯聯苯）、汽車廢輪胎和廢鉛蓄電池等。（參見：特殊廢物）

Special waste which is solid, i.e. a solid waste which requires special treatment as a result of its composition and problem potential. Examples: toxic filter residues, spent activated carbon, contaminated packaging material, contaminated electric equipment (PCB), used tyres from automobiles, used lead accumulators, etc. (See also *special wastes*)

solid waste 固體廢物 *guti feiwu*

是固態的廢物,它包括具有不同潛在危害和有害組分的特殊廢物。(參見:廢物)

Waste material which is solid. The term includes different hazard potentials and compositions. (See also *waste*)

solidification of wastes 廢物固化 *feiwu guhua*

參見:廢物處置的穩定化法。

See *stabilization processes in waste disposal*.

sorption processes 吸著過程 *xizhao guocheng*

在非常小的粒子(主要是分子)和土壤表面之間發生的吸附與解吸過程。另一方面,吸收作用涉及到將某一物質吸收於另一(流體)介質中,如將氣體溶解於液體中。(參見:吸收、吸附、解吸)

Adsorption and desorption are processes between very small particles (mainly molecules) and a solid surface. Absorption on the other hand deals with taking up a substance in another (fluid) medium, e.g. dissolving a gas in a liquid. (See also *absorption*; *adsorption*; *desorption*)

sound 聲音 *shengyin*

通過耳器官產生的感覺。從物理意義上講,聲音是聲源物質振動的結果。聲源物質的振動會使與振動部件相接觸的空氣產生壓力的震蕩,進而引起耳鼓膜震蕩從而產生聲感。(參見:噪聲)

The sensation produced through the organs of hearing. Sound in the physical sense is the result of a vibration of matter of a source, which causes an oscillation of pressure of air which is in contact with the vibrating part. This oscillation of air pressure in turn causes a vibration of the ear drum and the sensation of sound. (See also *noise*)

sound, impact of 聲音的影響 *shengyinde yingxiang*

聲音對現場所在地的人和目標的影響。

The effect sound has on persons and objects at the site where they are located.

sound emission 聲音發散 *shengyin fasan*

由某聲源如壓縮機發散的聲音，它是聲源的振動部件將其震盪傳播給與它相接觸的空氣而產生的。

The emission of sound from a source, e.g. a compressor. Sound emission is the result of vibrating parts of the source, parts which transmit their motion to the air in contact with them.

sound emission limits 聲音發散限值 *shengyin fasan xianzhi*

對露天設備如汽車、飛機和建造設備的*聲功率級*的限值。在有些國家的立法中規定有此類限值。對於工業設備和封閉聲源，其發散特徵必須在制定設備方案過程中加以評價，然後採取合適的控制措施。

Limiting values for the *sound power level* of open installations, e.g. vehicles, airplanes and construction machines. Limiting values are set in the national legislation of some countries. For industrial installations and sound sources enclosed therein, emission characteristics have to be evaluated in the planning process of the installations and suitable control measures have then to be integrated.

sound power level 聲功率級 *shenggonglüji*

聲源聲音發散的特徵值。它是聲音發散和影響預測的一個基本參數，可由離聲源一定距離處測定的*聲壓級*計算得到。

A characteristic value for the sound emission of a source. It is a fundamental quantity for sound emission and impact prognosis and is calculated from the *sound pressure level* measured at a given distance from the source.

sound pressure level 聲壓級 *shengyaji*

　　描述聲音物理強度的基本參數，以分貝表示。分貝（A聲級）是人耳正常可感知的聲壓的水平。1分貝之差是一般人耳剛能感知的聲壓。

　　The basic parameter for describing the physical intensity of sound. It is given in decibel (dB). Decibel (A) or dB(A), is the level of sound pressure which takes into account the normal perceptibility of the human ear. A difference of one dB(A) is just perceptible for the average human ear.

sound transmission 聲音的傳播 *shengyinde chuanbo*

　　從聲源到影響點聲音震蕩的傳播過程，它取決於傳播的途徑和傳播的條件。

　　The propagation or transmission of sound vibration from the source to the point of impact. The transmission of sound depends on the path of transmission and on the conditions found along the path.

sour water 酸性水 *suanxingshui*

　　原油煉製如蒸餾過程產生的酸性廢水。

　　An acidic effluent stream generated in refineries of crude oil, e.g. in distillation steps.

source air 源頭空氣 *yuantou kongqi*

　　在反應器或生產設備如密閉過濾、粉碎或乾燥設備的排氣口處直接收集的廢氣流，偶而也用"設備廢氣"這一術語。這種廢氣所含的污染物有中高濃度粉塵、溶劑和一些工藝氣體。其流量範圍一般在1,000–15,000立方米/小時之間。（參見：廢氣類別、工藝廢氣）

　　An off-gas stream, collected directly at the vent of a reactor or manufacturing unit, e.g. a closed filter, a mill, or a drying unit. The term 'apparatus off-gas' is also used occasionally. The pollutants in this off-gas stream include medium high concentrations of dust, solvents and some process gases. Flow volumes range typically from 1,000 to 15,000 m^3/hour. (See *off-gas classes*; *process gases*)

source reduction 源頭削減 *yuantouxuejian*

見：integral process development。

See *integral process development.*

special waste analysis 特殊廢物分析 *teshu feiwu fenxi*

見：analysis of special wastes; analysis plan in tiers。

See *analysis of special wastes; analysis plan in tiers.*

special waste incinerator 特殊廢物焚燒爐 *teshu feiwu fenshaolu*

適用於*特殊廢物*焚燒處理的任何焚燒爐。這類焚燒爐的主要特點是：(1)採用熱氧化法(在足夠高的溫度和足夠長的焚燒時間下)處理，破壞效率高；(2)具有相匹配的煙氣處理，能將煙氣中的所有污染物去除到所要求的程度，此外也還包括洗滌液的處理；(3)具有除去礦質化爐渣的系統，包括爐渣的處理；(4)具有節能和熱回收系統。

Any incinerator which is suitable for the incineration of *special wastes*. Key characteristics of such an incinerator are: high destruction efficiency by thermal oxidation (adequate temperature and incineration time); suitable flue gas treatment to remove all pollutants in the flue gas to required levels, including treatment of the scrubbing liquors; a system to remove the mineralized slag, including slag handling; and energy optimization and heat recovery.

special waste management 特殊廢物管理 *teshu feiwu guanli*

見：management of special wastes。

See *management of special wastes.*

special wastes 特殊廢物 *teshu feiwu*

需要進行特殊處理的*廢物*。其處理方法取決於廢物的物理性質(固體、半固體、污泥、淤漿、液體)、化學組成及其危害性。特殊廢物由於含有有毒的或具有*生態毒性的污染物*，因此在生態系統中會產生不良影響。它也常常稱為危險廢物、有毒廢物和化學廢物。

Any *waste* which needs special treatment for its disposal. The treatment depends on the physical properties (solid, semi-solid, sludges, pastes, liquids), the chemical composition of the waste and their hazards. Special wastes may have a detrimental effect in an ecosystem as a result of toxic or *ecotoxic contaminants*. Often, other terminology is used, such as hazardous wastes, toxic wastes, chemical wastes.

special wastes, definitions 特殊廢物，定義 *teshu feiwu, dingyi*

用來鑒別特殊廢物(常常也稱為危險廢物)以及把它們區別於一般廢物的判斷標準。其定義可以基於不同的原則。一個明確、簡要和可操作的定義，對於保證正確處理這類廢物是很重要的，另外從控制和實施上看也是必要的。

經常使用的判斷標準包括：廢物的源頭工藝過程；物理、化學或毒理學性質；廢物的化學組成；潛在危害；合乎需要或不合乎需要的處置方法。

不同國家制定的國家法規中所確定的特殊廢物的定義是很不相同的。聯合國環境規劃署廢物分類系統，是一個國際上公認的系統，它也是控制廢物越境轉移的巴塞爾公約的一部分內容。

因為大多數廢物是複雜的混合物，因此僅以分析為基礎的分類，可能會產生一些問題，而且工作量會很大。

Descriptions of criteria to identify special wastes (often also called hazardous wastes) and distinguish them from common wastes. Definitions may be based on different principles. A clear, concise and workable definition is important to make certain that such wastes are disposed of properly and also for reasons of controls and enforcement.

Groups of criteria which are often used are: the source process of the waste; physical, chemical or toxicological properties; the chemical composition of a waste; the presence of certain critical compounds; potential hazards; and desirable or undesirable modes of disposal.

Definitions in national legislation of different countries vary considerably. The *UNEP* waste classification system, as part of the *Basel Convention* on transboundary movement of wastes, is a system which is recognized internationally.

As most wastes are heterogeneous mixtures, a classification based only on analysis may cause problems and need much effort.

specific adsorption surface 比吸附表面 *bixifu biaomian*

　　吸附劑如活性碳的內活性表面。質量良好的活性碳具有1000米²/克以上的比表面。在*活性碳再生*時，其活性表面將獲得再生和重整。

　　The internal active surface of an adsorbent, e.g. activated carbon. Good activated carbon qualities have a specific surface of over 1,000 m²/g. On *regeneration of activated carbon*, this surface has to be regenerated and reformed.

specific costs 比（單位）費用 *bi (danwei) feiyong*

　　在成本管理中單位能源、單位原材料、一個加工小時或單位廢物處置的費用。

　　In cost management, the costs of one unit of energy, one unit of of raw material, one processing hour, or the disposal of one unit of waste.

specific disposal costs 比處置費用 *bichuzhi feiyong*

　　正確處置每單位（公斤或立方米）特定廢物的總費用，它用於把污染控制費用分攤到產品價格中。

　　The total costs for the correct disposal of one unit (kg or m³) of a particular waste stream. Specific disposal costs are used in the allocation of pollution control costs to product prices.

specific waste quantity 比廢物量 *bifeiwuliang*

　　生產每單位產品產生的廢物量，例如以公斤廢物/公斤產品表示。這一術語常用於*污染控制費用的分攤*以及生產過程的環境評價。（參見：排放因子）

　　Waste quantity per unit quantity of product, e.g. in kg of waste /kg of product. A term used in the *allocation of pollution control costs* and also in the environmental evaluation of production processes. (See also *emission factors*)

spectrophotometry 分光光度測定法 *fenguang guangdu cedingfa*

　　基於測定電磁輻射（在大多數情況下測定光）的各波長強度的*儀器分析方法*。它是將一束單色輻射（紅外光、可見光、紫外光；可用單色器或濾光器獲得) 通過被分析的樣品而進行測定的。其輻射強度的衰減服從比爾 —— 朗伯定律,可用來度量該樣品中物質的濃度。

Methods of *instrumental analysis* which are based on intensity measurements of individual wavelengths of electromagnetic radiation, in most cases of light. A more or less monochromatic beam of radiation (infrared, visible, ultraviolet light; obtained with a monochromator or with filters) is passed through the sample to be analyzed. The attenuation of the intensity is a measure of the concentration of a substance in the sample. This attenuation is governed by the Beer-Lambert Law.

spills 泄漏 *xielou*

由於事故或粗心大意造成的意外的液體排放或物料損失。

Unwanted discharges of liquids or losses of material, as the result of an accident or of negligence.

splitting of emulsions 破乳 *poru*

見:emulsion splitting。

See *emulsion splitting*.

spray deflector 噴淋折板 *penlin zheban*

見:deflector elements。

See *deflector elements*.

spray tower (scrubber) 噴淋塔(洗滌器) *penlinta (xidiqi)*

一種簡單的濕式空氣污染控制系統。運行時,將洗滌液在噴淋塔的上部用噴咀噴入塔中,而氣流則以並流或逆流方式通過噴淋區。由於沒有什麼特殊的措施來提高傳質效率,因此它的處理效率不高。在有些設計中安設有洗滌器的徑向噴淋設備。

A simple wet APC-system. Scrubbing liquor is sprayed into the spray tower through nozzles at the upper end. The gas stream is passing this spray either co-current or counter current. There are no special provisions to provide for an efficient mass transfer, hence the efficiency is not high. Some designs are equipped with an arrangement for the radial spraying of the scrubbing liquor.

SS 懸浮固體 *xuanfu quti*

見：suspended solids。

See *suspended solids*.

stabilization processes in waste treatment 廢物處理的穩定化方法 *feiwu chulide wendinghua fangfa*

使廢物變得比較穩定以利於最終處置的方法，包括重金屬組分的化學固定，或者將廢物固定在較穩定的基體(如混凝土基體)中，以便在*受控填埋場*中進行最終處置，從而能減輕浸出液造成的地下水污染。剩餘活性污泥穩定化包括好氧或厭氧處理，目的在於進行污泥的處置或作為肥料用於農業。

Processes which render wastes more suitable for final disposal. Stabilization processes may include chemical fixation of heavy metal components, or general stabilization of wastes in a more stable matrix e.g. in a concrete matrix, for disposal in a *controlled landfill*. In this case contamination of groundwater through leachate is reduced. Stabilization of excess activated biosludge for disposal and use in agriculture may involve aerobic or anaerobic treatment.

stack emissions 煙囪排放 *yancong paifang*

工廠*點源排放*的同義詞。

A synonym for *point source emissions* of a factory.

staff function 參謀職能 *canmou zhineng*

工廠中不直接涉及生產的職能機構。這些機構常常具有協調職能，它們負責收集和準備數據和資料，提供作為決策的依據。*控制*職能通常就是參謀職能。

A function or unit of the organization of a factory which is not directly associated with production. Staff units often have coordinating functions. They may collect and prepare data as a decision base. *Controlling* functions generally are staff functions.

staff responsibility 參謀職責 *canmou zhize*

參謀職能的責任。除非特別指出,參謀職能一般不具有指揮命令的直接權力。負有參謀職能的人員可以向負有指揮職能的經理報告,通過經理來施加影響。

The responsibility of a staff function. Unless specifically defined otherwise, staff functions in general have no direct power to mandate. Staff employees may exert their influence through the manager of the line function to whom they report.

stakeholders of a company 公司的利害關系者 *gongside lihai guanxizhe*

與公司有直接或間接關係並會受公司的運作和活動影響的人或機構。其有利害關係者(團體)包括職工、用戶、股東、供貨商、鄰近居民、市政當局、公共關係部門。現代管理的一個重要方面,是與所有的利害關係者(團體)保持良好關係。

Persons or institutions which have a direct or indirect interest in a company and are affected by the operation and the activities of the company. Stakeholder groups include the employees, the customers, the shareholders, the suppliers, the neighbors, the municipality, the public administration. One aspect of modern management is to maintain good relations with all stakeholder groups.

standard cost calculation 標準成本計算 *biaozhun chengben jisuan*

計算真實的產品生產成本的管理方法。標準單位成本是計算每一產品生產成本的基礎(參見:直接費用、間接費用)。應該按照生產過程的物料衡算來求得物料量。在成本管理中也應包括真實的污染控制費用要素。

A management procedure to calculate true product manufacturing costs. The bases for the calculation of the manufacturing costs for each product are standard unit costs (*direct costs*, and *indirect costs*). Quantities of materials should be based on the mass balance of the process. The procedure should also include true cost elements for pollution control.

standard deviation 標準偏差 *biaozhun piancha*

統計分析中使用的一項術語,它用於衡量一組重複測量的數據偏離其平均值的分散程度。標準偏差與測定次數有關,它可用於計算 *置信界限*以及評價結論的 *顯著水平*。

A term used in the statistical analysis of a set of repeated measurements. The standard deviation is a measure of the scatter of the set of observations around the mean value of the set. The standard deviation is related to the number of measurements in the set. It is used to calculate *confidence limits* and to evaluate the *significance level* of conclusions.

standard methods in analysis 標準分析方法 *biaozhun fenxi fangfa*

由國家和國際標準組織頒佈的分析方法。這些方法給出了能得到可靠、重複結果的分析規程,其結果可在不同實驗室之間進行比較。這些方法的重要來源有:美國材料試驗學會標準方法,美國環境保護局方法,聯邦德國水、廢水和污泥分析的標準方法(DIN標準方法)。此外,在國家環境法規的有關內容中也頒佈有標準的分析方法。

Methods for analyses published by national and international norming organizations. These methods give instructions for analyses which yield reliable and repeatable results which can be compared between different laboratories. Important sources of such methods are for instance: ASTM, the American Society for Testing of Materials, Washington, D.C., EPA Methods, published by the American Environmental Protection Agency, DIN Standard Methods for water, effluent and sludge analyses of the Federal Republic of Germany. In addition, there are methods published in the context of national environmental legislation.

standard unit costs 標準單位成本 *biaozhun danwei chengben*

*標準成本計算*中使用的那些費用要素:物料、能源、工時、設備費用的比單位成本,此外還應包括所有廢物處置過程的處置費用。

Specific unit costs for those elements which are used in the *standard cost calculation*: materials, energies, work-hours, equipment cost. Also included should be the disposal costs for all waste disposal processes.

standard unit costs for pollution control 污染控制的標準單位成本 *wuran kongzhide biaozhun danwei chengben*

正確處置廢物所產生的單位污染物的處置費用。比單位成本是按處置某一特定廢物的費用和所處理的廢物總量來計算的。在處置總費用中還應包括現場的處理與運輸。這些污染控制費用需按*污染者付費原則分攤到產品成本中去*。(參見：污染控制費用的分攤)

The specific costs per unit of pollutant resulting from the environmentally correct disposal of waste streams. Specific unit costs are calculated from the total costs of the disposal of a specific waste and the total quantity of a particular waste stream treated. The total costs should also include on-site handling and transport. These costs are needed for the allocation of pollution control costs to the products by the *polluter-pays-principle*. (See also *allocation of pollution control costs*)

standards by implication 蘊涵標準 *yunhan biaozhun*

基於審計者的知識與經驗，由環境審計結果產生的工作標準。

Performance standards which can be derived from the result of a professional environmental audit and which are based on the knowledge and the experience of the auditor.

state-of-the-art 現代工藝水平 *xiandai gongyi shuiping*

在一定時期用來完成某一技術工作的現代工藝技術。(參見：可獲得而經濟的最佳技術、最佳實用技術)

A definition and description of modern technology generally applied at a given time to carry out a certain technical task. (See also *best available technology economically achievable*; *best practicable technology*)

state-of-the-art in pollution control 污染控制中的現代工藝技術*wuran kongzhizhongde xiandai gongyi jishu*

應用"現代工藝水平"這一術語於污染控制技術上，表明該技術已成功地應用於許多實例中，並且證明是有效和可靠的。

The application of the term 'state-of-the-art' to pollution control technology.

This level of technology has been applied successfully in a number of cases and has proven its efficiency and reliability under practical conditions.

static chamber incinerator 靜態燃燒室焚燒爐 *jingtai ranshaoshi fenshaolu*

適用於處理含水或者有機物廢液的一種焚燒爐，其焚燒過程係在一個垂直圓柱形的焚燒室中進行。焚燒室襯有耐火磚或者具有雙壁結構 (*冷壁式焚燒爐*)。廢物噴入方式是很嚴格的，需用專門設計的汽化噴咀。在有些設計中，用水蒸汽來進行汽化。在焚燒爐頂部還安設有輔助火焰和燒咀，以維持使廢物完全分解所需的焚燒溫度。焚燒所需的空氣也是從焚燒爐頂部引入的。其他輔助設備包括熱回收和煙氣處理的設備。

這種焚燒爐可用於有毒母液或廢溶劑的焚燒處理。此外，它還能用於處理惡臭廢氣，在這種情況下可用惡臭廢氣來代替等體積的焚燒空氣。

An incinerator type which is suitable for aqueous or organic liquids. The incineration process takes place in a vertical cylindrical incineration chamber. This may be lined with fire brick or constructed as a double walled structure (*cold wall incinerator*). Waste injection is critical and requires specially designed vaporization nozzles. In some designs, steam is used for vaporization. A supporting flame and burner, also arranged at the top, serves to maintain the incineration temperature which is necessary for complete destruction of the waste material. Air for incineration is also introduced at the top. Auxiliary parts include heat recovery units and flue gas treatment.

This type of incinerator is in used for the disposal of aqueous, toxic mother liquors or for the incineration of waste solvents. In addition, it can be used to treat odorous off-gas streams, which then replace an equivalent volume of the incineration air.

static equilibrium 靜態平衡 *jingtai pingheng*

一旦達到平衡不發生變化或脈動的平衡狀態。例如，化學反應中的平衡狀態，或大多數物理體系中的平衡，如天平的平衡。

An equilibrium state which does not change or oscillate once it has been reached. Examples: the equilibrium state in chemical reactions or the equilibrium in most physical systems, e.g. the equilibrium of a balance.

statistics 統計 *tongji*

統計是數學上對實驗數據進行評估的一個特殊領域。如果在同樣條件下重複進行試驗許多次,觀察得到的結果將是不同的,它們將隨機地分佈在平均值的周圍。統計所涉及的內容之一,就是進行數據的分析、誤差的評估和數據顯著性的計算。在簡單情況下,計算出平均值、標準偏差和置信界限就足以滿足要求了;而對比較重要的場合,則需運用許多不同的統計程序和檢驗方法。將統計方法應用於環境數據,將可保證不會從一組數據得出不正確的結論。

Statistics is a special field of mathematics dealing with the evaluation of experimental data. Experience shows that if an experiment is repeated a number of times under the same conditions, the observed values will differ. They will be distributed at random about an average value. One aspect of statistics deals with the analysis of data, the evaluation of errors and with the calculation of the significance of data. For simple cases the calculation of the mean value, the standard deviation and the limits of confidence are sufficient. In more critical cases, many different statistical procedures and tests are available. The application of statistics to environmental data helps to make certain that no incorrect conclusions are drawn from a set of data.

STEL 短期接觸限值 *duanqi jiechu xianzhi*

Short term exposure limits的縮寫。它是與閾限值一起發佈的。短期接觸限值是從職業衛生上考慮在工作日期間不應接觸超過15分鐘的限值。它和閾限值之間的關係與物質的性質有關。(參見:閾限值)

Short term exposure limits, published together with the TLV values. Short term exposure limits are values for occupational health considerations which should not be exceeded for periods longer than 15 minutes during the work day. The relation between STEL and TLV value depends on the nature of the substance. (See also *TLV*)

sterilization 消毒（滅菌）*xiaodu (miejun)*

使某種物質（食物、水、醫療器械、房間）完全無致病生物。這可以使用殺菌劑、進行高溫處理、投加強氧化劑或通過微過濾來進行。為此，常常使用氯氣、臭氧和產氧物質。在技術部門可用微過濾來進行諸如水的滅菌或消毒。

Rendering a material (food, water, medical instruments, rooms) totally free of pathogenic organisms. Sterilization can be carried out either by the application of bactericides, by thermal processes, by the addition of strong oxidizing agents or by microfiltration. Chlorine, ozone and oxygen producing substances are used for his purpose. In the technical sector sterilization or disinfection e.g. of water, can achieved by microfiltration.

sterilization of sludge 污泥滅菌 *wuni miejun*

見：hygienization, thermal, of activated sludge。

See *hygienization, thermal, of activated sludge.*

Stockholm Conference 1972 1972年斯德哥爾摩會議 *yijiuqi'ernian sidege'ermo huiyi*

全世界範圍涉及環境問題的第一次國際會議。這次會議提高了人們對環境問題的認識，產生了聯合國的一個專門分支機構，即*聯合國環境規劃署*。斯德哥爾摩會議是由加拿大的斯特朗（Maurice Strong）先生的倡導下組織召開的。第二次會議於1992年在*里約熱內盧*舉行。

The first international conference dealing with environmental issues on a worldwide scale. It raised the awareness of environmental problems and led to the organization of a special branch of the United Nations, the *UNEP*. The conference was primarily organized through the initiative of Maurice Strong of Canada. A second conference was held in *Rio de Janeiro*, 1992

stormwater overflow 暴雨水分流排管 *baoyushui fenliu paiguan*

下水道系統中設在污水處理場之前起分流作用的可控溢流設施（堰），在大雨之後它可把大量的暴雨水（*逕流水*）分流出來，直接排入受*納水體*。

A controlled overflow device (a weir) to a branch-off in the sewer system before its input to the effluent treatment plant. The branch-off serves to divert excessively high input volumes of stormwater (*runoff water*) after heavy rains to a direct discharge to the *receiving waterbody*.

stratospheric ozone 平流層臭氧 *pingliuceng chouyang*

見：ozone, stratospheric。

See *ozone, stratospheric*.

stripping 氣提 *qiti*

揮發性物質從液相至氣相進行質量傳遞的單元過程。其效率取決於化合物的蒸氣壓與溶解度。(參見：空氣氣提)

A unit process for the mass transfer of a volatile substance from the liquid phase to the gas phase. The efficiency of stripping depends on the vapor pressure and the solubility of the compound. (See also *air stripping*)

stripping column 氣提塔 *qitita*

進行氣提過程的垂直塔式加工設備，其設計通常與精餾塔相似，有液體與氣提氣的進料與出料。氣提塔可以裝填有*規填料*。(參見：塔填料)

A vertical processing column to carry out a stripping process. The design is generally similar to a column for rectification, with inputs and outputs for the liquid and the stripping gas. The column may be filled with a *structured packing*. (See also *column packing*)

stripping with steam 蒸汽汽提 *zhengqi qiti*

應用於工藝過程末端的預處理技術。它通過吹入蒸汽使廢水中的*揮發性組分*從水溶液中分離出來。它通常以並流操作方式在板式塔或填料塔中進行，並可將蒸汽中提出的物質加以回收。蒸汽汽提的效率取決於揮發性物質在水中的溶解度和其蒸汽壓。

A pretreatment technology, to be applied at the end of a process. Volatile components (*VOC*) of an effluent can be separated from the aqueous solution by blowing steam through the solution. The process is usually carried out in a plate column or a filled column in a countercurrent process. The purged material in the steam can be recovered. The efficiency of steam stripping depends on the solubility in water and the vapor pressure of the volatile component.

structurally packed column 有規填料塔 *yougui tianliaota*

 裝填有規則形狀載體材料的填料塔，如生物反應器或蒸餾塔。其載體可以由塑料(如聚氯乙烯、高密度聚乙烯)、鋼材或任何其他適合的材料制成。有規填料在市場上有供應。(參見：無規填料塔)

 A column, e.g. a bioreactor or a distillation column, which is packed with a regularly shaped, structured support material. The support may be made of plastic (e.g. PVC, HDPE), steel or any other material which is suitable for the desired application. Structured packings are commercially available. (See also *randomly packed column*)

structured packing 有規填料 *yougui tianliao*

 化工工藝中加工處理塔的填料，它由不同設計的有規材料構成並裝填在塔內。在市場上有不同類型和材料的有規填料供應。

 The packing of a processing column in chemical technology, made of structured elements of various designs and mounted inside the column. Structured packings of different types and materials are available commercially.

subsidies or tax reductions 補貼或減稅 *butie huo jianshui*

 促進某些活動如環境友好產品或清潔生產工藝的經濟刺激手段。(參見：獎勵、經濟刺激)

 Economic incentives to promote certain activities, e.g. environmentally friendly products or clean manufacturing processes. (See also *incentive rewards*; *economic incentives*)

suggestion box 建議箱 *jianyixiang*

 工廠中用來收集職工改進工作建議的信箱。這種簡單易行的措施是工廠*建議系統*的一部分，目的在於使職工參與到環境活動中去。因此，它是人事管理的一種手段。其建議可以以環境問題為重點，或者也可涉及任何其他需要改進的問題。

A neutral letter box within a factory to collect suggestions for improvements by the employees of all levels. This simple instrument is part of a *suggestion system* of a site which serves to integrate personnel into the environmental activities. It is thus an instrument of personnel management. Suggestions may focus on environmental aspects or any other improvement.

suggestion system (personnel management) 建議制度(人事管理) *jianyi zhidu (renshi guanli)*

作為人事管理一個組成部分的一項制度，其目的是廣泛徵求所有職工的建議，並在問題解決的過程中充分考慮這些建議。這一制度可應用於各種不同的管理領域和技術操作中。它鼓勵職工找出需要改進的問題，並提出如何進行改進的建議。(參見：建議箱)

A system used as an element of personnel management. The objective of a suggestion system is to solicit personal know-how of employees at all levels and to integrate it into the problem solving process. The system can be applied to very different areas of management and technical operations. Personnel is actively asked to identify areas which need improvement and also to make suggestions on how to improve situations. (See *suggestion box*)

sum parameters 綜合指標 *zonghe zhibiao*

見：group parameters。

See *group parameters*.

summer smog 夏季煙霧 *xiaji yanwu*

見：photochemical smog。

See *photochemical smog*.

supercritical fluid chromatography 超臨界流體色譜法 *chaolinjie liuti sepufa*

用某種氣體如二氧化碳於高壓下作為洗脫劑的液相色譜法技術。它可以成功地用於分析高極性物質或者是在一般色譜條件下不穩定和不能分離的物質。

Liquid chromatography, in which a gas, for instance carbon dioxide, is used under high pressure as the eluant. This technique can be successful in the analysis of very polar substances, or substances which are unstable and cannot be separated under normal conditions of chromatography.

supercritical fluid extraction 超臨界流體萃取 *chaolinjie liuti cuiqu*

用超臨界氣體或氣體/溶劑混合物進行萃取的技術。在用*高壓液相色譜法*或氣相色譜法分析土壤中的污染物時，可用超臨界流體萃取技術作為富集手段。其操作溫度和壓力取決於所使用的氣體或氣體/溶劑混合物的種類。

An extraction technique with supercritical gases or gas/solvent mixtures. The technique is used for instance in the enrichment step in *HPLC* analysis or GC analysis of contaminants in soil. Pressure and temperature depend on the gas or gas/solvent mixture which is used.

SUPERFUND program 超級基金計劃 *chaoji jijin jihua*

在美國為了清理與淨化老的填埋場以及與之有關的已污染的地下水體而實施的一項計劃。它是基於1980年的立法（綜合環境響應、賠償和責任法，CERCLA）和1986年的修正案（超級基金修正案和再授權法，SARA）而制定的。它包括建立全國污染場地的清單，對每一場地進行評價，確定重點場地，以及進行場地的清理與淨化活動。該法律還規定了經濟責任問題和向政府主管部門報告的義務，由美國環境保護局負責管理和實施。

"超級基金"這一術語源於總體費用極高，"需要超級基金"這一事實。1995年的估算為50-100億美元。由於其費用十分巨大，1996年正在對法律與程序進行重新審查。

A program which is carried out in the USA with the objective to clean up old landfills and associated polluted groundwater bodies. The program is based on the legislation of 1980 (the Comprehensive Environmental Response, Compensation, and Liability Act (CERCLA)), and amendments of 1986, (the Superfund

Amendments and Reauthorization Act (SARA)). It includes the establishment of a nationwide inventory of contaminated sites, individual assessment of each site, the setting of priorities and clean up activities at the sites. The law also defines the question of financial liability and the obligations to notify the authorities. The law is administered by the American Environmental Protection Agency, EPA.

The term 'SUPERFUND' is derived from the fact that the overall costs are extremely high 'needing a super fund'. Estimates in 1995 are in the order of between five and ten billion (thousand million) of US$. In view of these immense costs, legislation and procedures are being reviewed in 1996.

support fuel burner 輔助燃料燒咀 *fuzhu ranliao shaozui*

安裝在特殊廢物焚燒爐中的燒咀(一個或若干個),如果廢物的熱值太低,則需用輔助火焰來達到和維持焚燒溫度。為了完全燃燒,溫度最好高於1,000°C。燒咀可安設在主焚燒室中,或者也可安設在後燃燒室中,可以用礦物油、天然氣或廢溶劑作為輔助燃料。(參見:焚燒爐、回轉窯焚燒爐)

One or several burners installed in an incinerator for special wastes. The support flame is needed to reach and maintain the incineration temperature if the heat content of the waste material is too low. For complete combustion, a temperature of over 1,000 degree C is recommended. A burner may be installed in the main incineration chamber or in the post combustion chamber. Mineral oil, natural gas or waste solvents may be used as support fuel. (See also *incinerator*; *rotary kiln incinerator*)

suppression in ion chromatography 離子色譜法中的抑制作用 *lizi sepufazhongde yizhi zuoyong*

見:ion chromatography, chemical suppression。
See *ion chromatography, chemical suppression.*

surface aerator 表面曝氣機 *biaomian baoqiji*

安裝在廢水生物處理淺平式曝氣池混合液表面上方的具有垂直旋轉

385

軸的渦輪機，其葉輪位於水面上轉動，以把空氣引入到液體中。表面曝氣機的效率以單位時間單位能耗所輸入的氧量表示。

A turbine-like element with a vertical axis of rotation, mounted above the surface of the mixed liquor in the flat aeration basin of the biology stage of an effluent treatment plant. The turbine-wheel rotates just below the still water level and serves to introduce air into the liquid. The efficiency of surface aerators is characterized by the amount of oxygen input per unit of time and energy input.

surface regeneration 表面再生 *biaomian zaisheng*

見：regeneration of activated carbon; specific adsorption surface。

See *regeneration of activated carbon; specific adsorption surface.*

suspended solids (SS) 懸浮固體（懸浮物）*xuanfu guti (xuanfuwu)*

用以表示水和廢水中不溶且高度分散的污染物的一項綜合指標。懸浮固體可以是無機物或有機物，它們會使污染水體產生濁度。

Group parameter for a group of insoluble finely dispersed pollutants in water and effluents. Suspended solids are responsible for the turbidity of a polluted water-body. Suspended solids may be inorganic or organic.

suspended solids in effluents, measurement 廢水懸浮物的測定 *feishui xuanfuwude ceding*

廢水懸浮物的測定通常有兩種方法。一種是簡單的常規方法，即將一升廢水置於錐形瓶（英霍夫漏斗）中至少24小時，然後測量英霍夫漏斗底端處沉降物的體積，以毫升表示。另一種較定量的方法，是把廢水通過孔徑為0.45微米的過濾器，將過濾殘留物烘乾後稱重。

懸浮物測定是廢水處理場運行控制的一部分（見：*污泥濃度和污泥揮發份濃度*），它也是廢水處理場排放廢水總控制指標的一項內容。

Two methods are generally used for the measurement of suspended solids (SS) in effluents. In a simple and routine method, one liter of the effluent is kept in a conical flask (Imhoff funnel) for at least 24 hours. The volume of the settled matter can be measured at the tip at the bottom of the Imhoff funnel and is reported in milliliters. In a more quantitative method, the sample is passed through a 0.45 micrometer filter. The residue is dried and weighed.

The measurement of SS is part of the operation control of an effluent plant (*MLSS*, *MLVSS*) and of the general control of the effluent discharged from an effluent treatment plant.

sustainable development (SD) 可持續發展 *kechixu fazhan*

1992年聯合國於里約熱內盧召開的國際會議上簽署制定的一項政策，其定義為：一個國家（或一個城市）的發展不僅應滿足當代人的需要，而且應不會損害後一代人的潛在需要。可持續發展包括充分利用不可再生資源。由於能源是一種重要資源，因此可持續的能源政策是可持續發展的一個重要組成部分。清潔的環境作為一種不可再生資源，是後代人生存之根本，因此解決環境污染問題不能無限期地拖延下去。

A policy endorsed at the international Conference of the United Nations in Rio de Janeiro in 1992, or Rio Conference. Sustainable development (SD) can be defined as: development of a country (or a city) in such a way that the needs of present generations are met without compromising the ability of future generations to meet their own needs. SD includes the use of non-renewable resources in a thoughtful manner. Energy is an important resource and a sustainable energy policy is thus an important element of SD. As a non-renewable resource, a clean environment is the base for future generations, hence solving environmental pollution problems may not be postponed indefinitely.

SVI 污泥體積指數 *wuni tiji zhishu*

見：sludge volume index。

See *sludge volume index*.

symbiosis 共生 *gongsheng*

生態系統中不同生物群體之間的一種特殊類型關係。在共生的相互作用中，雙方彼此存在和獲益。地衣就是藻類和真菌的共生現象。共生作用的一個重要例子是豆科根瘤中的根瘤菌從空氣中固定氮的過程。在此過程中，氮被植物利用，而細菌又依賴植物而生存。

A specific type of relationship between populations of different organisms in an ecosystem. In symbiotic interactions both partners depend on each other and both profit from each other. Lichens are a symbiosis of algae and fungi. An important example of a symbiosis is the process of fixation of nitrogen from air by rhizobium bacteria in the root nodules of leguminosae. Nitrogen is made available to the plant, while the bacterium depends on the plant for living.

synecology 群落生態學 *qunluo shengtaixue*

生態學中研究某群落中單個物種的相互作用及該群落對它們生境的影響的領域。（參見：生態學、個體生態學）

Ecology focusing on the interactions of single species in a community and on the influence of the community on their habitat. (See also *ecology*; *autecology*)

system boundary 系統邊界 *xitong bianjie*

對調查研究的系統所限定的範圍（或界線）。在對若干系統進行比較時，例如通過*生態平衡*來比較兩個產品時，系統邊界是很重要的。在*環境審計*中，系統邊界由所調查工廠的界區確定。

The boundary (or borderline) which limits a system under investigation. System boundaries are important in the comparison of systems, for instance in the comparison of two products through *ecobalances*. In an environmental *audit*, the system boundary is given by the borderline of the factory under investigation.

systematic error 系統誤差 *xitong wucha*

重複測量或測定時出現的試驗誤差與波動，它是由於整個分析過程，包括從樣品的獲得到結果的評估與解釋上存在某些不足而產生的。仔細研究全部的分析過程，尋找出產生誤差的一切可能的原因，通常就能鑒別出系統誤差，並加以消除。（參見：偶然誤差、準確度、精密度、可靠性）

Experimental errors and fluctuations which occur when measurements or determinations are repeated. Systematic errors have their origin in certain inadequacies of the whole analytical procedure, from the drawing of the sample to the evaluation and interpretation of the results. Careful investigation of the complete analytical procedure and a search for all possible causes usually leads to the identification and subsequent elimination of systematic errors. (See also *random error*; *accuracy*; *precision*; *reliability*)

T

tank storage emissions 儲罐（無組織）排放 *chuguan (wuzuzhi) paifang*

由儲罐中儲存的揮發性液體引起的向大氣的排放。這是由於罐區在正常操作時罐中的儲存物蒸發所造成的；並不僅限於泄漏。大氣壓和溫度的變化會導致儲罐的呼吸，從而造成向大氣的排放。這種呼吸排放可採用壓力儲罐或者具有氣體洗滌設備的排氣筒來加以控制。

在裝罐時也會產生向大氣的排放或損失。例如在單管輸送系統中，儲罐中飽和氣體的空間為加入罐中的液體所取代，會導致氣體排入到大氣中。一般可採用*氣體平衡管系統*來減少這類排放。

Emissions into the atmosphere associated with the storage of volatile liquids in tanks. Such emissions result from evaporation of the tank contents during normal operation of a tank farm; they are not limited to spills. Changes of atmospheric pressure and temperature lead to a breathing of the tank and as a result to emissions. These breathing emissions can be kept under control by pressurized tanks or vents equipped with gas scrubbing units.

Losses or emissions occur also when the tank is filled. With a single pipe transfer system, the saturated gas volume in the tank is displaced by the liquid which is being filled into the tank and is discharged to the atmosphere. To reduce such emissions, *gas balancing pipes* are generally applied.

target organism 靶生物 *bashengwu*

受外來物質影響的生物體。它是毒性試驗中物質作用的靶。（見：對魚的毒性、對蝦的毒性、廢水對無脊椎動物的毒性）

The organism which is affected by a foreign substance. It is thus a target for the substance in a toxicity test. (See also *toxicity versus fish*; *toxicity versus shrimp*; *effluent toxicity with invertebrates*)

tax 稅 *shui*

向國家稅務部門繳納的費用。它一般用來籌集國家所需的經費，包括國務活動的開支。有許多不同的稅收體系，它們是由政治因素決定的。而且收稅的多少並不直接與所受到的服務有關。

A payment made to the tax authority of a state, generally to raise the money needed by the state to cover the expenses of its activities. Many different systems of taxation are applied. The systems are the result of a political process. The magnitudes of taxes are not directly related to a service received.

TC 總碳 *zongtan*

Total carbon的縮寫。總碳是廢水水樣中無機碳和有機碳的總量。（參見：總有機碳）

Total carbon, i.e. the total content of inorganic and organic carbon in an effluent sample. (See also *TOC*)

technologies for pollution control 污染控制技術 *wuran kongzhi jishu*

見：pollution control technology。

See *pollution control technology*.

terrestrial ecosystems 陸地生態系統 *ludi shengtai xitong*

岩石圈生態系統，包括各種不同的群落生境及其生物群落。

Ecosystems of the lithosphere, including a large variety of very diverse biotopes and their biocenoses.

tertiary treatment of effluents 廢水三級處理 *feishui sanji chuli*

廢水在一級（機械、物理、化學）處理和二級（生物）處理後所進行的高級處理。廢水三級處理有不同目的，如用*硝化/反硝化*法去除含氮化合物（如氨、亞硝酸鹽和硝酸鹽），用沉澱和過濾法去除會造成*受納水體富營養化*的磷酸鹽。

它還可以用來減少會引起濁度問題的廢水中殘留的懸浮物。三級處理還包括最終淨化塘或用不同方法進行補充生物處理和過濾處理的任何其他設備。

Treatment steps for advanced effluent treatment, in addition to primary (mechanical/physical/chemical) treatment and secondary (biological) treatment.

Tertiary treatment is applied with different objectives. It may serve to remove nitrogenous compounds (e.g. ammonia, nitrite, nitrate) with the *nitrification/ denitrification* process. Phosphates, which add to the *eutrophication* of a *receiving waterbody*, can be removed by precipitation and subsequent filtration.

Tertiary treatment may also serve to reduce the residual load of suspended solids in the effluent, which would add to the turbidity. Additional process steps for tertiary treatment include also polishing ponds or any other equipment for additional biological treatment and efficient filtration by different methods.

test fish 試驗魚 *shiyanyu*

見：toxicity versus fish。

See *toxicity versus fish*.

textile effluents 紡織廢水 *fangzhi feihui*

見：dye house effluents。

See *dye house effluents*.

theoretical oxygen demand 理論需氧量 *lilun xuyangliang*

假定某化合物的所有元素被全部氧化而由其化學組成計算得到的需氧量。對於鹵代化合物，這可能會造成誤差。計算廢水的理論需氧量，只有在定量地確知廢水的所有組成成分的情況下才有可能，但這種情況是極少見的。（參見：總需氧量）

The oxygen demand calculated from the chemical composition of a compound, assuming that all elements are totally oxidized. For halogenated compounds, this may lead to uncertainties. The calculation of the theoretical oxygen demand of effluents is only possible if all components are known quantitatively, which is rarely the case. (See also *total oxygen demand*)

theoretical plate 理論板 *lilunban*

表示分離塔或精餾塔的分離效率的指標或參數，它與板式精餾塔內理想的盤板的板數有關。這一參數也可應用於其他類型的分離塔，如填料塔。

An indicator or parameter to describe the separation efficiency of a separation or rectification column. It is related to the number of ideal tray-like plates inside a plate rectification column. The indicator number is also applied to other types of separation columns, such as *packed columns*.

thermal hygienization of activated sludge 活性污泥的熱衛生處理 *huoxing wujide reweisheng chuli*

見：hygienization, thermal, of activated sludge。

See *hygienization, thermal, of activated sludge*.

thermal oxidation systems 熱氧化系統 *reyanghua xitong*

見：combustion。

See *combustion*.

thermodesorption 熱解吸 *rejiexi*

用熱氣體(如熱氮氣或熱蒸汽)將被吸附的化合物從吸附狀態下解吸出來。

Desorption of an adsorbed compound from the adsorbed state by a hot gas, e.g. hot nitrogen or hot steam.

thin layer chromatography (TLC) 薄層色譜法 *boceng sepufa*

在塗有一薄層吸附劑的玻璃板或鋁片表面上進行的色譜分離方法。吸附劑應依被分離的化合物的性質而定。其固定相吸附劑常常含有硅膠、硅藻土或礬土，並被固定在支承板上。測定時，加入一定量的樣品混合物作為薄層色譜板上的色斑。用純溶劑或混合溶劑作為流動相。為了防止溶劑蒸發，在一密閉的玻璃槽中顯現色譜。可以採用許多不同的檢測方法來鑒別所要分離的物質。薄層色譜法可用來研究諸如染料合成中所生成的非揮發性化合物和副產物的混合物。在一定限制範圍內，它也可用於痕量雜質的半定量分析。

A chromatographic separation procedure on a thin film of an adsorbent on a glass plate or an aluminium sheet. The type of adsorbent is specific for the type of compounds to be separated. The stationary adsorbent often contains silica gel, diatomaceous earth or alumina. This is fixed on a support plate. A defined quantity of the sample mixture is added as a spot onto the TLC plate. Pure solvents or solvent mixtures are used as mobile phase. The chromatogram is developed in a closed glass tank, in order to prevent evaporation of the solvent. Many different detection methods are used to identify the separated substances. Thin layer chromatography is used to investigate mixtures of mainly non-volatile compounds and by-products, resulting for instance in the synthesis of dyestuffs. Within certain limits, the method can be used for semi-quantitative analysis of trace impurities.

threshold limit value 閾限值 *yuxianzhi*

見：TLV。

See *TLV*.

threshold odor concentration 臭閾濃度 *chouyu nongdu*

見：odor index。

See *odor index*.

tiers analysis 層次分析 *cengci fenxi*

見：analysis plan in tiers。

See *analysis plan in tiers*.

time average sample 時間平均樣 *shijian pingjunyang*

以相等的時間間隔進行廢水的取樣，例如，每小時取一次樣品。由於排放量可能發生變化，所以平均樣並不一定代表實際的排放情況。（參見：瞬時樣、流量平均樣）

Sampling of a waste water stream at equal time intervals, for instance one sample every hour. The time average sample does not necessarily represent the true discharge, as the discharge volume may fluctuate. (See also *grab sample*; *flow average sample*)

time horizon dilemma 時間範疇上的難題 *shijian fanchoushangde nanti*

環境保護管理中遇到的難題。它產生的原因是由於環境保護的費用是即時費用，但其效益卻在相當長時間後而且可能會在另一地方顯現出來。另一方面，商業利潤一般是短期利潤。因此，經理們在花錢用於環境保護方面時會處於困難境地，因為環境保護的投資看來是不會立即得到回報的，而把同樣的資金分配用於商業活動將會立即獲得利潤。作為解決這一難題的辦法，要求經理們必須具有長遠和超前觀點，並基於長遠的考慮來進行評估和決策。（參見：費用－效益分析）

A dilemma encountered in environmental protection management. The dilemma is caused by the fact that the *costs of environmental protection* are immediate costs, while the benefits manifest themselves only after a considerable time, and possibly in another place. On the other hand, business returns are generally short term returns. The manager therefore is in a dilemma to spend money for an environmental protection activity for which there seems to be no immediate return, while the allocation of the same funds to business would generate immediate profits. As a solution to this dilemma, the manager is asked to apply long term vision and leadership and bases his evaluation and decision on long term considerations. (See also *cost-benefit analysis*)

TLC 薄層色譜法 *baoceng sepufa*

見：thin layer chromatography。

See *thin layer chromatography*.

TLV 閾限值 *yuxianzhi*

Threshold limit value的縮寫。在*職業衛生*管理中對所使用的化學物質和物理介質規定的閾限值。它給出了工作場所空氣中物質的限制濃度，在該濃度下一般職工在8小時工作日期間能安全地接觸而不會產生任何不良的影響。美國政府工業衛生工作者會議出版了閾限值表。它們還包括了一些補充數據和資料，如一些特殊案例和一些具有特殊性質的化合物，例如致癌物。（參見：最大容許濃度、短期接觸限值）

Threshold limit values (TLVs) for chemical substances and physical agents for use in *occupational health* management. The TLVs give the limiting concentrations of substances in the work atmosphere to which the average employee can safely be exposed during an 8-hour day at work without any adverse effect. TLV tables are published by the American Conference of Governmental Industrial Hygienists. They also include additional information and data for special cases and for compounds with special characteristics, e.g. substances causing cancer. (See also *MAK*; *STEL*)

TOC 總有機碳 *zongyoujitan*

　　Total organic carbon的縮寫。廢水分析中使用的一項綜合指標。在分析中必須區別有機化合物的碳與無機化合物如碳酸鹽的碳。（參見：總有機碳/總碳分析）

　　A group parameter used in the analysis of effluents. In the assessment one has to distinguish between carbon from organic compounds and inorganic carbon, e.g. in carbonates. (See also *TOC/ TC analysis*)

TOC analysis 總有機碳分析 *zongyoujitan fenxi*

　　廢水水樣中總有機碳含量的分析。詳見*總有機碳/總碳分析*。分析時還必須小心考慮無機碳。然而，不是所有的儀器都能進行無機碳分析的。

　　The analysis of the total organic carbon content of a sample of effluent. For details see *TOC/TC analysis*. Care has to be taken to account also for inorganic carbon. Not all instruments are equipped to do this .

TOC load 總有機碳負荷 *zongyoujitan fuhe*

　　單位時間內承受的廢水總有機碳量。

　　The mass flow per time of total organic carbon in an effluent.

TOC/TC analysis 總有機碳/總碳分析 *zongyoujitan/zongtan fenxi*

　　水樣中總碳和總有機碳的測定。其測定儀器大多裝配有兩根分離柱，一根裝填酸化載體（多磷酸），用以測定無機二氧化碳。另一根裝填氧化催化劑，用以在$700-900{}^{\circ}C$下分解有機碳化合物。在兩根柱子的出口處分別測定其二氧化碳，然後由它們的差值計算得到總有機碳含量。應該注意，市場上銷售的儀器並不都具有區別無機碳和有機碳的功能。

　　Measurement of total carbon and total organic carbon in a water sample. Most instruments are equipped with two separate columns. One column holds an acidified support (polyphosphoric acid) and is used for determining the inorganic carbon dioxide. The second column is filled with an oxidation catalyst and used to decompose the organic carbon compounds by combustion at 700-900°C. Carbon dioxide is measured at the outlet of both columns and the total organic carbon content (TOC) is then calculated from the difference of the two values. It is to be noted that not all commercially available instruments distinguish between inorganic and organic carbon.

TOD 總需氧量 *zongxuyangliang*

    見：total oxygen demand。

    See *total oxygen demand.*

total carbon 總碳 *zongtan*

    見：TC。

    See *TC.*

total organic carbon 總有機碳 *zongyoujitan*

    見：TOC。

    See *TOC.*

total organic halogen compounds 總有機鹵素化合物 *zhongyoujilusu huahewu*

    見：TOX。

    See *TOX.*

total oxygen demand (TOD) 總需氧量 *zongxuyangliang*

    表示廢水特性的一項*綜合指標*。它用熱燃燒法氧化水樣中所有雜質所需的氧量來度量。總需氧量的測定是在裝填有催化劑的燃燒柱中進行的，燃燒溫度為900°C。測定時，將廢水水樣注射入燃燒柱前的系統中。該儀器中配備有氧敏感檢測器，用以測定燃燒柱前後兩點 N_2/O_2 流中的氧濃度。根據此兩點之間氧濃度的降低量，可計算出總需氧量。因為總需氧量值包括所消耗的全部氧量，因此它總是比由總有機碳所計算得出的耗氧量大。

A *group parameter* for the characterization of effluents. It is a measure of the amount of oxygen used to oxidize by thermal combustion all impurities in a water sample. This is achieved in a combustion column, containing a catalyst. The combustion temperature is 900°C. The effluent sample is injected into the system before the combustion column. The instrument is equipped with oxygen sensitive detectors. The oxygen concentration is measured in the N_2/O_2 stream before and after the combustion column. The decrease between these two points is measured and used to calculate the TOD value. Since the TOD value includes the total oxygen con- sumption, it is always higher than the oxygen consumption calculated from the TOC value alone.

total product manufacturing costs 產品生產總成本 *chanping shengchan zongchengben*

產品的 *直接費用* 和生產部門的一部分 *間接費用* 的總和。

The total of the *direct costs* of a product and a fraction of the *indirect costs* of a production department.

total suspended solids (TSS) 總懸浮固體 *zong xuanfu guti*

見：suspended solids。

See *suspended solids*.

tower bioreactor 塔式生物反應器 *tashi shengwu fanyingqi*

一種特殊設計的廢水生物處理反應器，其曝氣池建成為較高的塔或槽，空氣或純氧由底部引入。總的來說，這種塔式生物反應器具有充氧效率高、佔地面積較小的優點，而且易發現曝氣池的泄漏情況。

A special design of the biological stage of an effluent treatment plant. The aeration basin is constructed as a relatively tall tower or tank; air or pure oxygen is introduced at the bottom. Overall, this results in a better efficiency of oxygenation. In addition, less space is needed for this construction and the basins can be inspected for leaks.

TOX 總有機鹵素化合物 *zhongyoujilusu huahewu*

Total organic halogen compounds的縮寫。它是水污染控制中使用的一項*綜合指標*，其值是將所有鹵素化合物通過燃燒轉化為鹵化物，再用銀鹽法分析測定鹵化物離子而得到的。（參見：可吸附的有機鹵素化合物、可提取的有機鹵素化合物、可汽提的有機鹵素化合物）

Total organic halogen compounds. A *group parameter* used in the field of water pollution control. The TOX value is analyzed by converting all halogen compounds to halide (through combustion), followed by argentometric determination of halide ion. (See also *AOX*, *EOX*, *POX*)

toxic waste 有毒廢物 *youdu feiwu*

一類需用特殊方法進行處置的有毒的廢物。然而，毒性的定義往往是不明確和含混的，因此偏向於稱之為*特殊廢物*。

A type of waste which is toxic and which therefore needs to be disposed of in a special way. The definition of toxicity is often not clear and not unambiguous. A preferred term is *special waste*.

toxicant 毒物 *duwu*

對*靶生物*具有毒害作用的物質。

A substance which exerts a toxic effect on *target organisms*.

toxicity, aquatic, of single substances 單一物質的水生生物毒性 *danyi wuzhide shuisheng shengwu duxing*

水中某單一物質的一種特性，可用標準方法來測定*半致死濃度*和*半作用濃度*的一類參數。但一種複雜廢水的毒性，是廢水中所存在的各種物質的毒性的綜合作用結果。由於毒性是非加成性的，因此不可以將廢水中各種化合物的毒性直接相加成來得到該廢水的總的毒性。

A characteristic property of a single substance in water. Parameters such as *LC-50* and *EC-50* are determined by standardized methods. The toxicity of a complex effluent is the result of the toxicity of all individual substances present. Since toxicity is not an additive property, the toxicity of individual compounds may not be added directly to give the overall toxicity of an effluent.

toxicity of effluents 廢水的毒性 *feishuide duxing*

見：effluent toxicity。

See *effluent toxicity*.

toxicity versus fish 對魚的毒性 *duiyude duxing*

污染物或含污染物的廢水的一種性質。某物質對魚的毒性與魚的種類和試驗條件有關。它用LC-50表示，即在標準條件下接觸某一濃度物質的試驗魚中有50%存活時該物質的濃度。某一經處理後的廢水的毒性，可以用稀釋倍數表示，即用新鮮充氧水進行稀釋，試驗魚中有50%存活時的稀釋倍數。在某些條件下，魚的毒性試驗可以用其他水生物的毒性試驗如*對蝦的毒性*來代替。

A property of a pollutant or of an aqueous effluent containing toxic pollutants. The toxicity of a substance versus fish depends on the species and on the test conditions. It is reported as LC50, that is the concentration of a substance at which under standardized conditions 50% of the exposed fish survive. The toxicity of a treated effluent can be reported as a dilution factor, i.e. that dilution with fresh oxygenated water, at which 50% of the exposed fish survive. Tests for fish toxicity can under certain conditions be replaced by tests for other water organisms, e.g. *toxicity versus shrimp.*

toxicity versus shrimp 對蝦的毒性 *duixiade duxing*

在評價廢水或單一物質的水生生物毒性時，可用某些種類的蝦代替魚作為試驗生物。試驗結果要註明蝦的種類，如長額蝦、褐蝦、海水蝦。

這些試驗通常是長期試驗，它需延續到蝦的整個生命周期，即從破卵開始直到死亡為止，評價其慢性效應。但用其他無脊椎動物作為*靶生物*得到的毒性數據，與用魚得到的毒性數據，它們之間沒有直接的相關關係。

Certain types of shrimp may be used as test organisms as an alternative to fish for the evaluation of aquatic toxicity of effluents and of single substances. Test results have been reported with: Pandalus danae; Crangon crangon (brown shrimp); Artemia salina (brine shrimp).

These tests are often long term tests to evaluate chronic effects. They may extend over the full life cycle of the shrimp, from hatching to death. As with other invertebrate *target organisms*, there is no direct correlation with data from fish.

toxicology 毒理學 *dulixue*

生命科學的一個分支學科，它研究與說明作用於*靶生物*的毒效應。靶生物可以是人、哺乳動物、鳥、昆虫、魚、浮游生物、植物或微生物。毒理學可幫助人們了解生態系統中化合物的作用。

A discipline within the broader range of life sciences. Toxicology investigates and interprets toxic effects upon *target organisms*. Target organisms may be very different: man, mammals, birds, insects, fish, plankton, plants or microorganisms. Toxicology adds to the understanding of the effects of compounds in eco-systems.

toxicology, aquatic 水生生物毒理學 *shuisheng shengwu dulixue*

研究水生生態系統中的生物如魚、*食物鏈*生物、浮游生物、藻類等的毒理學，它可幫助人們了解*受納水體*中各物質或複雜廢水的毒害作用。

Toxicology dealing with organisms in aquatic ecosystems, such as fish, organisms of the *food chain*, plankton, algae, etc. Aquatic toxicology helps to understand toxic effects of individual substances or complex effluents in the *receiving waterbodies*.

trading of emission rights 排放權交易 *paifangquan jiaoyi*

基於*經濟刺激*，在環境立法中運用的經濟手段之一。（參見：排放權）

A legal economic instrument in environmental legislation based on *economic incentives*. (See also *emission rights*)

traffic emissions 汽車尾氣排放 *qiche weiqi paifang*

汽車尾氣排放是環境保護重要問題之一，它是構成市區空氣污染的重要部分。其主要污染物為一氧化碳、揮發性有機化合物和氮氧化物，它們是形成*夏季煙霧*的主要原因。鉛也與此有關，其量取決於所用汽油的類別。柴油機還會排放煙灰和二氧化硫（取決於柴油的質量）。現代化的汽車安裝了 *催化轉化器* 以減少這些污染物的排放。

此外，二氧化碳的排放還與 *溫室效應* 有關。汽車排放也還包括道路粉塵和磨蝕粉塵。汽車噪聲污染也很重要，但沒有受到足夠的重視。

Emissions from motor vehicle traffic constitute one of the important problem areas in environmental protection. These emissions constitute a major fraction of air pollution in urban areas. The main pollutants are carbon monoxide, VOC, NO_x. These emissions contribute to the formation of *summer smog*. Lead is also relevant. Its quantity depends on the type of gasoline used. Diesel engines also produce emissions of soot and sulfur dioxide, depending on the quality of the diesel oil. Modern vehicles are equipped with *catalytic converters* for reducing these emissions.

In addition, carbon dioxide emissions are relevant in the context of the *greenhouse effect*. Traffic emissions also include general dust from roads and from abrasion. Noise emissions are also significant but have received less attention.

transport phenomena 傳輸現象 *chuanshu xianxiang*

在物理化學、物理學、生產工藝學和有關科學中研究和應用的一種現象，它可在不同相之間和在同一相之內發生，對於質量或能量傳輸過程是非常重要的。能量和熱量的傳輸對於工藝過程能源利用的優化（熱交換）和安全生產具有重要作用。當物質通過一 *相界面* 傳輸時，如在萃取過程和廢氣 *洗滌* 過程，總是會發生質量傳遞的。

A type of phenomena studied and applied in physical chemistry, physics, processing technology and sciences related to these. Transport phenomena are important in all cases where a transport of mass or energy takes place. This can be between different units or phases as well as within a single phase. The transport of energy or heat is important in the optimization of energy aspects of processes (*heat exchange*) and in safety considerations. Mass transfer always takes place when a substance is transported across a *phase boundary*, e.g. in extraction processes and in off-gas *scrubbers*.

trickling towers (effluents) 塔式滴濾池（廢水） *tashi delüchi (feishui)*

廢水生物處理的一種塔式設備，在塔內裝填有作為活性 *生物膜* 載體的有規填料，因此塔式滴濾池是一種在*載體上*生長有微生物的生物反應器。運行時，從塔底部通入空氣向上逆流通過滴濾池進行曝氣。與活性污泥法一樣，它也需對廢水進行預處理，但它所產生的剩餘污泥量較少。塔式滴濾池可應用於處理城市污水。（參見：活性污泥處理、生物降解）

A unit for biological treatment of effluents. The active unit consists of a tower or column with a structured filling, acting as a support for the activated *biofilm* or biomass. A trickling tower is thus a *biology on a support*. The effluent trickles at a low flow rate through the tower. Air for aeration is passed countercurrent from the bottom up. Pretreatment of the effluent is needed as in other activated sludge technologies. The amount of excess sludge to be separated after the biological treatment is smaller. Trickling towers may be applied for the treatment of municipal sewage. (See also *activated sludge treatment*; *biodegradation*)

trophic 有關營養的 *youguan yingyangde*

關於食物或營養的狀況。

With respect to food or nourishment.

trophic level 營養級 *yingyangji*

*食物鏈*中生物體的各別系統的水平。每一營養級的生物體是下一營養級生物體的捕食者，同時又是上一營養級生物體的被捕食者。例如，在水生態系統中，由最低營養級至最高營養級的食物鏈是：微生物和藻類－浮游動物－魚－食肉魚－水禽－人。（參見：捕食者與被捕食者的關係）

The individual system levels of the organisms in a *food chain*. The organism of each trophic level is a predator of the ones at the next lower level and at the same time the prey of the next higher one. Examples in an aquatic ecosystem, from the lowest to the highest level: microorganisms and algae – zooplankton – fish – predatory fish – waterfowl and/or – man. (See also *predator-prey relationship*)

true total product costs 產品的真實總成本 *chanpingde zhenshi zongchengben*

把按照廢物的真實比排放量和處置價格計算得到的污染控制費用包

括在内的產品成本。真實的產品成本是按現代的*成本計算方法*計算的，它包括了*污染控制費用的分攤計算*。

Product costs which fully include elements of pollution control based on true specific emission quantities and disposal prices. True product costs are the result of a modern *cost calculation procedure* including the *allocation of pollution control costs*.

TSS *總懸浮固體 zongxuanfu guti*

Total suspended solids的縮寫。（參見：suspended solids）

Total suspended solids. (See also *suspended solids*)

turnover rate *周轉率 zhouzhuanlü*

變化率或變化速度。當應用於連續生物反應器時，指在一定時間內單位容積反應器所能處理的廢水量或負荷（例如，每天每立方米反應器體積處理的公斤BOD_5量）。

Rate or velocity of change. When applied to a continuous biological reactor, the quantity or load of effluent which can be treated per unit volume of reactor in a given time (e.g. kg of BOD_5 per reactor volume per day).

U

UASB treatment 升流式厭氧污泥床處理 *shengliushi yanyang wunichuang chuli*

見：upflow anaerobic sludge blanket treatment。

See *upflow anaerobic sludge blanket treatment.*

ultrafiltration 超濾 *chaolü*

分離1-100毫微米粒子的過濾過程，它是介於微過濾和逆滲透或毫微米級過濾之間的一種過濾技術。超濾必須在3-4巴的壓力下進行。其應用包括：從牛奶製取乳清，水污染控制中機械行業切削油乳液的分離，以及在化工行業中的各種應用。（參見：過濾、微過濾、逆滲透、毫微米級過濾）

A filtration process to separate particles in the range of 1 nm to 100 nm. Ultrafiltration is therefore a technology between microfiltration and reversed osmosis or nanofiltration. Ultrafiltration has to be carried out under pressures of 3 to 4 bar. Applications include the processing of whey from milk, in the field of water pollution control, the separation of emulsions of drilling and cutting oils from the machine and mechanical industry, and different applications in the chemical industry. (See also *filtration*; *microfiltration*; *reversed osmosis*; *nanofiltration*)

ultrasonic mass flow meter 超聲波質量流量計 *chaoshengbo zhiliang liuliangji*

見：mass flow meter, ultrasonic。

See *mass flow meter, ultrasonic.*

underground waste deposits 廢物的地下堆存 *feiwude dixia duicun*

將特殊廢物堆存於地下如石灰石岩洞中的可能性。對於這種廢物的地下堆存需仔細調查了解*地下水*和*喀斯特水*的情況。喀斯特水體可能將廢物堆放場區與飲用水資源相連，因此可能會導致飲用水污染，而這種環境風險是不能接受的。

A possibility of depositing special wastes underground, e.g. in caverns in limestone formations. Such a deposit requires a careful investigation of the *groundwater* and *karst water* situation. Karst waterbodies may connect the waste deposit area with resources of drinking water and may therefore potentially lead to drinking water contamination. The environmental risk in such cases may not be acceptable.

underground waste deposits in salt mines 地下鹽礦中廢物堆存 *dixia yankuangzhong feiwu duichun*

　　一種非常特殊的廢物受控堆存方法。在通過適當控制、分析和審批程序後，將廢物裝桶堆存在開採出岩鹽或鉀鹽後留下的地下岩洞中。一個聞名的地下鹽礦廢物堆存場，位於德國的赫法－紐羅德（Herfa-Neurode）附近，其岩洞位於地下大約600米深處，而且是完全乾燥的（岩鹽洞！），廢物不會進入水圈及污染水圈的任何部分。從環境風險管理的觀點看，在這種完全乾燥的地下鹽礦中進行廢物的地下堆存是可以接受的。

A very special type of a controlled deposit. Waste, after appropriate control, analysis and permitting procedure, is deposited in drums in underground caverns left empty after extraction of rock salt or potassium salt. A well-known deposit of this type is situated in the Federal Republic of Germany near Herfa-Neurode. The caverns are about 600 meters below ground. They are completely dry (rock salt caverns!) and the wastes cannot get into, and contaminate, any part of the hydrosphere. Under these totally dry conditions, such deposits are acceptable from the point of view of environmental risk management.

UNEP 聯合國環境規劃署 *lianheguo huanjing guihuashu*

　　聯合國中負責環境保護工作的機構，其總部設在肯尼亞的內羅比，在墨西哥城、麥納麥（巴林首都）、曼谷、日內瓦、紐約、華盛頓特區和巴黎設有辦事處。聯合國環境規劃署的任務是提高全世界人們的環境意識，以及與聯合國其他機構協調進行環境保護工作。其活動與計劃涉及所有環境問題。它所發佈的與環境有關的文件資料，可通過它設在世界各地的辦事處獲得。

The Environmental Protection Program of the United Nations Organization. The central office of UNEP is in Nairobi, Kenya, with branch offices in Mexico City, Manama (Bahrain), Bangkok, Geneva, New York, Washington, D.C., and Paris. The function of UNEP is to raise the level of environmental awareness worldwide and to coordinate EP work within other parts of the United Nations Organization. The activities and programs of UNEP cover all environmental segments. Relevant documents on environment from UNEP are available through its branch offices.

UNEP Convention on waste transports 聯合國環境規劃署廢物轉運公約
lianheguo huanjing guihuashu feiwu zhuanyun gongyue

見：1989年巴塞爾公約。

See *Basel Convention of 1989.*

UNEP waste classification system 聯合國環境規劃署廢物分類系統 *liangheguo huanjing guihuashu feiwu fenlei xitong*

為1989年聯合國環境規劃署廢物轉運公約（也稱為1989年巴塞爾公約）制定的廢物分類系統。

根據這一系統，可基於以下方式將廢物分成若干類和子類：（1）按照廢物流的源頭工藝來分類（Y1至Y18）；（2）按照存在的某些組分來分類（Y19至Y45）；（3）按照某些特殊的考慮來分類（Y46和Y47）；（4）按照特殊危害性和/或潛在危險來分類（H1至H12）。這一分類系統的使用與限制廢物的越境轉移有關。（參見：1989年巴塞爾公約）

A waste classification system which was developed for the 1989 UNEP Convention on waste transports, also known as the 'Basel Convention of 1989'.

Following this system, wastes are classified into a number of categories and subcategories, based on (a) the source process of the waste stream (categories Y1 to Y18); (b) the presence of certain constituents, (categories Y19 to Y45); (c) special considerations (Y46 and Y47); (d) specified hazardous properties and/or danger potentials. (codes H1 to H12). The classification system is used in conjunction with restrictions on the transboundary movement of wastes. (See also *Basel Convention of 1989*)

UNOX System 優諾克斯系統 *younuokesi xitong*

由聯合碳化物公司開發的應用純氧技術的一種特殊的廢水生物處理系統。在該系統中，曝氣池上方的空間被分隔成4個獨立的小間。曝氣

池加蓋密閉，因此充氧效率較高，空氣污染較輕。（參見：林多克斯系統、純氧曝氣系統）

A special application of biological effluent treatment with technically pure oxygen developed by Union Carbide Corp. In this system, the space above the oxygenation basin is also subdivided into four separate compartments. The aeration basins are closed and therefore lead to higher oxygenation efficiency and less air pollution. (See also *LINDOX*; *aeration systems with oxygen*)

upflow anaerobic sludge blanket (UASB) treatment 升流式厭氧污泥床處理
shengliushi yanyang wuni chuang chuli

基於*厭氧流化床生物處理*的*生物反應器*。運行時，廢水從底部向上流經反應器。其厭氧污泥可能形成為顆粒狀。在某些實際應用中，"污泥床"係形成在以無機惰性載體為核心的周圍。這一系統可應用於處理食品和化工廢水。

A *bioreactor* which is based on anaerobic *fluidized bed biology*. The waste water flows through the reactor from bottom up. The anaerobic sludge may have the form of granules. In some cases, a 'sludge blanket' is formed around an inorganic inert support nucleus. The system may be applied in food and chemical industry.

UV radiation 紫外輻射 *ziwai fushe*

波長在100-400毫微米之間的電磁輻射。地球上的紫外輻射可分類為：UV-A，波長為320-400毫微米；UV-B，波長280-300毫微米；UV-C，波長200-280毫微米。UV-B對於臭氧層耗竭和皮膚損害問題具有重要作用。（參見：平流層臭氧、多勃遜單位）

Electromagnetic radiation with a wavelength between an upper limit of about 400 nm and a lower limit of much below 100 nm. UV radiation on earth is commonly classified into UV-A with a wavelength of 320 to 400 nm, UV-B with a wavelength of 280–300 nm and UV-C with a wavelength of 200 to 280 nm. UV-B is important in the context of ozone depletion and the skin damage problem. (See also *ozone, stratospheric*; *Dobson Units*)

V

Valdez Principles 瓦爾德之原則 *wa 'erdezhi yuanze*

見：CERES Principles。

See *CERES Principles*.

validity of a conclusion 結論的有效性 *jielunde youxiaoxing*

見：顯著水平。

See *significance level*.

values 價值 *jiazhi*

在討論有關管理、*生活質量和費用-效益*問題時使用的術語，指人們非常珍惜和重視的事物或條件。

A term used in the context of management, *quality of life* and also *cost-benefit* discussions. The things or conditions which man cherishes and values highly.

venturi canal 文丘里管道 *wenqiuli guandao*

沿氣體或液體介質流動方向寬度逐漸變小的管道。這種管道的形狀是根據流體動力學或空氣動力學原理設計的。由於管道逐漸變窄，使得介質的速度逐漸增大，而在縮頸處達到最大。如果在這一點改變流動方向，介質就會受到很大的離心力。在污染控制中，它已實際應用於各種空氣污染控制裝置以及測定氣體或液體介質流速的儀器設備中。

A canal with a width which is getting continuously smaller in the flow direction of a medium. The medium can be a gas or a liquid. In detail, the canal is shaped according to hydrodynamic or aerodynamic principles. Because of the narrowing, the velocity of the medium is increased and is highest at the constriction. If the flow is deflected at this point, the medium is subjected to high centrifugal forces. Practical applications in pollution control include various APC units and devices to measure flow rates of a gas or liquid medium.

venturi type scrubbers 文丘里型洗滌器 *wenqiulixing xidiqi*

　　結構為*文丘里管道*的空氣污染控制洗滌設備。在文丘里管道最小直徑處，由於節流作用，氣流被加速，同時在該處通過特殊噴咀噴入洗滌液，因此氣液之間發生密切接觸，加強了質量傳遞。在有些系統中，後來又使氣流轉向，氣流的高速轉向將產生離心力，使噴淋的小霧滴得到分離。

　　這一過程必須由風機提供推進氣流所需的能量。文丘里洗滌器的設計是很不相同的，其*壓降*變化很大。它們可應用於許多領域，詳細設計見*環形噴射器*、*X型分離器*、*噴射流洗滌器*。

　　APC scrubbing units structured as a *venturi canal*. As a result of the restriction at the smallest diameter of the venturi canal, the gas flow is accelerated. The scrubbing liquid is injected at this narrowest point through special nozzles. As a result, there is an intimate contact between liquid and gas. This contact is enhancing the mass transfer. In some systems, the gas stream is then deflected. Deflection at high speed leads to centrifugal forces which, in turn, lead to a separation of spray and mist droplets.

　　The energy needed for this process must be provided by the fans propelling the gas. The detail designs of venturi scrubbers and also their *pressure drops* vary considerably. Applications are in many fields. For some detail designs, see *ring-jet unit*; *X-separator*; *jet scrubber*.

VOC 揮發性有機化合物 *huifaxing youji huahewu*

　　一項*綜合指標*。(參見：揮發性有機化合物；廢水中的揮發性有機碳)

　　A *group parameter*. (See also *volatile organic compounds*; *volatile organic carbon in effluents*)

volatile organic carbon (VOC) in effluents 廢水中的揮發性有機碳
feishuizhongde huifaxing youjitan

　　廢水中的揮發性有機污染物(表示為碳)。它主要來自有機溶劑，並可用汽提方法從廢水中去除。揮發性有機碳可用*總有機碳*分析的現代儀器進行測定。此類儀器中配置有水溶液汽提和揮發性有機碳測定的設備。在有些廢水法規中規定了揮發性有機碳濃度的限值。揮發性組份既可能是空氣污染源,也可能是廢水處理場產生惡臭的原因。

The fraction of volatile organic pollutants (expressed as carbon) in an effluent. It is primarily resulted from organic solvents which can be removed from the effluent by stripping. Its measurement can be carried out in modern instruments for the analysis of *TOC*. The instrument is equipped with a device for purging the aqueous solution and for measuring the volatile part of the organic carbon. VOC concentration limits are included in some waste water legislation. Volatile components may be a source of air pollution and also the cause of bad odors of effluent treatment plants.

volatile organic compounds (VOC) 揮發性有機化合物 *huifaxing youji huahewu*

在常溫、常壓下能轉變成蒸氣相,即具有揮發性的有機化合物。這類化合物與空氣污染控制有關,它們會造成工藝廢尾氣*排放*,使得在許多不同排放源的*廢氣*中含有揮發性有機化合物。有些揮發性有機化合物是天然存在的,如萜烯、甲烷和植物香料。廢水中原來也可能含有揮發性有機化合物,但在廢水處理場它們會從廢水中被氣提出來而逸至大氣。

Organic compounds which are volatile, i.e. are transformable into the state of vapor under normal conditions of temperature and pressure. Such compounds are relevant in connection with air pollution control. VOCs contribute to process *emissions* and are contained in *off-gas* streams of many different sources. Some VOCs have a natural origin, e.g. terpenes, methane and plant fragrances. Volatile organic compounds may initially also be contained in effluents from where they are stripped in effluent treatment plants and given off to the atmosphere.

volume/volume ratio 體積/體積比 *tiji/tiji bi*

以體積比表示的濃度單位,如氣流中20%v/v的二氧化碳。

A concentration unit given as a volume related to another volume, e.g. 20% v/v CO_2 in a gas stream.

411

VOX 揮發性有機鹵素化合物 *huifaxing youji lusu huahewu*

Volatile organic halogen compounds 的縮寫。其較老的名稱為 *可氣提的有機鹵素化合物*。

Volatile organic halogen compounds, an older name for *POX, purgeable organic halogen compounds.*

W

waste 廢物 *feiwu*

　　現物主不再想要而且必須進行妥當處置的東西。廢物對環境具有不利的影響，因此對其處置必須加以控制。

　　A material or an object which the present owner does not want any more and which must be disposed of in a proper manner. Wastes have a negative commercial value and their disposal must therefore be controlled.

waste classification 廢物分類 *feiwu fenlei*

　　對不同來源與不同性質的廢物可以加以分類，其主要類別有：*城市和工業垃圾、建築廢物和瓦礫、農業廢物、工業特殊廢物*。世界各國採用了許多不同的廢物分類系統，它們按不同的判據即根據廢物的來源、物理狀態、廢物性質、組成、重要元素和化合物的痕量含量、火災與爆炸危害來區分廢物。每一分類系統有它本身的優缺點。實用的分類系統的優點，在於它考慮了廢物常常是含有各種不同組份的複雜混合物，而且它們的分析也是不明確的這一事實。*聯合國環境規劃署*已提出了一個國際上採用的分類系統。

　　Wastes of different sources and different nature can be classified. The main groups include: *municipal and industrial trash, construction waste* and debris, waste from agriculture, *special wastes* from industrial operations and from small trade. Numerous classification systems are used worldwide. They use different criteria to differentiate wastes: the waste source, the physical state, waste properties, composition, trace levels of critical elements and compounds, and fire or explosion hazards. Each system has its advantages and disadvantages. Pragmatic systems have the advantage that they take into consideration the fact that wastes are often heterogeneous mixtures of widely different components and their analysis is not clearcut. The *UNEP* has developed a system which is applied internationally.

waste data sheet 廢物數據卡 *feiwu shujuka*

　　用於*廢物管理*的文件資料。*廢物多聯單*是用來跟蹤每批廢物的行蹤的，而廢物數據卡則不同，它給出了每一種廢物的信息數據與資料。

廢物數據卡包括廢物產生源的管理數據，如生產現場、工廠、建築物、責任、費用分攤等；源工藝過程以及單位產品產生的廢物量；許可的處置方法、設施和處置的編碼（如再利用、焚燒、安全堆存）；廢物的物理性質，如液體、泥漿、污泥、臭味、外觀；廢物的組成，包括關鍵組分的數據；危害性（如毒性、安全方面的危害、處理、運輸）；日期和簽字。

廢物數據卡是針對每一種具體廢物的；如果數據有變化，例如工藝過程改變使得廢物數據相應改變，則應加以修正。

A document used in connection with *waste management*. While the *waste manifest* is used to track the movement of each lot of waste, the waste data sheet includes information for each separate type of waste.

Information elements are: administrative data on the source: site, factory, building, responsibilities, cost allocation, etc.; the source process and standard quantities of the waste per quantity of product; permitted disposal methods, facilities and disposal code numbers (e.g. reuse, incineration, secure deposit); physical properties of the waste: liquid, slurry, sludge, odor, appearance; composition, including data on critical components; hazards (toxicity, safety hazards, handling, transport); date and signatures.

The waste data sheet is valid for one specific waste; it must be re-edited if the data has changed, e.g. as a result of process changes.

waste deposits 廢物堆存 *feiwu duicun*

見：controlled landfill; dumps。

See *controlled landfill*; *dumps*.

waste disposal 廢物處置 *feiwu chuzhi*

在一般意義上指各種廢物包括固體、液體或氣體廢物的處置，但這一術語往往更專門用於指固體廢物的處置。（參見：空氣污染控制、水污染控制）

In the very general meaning: the disposal of all types of wastes, solid, liquid or gaseous. The term is often used more specifically for the disposal of solid wastes. (See also *air pollution control*; *water pollution control*)

waste exchange 廢物交換 *feiwu jiaohuan*

促進廢物交換與交易的機構。其目的是向廢物的可能用戶提供有關另一工廠具有的廢物的信息。因此，廢物交換應建立廢物及其來源的數據庫，並向可能的用戶提供這方面的數據。它可由行業協會按非盈利原則負責運作，或者也可按商業運作收取少量服務費。

An office or an institution to promote and facilitate the exchange and trade of waste materials. The objective is to inform potential users of wastes of the existence of wastes in another factory. The waste exchange thus keeps a database on wastes and their sources and makes this data available to potential users. A waste exchange may be operated on a non-profit base by a trade association, or it may by itself be a commercial operation, charging a small fee for the services.

waste incinerators 廢物焚燒爐 *feiwu fenshaolu*

見：回轉窯焚燒爐、爐箅式焚燒爐、流化床焚燒爐、多爐床焚燒爐、複合技術焚燒爐、冷壁式焚燒爐。

See *rotary kiln incinerator* , *grate incinerator*, *fluidized bed incinerator*, *multiple hearth incinerator*, *hybrid technology incinerator*, *cold wall incinerator*.

waste management 廢物管理 *feiwu guanli*

對*特殊廢物*和*城市廢物*處置的管理，包括採取技術與行政措施以保證各種廢物在裝卸（勞動衛生）、運輸和處置時不會產生任何問題。廢物造成對土壤的污染和對地下水與水體的污染，是需要從長遠考慮來解決的潛在問題。而對它所需進行管理的程度，依所處置的廢物的類型而定。（參見：特殊廢物管理、城市廢物管理）

Waste management, applied to the disposal of *special wastes* and *municipal wastes*. The activities include technical and administrative measures which guarantee a disposal of all wastes without causing any problems during handling, (work hygiene), transport, and disposal.

Ground contamination and contamination of groundwater and waterbodies are potential problems which need long term considerations. The level of management attention required depends on the type of wastes to be disposed. (See also *management of special wastes*; *municipal waste management*)

waste manifest 廢物多聯單 *feiwu duoliandan*

特殊廢物管理中作為*廢物跟蹤程序*所採用的文件資料。廢物多聯單係用來記載廢物從產生地轉移到最終處置地的過程。多聯單中含有廢物特徵、廢物編碼、廢物產生者、廢物運輸者、廢物量和最終處置等內容。各個處置環節的負責人要分別在單上簽字以證實廢物已得到正確的處理和處置。多聯單中的每一聯是分別對廢物產生者、運輸者、處置場和政府主管部門的。廢物多聯單制度通常是根據國家法律的要求而實行的。(參見：從搖籃到墳墓原則、廢物數據卡)

A document which is used as part of a *waste tracking procedure* for the management of special wastes. It serves to document the movement of wastes from the point of generation to the point of final disposal. The form includes information on waste characteristics, waste codes, waste generator, waste transporter, quantities and final disposal. Persons responsible along the chain of disposal confirm proper handling and disposal by their signature. Separate copies are for the waste generator, the transporter, the disposal site and the authorities. Waste manifests are generally based on national legislation. (See also *cradle-to-grave principle*; *waste data sheet*)

waste manifest procedure 廢物多聯單管理方法 *feiwu duoliangdan guanli fanfa*

應用"從搖籃到墳墓"的管理原則對廢物進行全過程控制的管理辦法。它用特殊設計的表格即廢物多聯單來跟蹤廢物，從廢物產生者處開始，沿著廢物處置的途徑進行控制，直到保證它們得到正確處置為止。廢物多聯單通常有幾聯，分別給(1)廢物產生者，(2)運輸者，(3)最終處置者，(4)政府主管部門。(參見：從搖籃到墳墓的控制)

A management procedure used to exert total control over wastes applying the 'cradle-to-grave' principle. A specially designed form, the waste manifest, is used to track wastes starting at the waste generator, carry out controls along their disposal route, and to certify correct disposal. Waste manifest forms generally include several copies for (a) the waste generator, (b) the transporter, (c) the final disposal site and (d) the authorities. (See also *cradle-to-grave control*)

waste minimization 廢物最少量化 *feiwu zushaolianghua*

見：integral process development。

See *integral process development*.

waste reduction at the source 廢物源頭削減 *feiwu yuantou xuejian*

在廢物產生源減少廢物量的措施，包括通過*完整的工藝過程開發*來優化生產過程，以及將廢物再用於其他生產過程。它的另一個目的是減少廢物可能產生的問題，以利於進行廢物的處置。（參見：循環利用、工藝措施）

Activities which serve to reduce the quantity of wastes at the source of their generation. These activities include the optimization of processes through *integral process development* and the reuse of wastes in other processes. Another similar objective is to reduce the problem potential of wastes in order to facilitate their disposal. (See also *recycling*; *in-process measures*)

waste tracking procedure 廢物跟蹤程序 *feiwu genzhong chengxu*

*特殊廢物處置管理*中的管理程序，它從生產過程中產生廢物那時開始，一直跟蹤到廢物在處置場進行最終處置時為止。現代的廢物跟蹤系統包括：*廢物數據庫*、廢物處理的明確責任、跟蹤*廢物*運輸與轉移的*多聯單*、以及廢物最終處置者確認對廢物進行了正確處置。

A management procedure in the context of the *management of special wastes disposal*. The tracking procedure serves to follow a particular lot of waste from the moment it is generated in production, to the moment when it is finally disposed of in a waste disposal facility. A modern waste tracking system includes a *waste database*, clearly assigned responsibilities for handling waste, a *waste manifest* (a set of forms) to follow transports and movement of a lot, and a confirmation from the final waste disposer for correct disposal.

waste water treatment plant 廢水處理場 *feishui chulichang*

見：effluent treatment plant。

See *effluent treatment plant*.

wastes, immobilized 固定化廢物 *gudinghua feiwu*

通過處理使所有有害組分被化學固定或被包封在無反應性的固體基塊中的廢物。因此，採用固定化技術可改善*廢物的浸出性*。已提出了一些固定化的方法，如用水泥將廢物固定化。

Wastes which have been treated in such a way that all critical components are bound chemically, or are encapsulated and enclosed in a solid non-reactive matrix. The process thus improves *leaching properties of wastes* (see *wastes, leaching properties*). Several procedures have been proposed for immobilization, e.g. as immobilization in cement.

wastes, leaching properties 廢物，浸出性 *feiwu, jinchuxing*

廢物可能含有水溶性組分，如無機鹽或某些有機物。當與水接觸時，這些組分會被溶解和浸出。其浸出液會滲透過土壤，導致地下水污染的潛在風險。由於浸出過程會受雨水的酸度和微生物增溶過程的影響，因此對擬填埋處置的廢物，應進行*浸出液測試*，並對浸出液進行分析測定。(參見：填埋場中浸出液的控制和處理)

Wastes may contain water-soluble components such as inorganic salts or certain organic substances. In contact with water, these components are dissolved and will leach out. This leachate may penetrate the ground and there is a potential risk of groundwater contamination. Leaching is influenced by the acidity of rain water and by microbiological solubilization processes. Wastes for deposits should therefore be tested and pass a *leachate test*. (See also *leachate control and treatment in landfill*)

wastes, non-reactive 非反應性廢物 *feifanyingxing feiwu*

在受控填埋場填埋體內所處的條件下不會進一步降解或分解的廢物。因此，無論在氣相(*填埋氣體*)或*浸出液*中都不會存在有害組分。對於非反應性廢物的處置，其填埋堆存的技術和環境要求較低。

Wastes which do not degrade or decompose further under the conditions found inside the body of a controlled landfill. As a result, no critical components are set free, neither in the gas phase (*landfill gas*), nor in the *leachate*. For the disposal of non-reactive wastes, the technical and environmental requirements on a waste deposit are lower.

wastes database 廢物數據庫 *feiwu shujuku*

用於支持和監督工廠中廢物管理的數據庫，它包括廢物特性、廢物來源、廢物最終處置數據。對於普通垃圾，只需有一般性的數據就足夠了；而對於特殊廢物，需要有關它們的性質、風險和最終處置的詳盡數據。廢物數據庫是*廢物跟蹤程序*的一個重要部分。（參見：從搖藍到墳墓的原則）。

A database to support and supervise management of wastes in a factory. It includes as data elements the characteristics of wastes, their sources, and data on their final disposal. General data is sufficient for common trash, while more details on properties, risks, and final disposal are needed for special wastes. The waste database forms an important part of a *waste tracking procedure*. (See also *cradle-to-grave principle*)

water 水 *shui*

分子式H_2O，它是一切生物生命中最基本和重要的物質，是眾多應用中必不可缺的資源。水除了用作為資源外，在哲學、藝術、宗教和文化上總是把它看作是一種象徵和基本要素。

水大約佔據地球表面三分之二的面積，估計其總量約為13.86億立方公里，其中97.5%在海洋，其餘在不同水體中以及存在於地下水、冰川和極地冰冠、大氣和土壤的水分中。此外，在生物體中也含有很小部分水。估計地球上淡水含量只有3,500萬立方公里。

地球上水循環包括蒸發、大氣中水的轉移、降水、河流水流和最終歸流海洋等過程。

水在地球生態圈中起著極為重要的作用，它是水圈各生態系統的主要組成部分，因此也是許多水生生物群落生境的重要組成部分。它的重要作用還包括對全球氣候和局地氣候的調節作用，風化和增溶植物營養物的作用，以及作為一切生命過程包括農業和工業應用的一種基本資源。同時，河流、湖泊和海洋也是食物輸送的途徑，是蛋白質（魚）的來源。

雖然水的總量是不變的，但由於各種各樣的污染源造成的水污染，使得清潔水正在日益減少。基於這一原因和水對生命的重要作用，防止各類水體進一步污染是*可持續發展*最重要的內容之一。

The molecule H$_2$O, a substance of fundamental importance in life of all organisms and an indispensable resource for a multitude of uses. In addition to its use as a resource, water has always been regarded as a symbol and fundamental element in philosophy, art, religion and culture.

Water covers about two-thirds of the surface of the earth. The total quantity of water is estimated at about 1386 million cubic kilometers; of this, about 97.5 % is contained in the oceans, the rest in different waterbodies, in groundwater, in the ice of glaciers and the polar ice caps, as humidity in the atmosphere and in the soil. A small fraction is contained in living organisms. The quantity of sweetwater is estimated at only about 35 million cubic kilometers.

The water mass flow goes through a global cycle including evaporation, transport in the atmosphere, precipitation, flow of rivers, and drainage to the oceans.

Water fulfills vital functions in the global ecosphere. It is the main component of different ecosystems of the hydrosphere and thus an important component of many aquatic biotopes. Important functions include also the regulating effect on the global and local climates, erosion and solubilization of plant nutrients, the role as a basic resource for all processes of life, including agricultural and industrial uses. Rivers, lakes and the oceans are at the same time a mean of transport and a source of proteins (fish) for food.

While the total quantity of water does not change, the quantity of clean water is decreasing as a result of many different sources of pollution. As a result of this, and because of the central role water plays in life, activities to prevent further pollution of all types of waterbodies are of utmost importance for *sustainable development*.

water, permeating 滲透水 *shentoushui*

見：permeate。

See *permeate*.

water database 水數據庫 *shui shujuku*

見：effluents inventory。

See *effluents inventory*.

water management basin 水域管理 *shuiyu guanli*

一國或跨國*水污染控制*管理的內容之一。水域管理的概念是基於水管理區域的定義，即指河流的流域或跨越若干不同國家邊界的水體。在歐洲，以法國為例，"河域管理局"管理的範圍跨越了法國許多行政區的邊界；在英國，也設有類似的行政機構。在國際上，有一些國際委員會在積極從事保護國際水體的工作，如萊茵河、日內瓦湖、康斯坦斯湖、地中海、北海等的保護。

A management element related to a national or supranational *water pollution control* concept. The concept of the water management basin is based on the definition of water management areas, which are identical with the boundaries of river catchment basins or waterbodies bordering several different countries. In Europe, on a national level, examples are found in France, the 'Agence de bassins' which transcends boundaries of the French 'Departments', as well as similar administrative institutions in England. On the international level, there are a number of active international committees. They all deal with the protection of international waterbodies, e.g. the Rhine River, Lake Geneva, Lake Constance, the Mediterranean Sea, the Baltic Sea, the North Sea and others.

water pollution control (WPC) 水污染控制 *shuiwuran kongzhi*

保護水體不受污染或*恢復*已污染水體所採取的一切行動和措施。水污染控制有賴於*完整的工藝過程開發*（源頭削減）和*水污染控制技術*。這些措施必須得到行政和管理部門的支持。另外，提高環保意識從而改變行為方式，對水污染控制也具有重要作用。

The total of all activities and measures taken to protect waterbodies from pollution, or, to *remediate* existing pollution.

Water pollution control rests on *integral process development* (source reduction) and *water pollution control technology*. These measures must be supported by administrative and management activities. Changes of behavior also play an important role.

water pollution control concept 水污染控制觀念 *shuiwuran kongzhi guannian*

企業實施水污染控制工作的總體觀念，它包括標準、管理體系和技術措施。管理方面的主要內容有：充分重視環境保護工作（包括水污染控制工作）；確定正常運行和事故性泄漏情況下的排放目標；建立包括所有廢水產生源和排放情況的數據庫；把水污染控制納入到計劃工作內容中；配備管理工作所必需的儀器設備，以進行監督和控制；對新工藝過程、新裝置、新設施和新設備，通過影響評價進行有效的技術管理。

技術措施包括：通過改進工藝、源頭削減和循環回收利用，從工藝過程本身解決水污染問題；將不同類別的廢水區分開來；對不同類別的廢水進行專門的預處理（工藝末端過程處理）；和最終處理（末端處理）。

An overall concept for water pollution control activities of an enterprise. Such a concept includes normative elements, management systems and technical measures.

The main elements on the management level are: giving adequate weight to environmental protection, including water pollution control activities; defining discharge goals for normal operations as well as for accidental spills; compiling a database covering all sources and all discharges of effluents; integrating water pollution control in the planning process; setting up the necessary management instruments to implement planned measures and to carry out controls on performance; and efficient technology management with impact assessments for new processes, facilities, units and equipment.

Technical measures include: solutions at the level of the process through processes improvements, source reduction and recycling; separation of different types of effluents; specific pretreatments for different types of effluents (end-of-process treatment) , and final treatments (end-of-pipe treatment).

water pollution control technology 水污染控制技術 *shuiwuran kongzhi jishu*

用於控制水污染的技術，它主要是開發*完整的工藝過程*和採取*工藝措施*，從源頭削減污染負荷。此外，還有各種不同的廢水處理技術，可應用於末端工藝過程處理以及作為末端處理措施。這些技術可用來處理污染的母液和廢水。

Technologies which keep the pollution of water under control. Of prime importance are technologies for reducing the pollutant load at its source through *integral process development* and *in-process measures*. In addition, various effluent treatment technologies are available, to be applied at the *end-of-process*, and *end-of-pipe measures*. All serve to treat contaminated mother liquors and effluents.

water quality goals 水質目標 *shuizi mubiao*

按污染物濃度制定的水體質量目標。對於使用目的不同的水體，如用作飲用水、娛樂用水或船舶用水，水質目標通常是不同的。如同環境空氣目標一樣，當污染是由許多不同源引起時，質量目標與排放限值之間的關係是複雜的。（參見：質量目標）

Quality goals defined in terms of concentrations of pollutants for waterbodies. Water quality goals are usually different for waterbodies which are used for different purposes, e.g. for drinking water, for recreation or for shipping. Similar to ambient air goals, the relationship between these goals and emissions limits is complicated when a number of different sources contribute to the pollution. (See also *quality goals*)

WBCSD 世界可持續發展工商委員會 *shijie kechixu fazhan gongshang weiyuanhui*

見：World Business Council for Sustainable Development。
See *World Business Council for Sustainable Development*.

wet air oxidation, high pressure 高壓濕式空氣氧化 *gaoya shishi kongqi yanghua*

一種很有效的廢水處理技術。它在壓力100-200巴、溫度200-300°C下於液相中用氧來氧化高濃度廢水中的污染物。其反應器可設計為泡罩塔形式。廢水停留時間約為1-6小時。為防止設備腐蝕，反應器需用耐腐蝕材料如不銹鋼、哈斯特合金（商品名）或鈦製成或襯裏。處理時可使用少量催化劑，而且在最佳條件下，氧化反應熱幾乎可足以維持反應溫

423

度。此外，還需要有一些輔助設施來處理廢氣和反應後的含低分子量易
生物降解物質的廢水。濕式空氣氧化是處理各種不可生物降解污染物的
一種高新技術，其基建投資大，但考慮到它具有很好的處理效果，因此
在經濟上還是可行的。（參見：廢水低壓氧化）

A very powerful technology for the treatment of effluents. Pollutants in relatively highly concentrated solutions are oxidized at a pressure of about 100–200 bar
and at a temperature of 200–300℃ in the liquid aqueous phase by oxygen. The
reactor is designed e.g. as a *bubble column*. The residence time of the effluent is in
the order of one to six hours. To prevent corrosion, this reactor is constructed from,
or lined with, resistant material, such as stainless steel, Hastalloy (trade name) or
titanium. A small amount of a catalyst may be used. Under optimal conditions, the
heat of oxidation is almost sufficient to maintain the reaction temperature. Auxiliary
facilities are needed to treat the off-gases and the treated mother liquors which may
contain some low molecular weight easily biodegradable substances.

Wet air oxidation is a high tech method for the treatment of various non-
biodegradable pollutants. It requires a considerable capital investment but, con-
sidering its efficiency, is economically viable. (See also *low pressure effluent
oxidation*)

wet APC-systems 濕式空氣污染控制系統 *shishi kongqi wuran kongzhi xitong*
見：APC-systems, wet; packed column scrubber。
See *APC-systems, wet; packed column scrubber.*

wet electrostatic precipitator 濕式靜電除塵器 *shishi jingdian chuchenqi*
一種特殊設計的用於去除生產工藝過程排放的微細霧滴和塵粒的 *靜
電除塵器*。在濕式靜電除塵器中，氣流可能是冷而潮濕的，可以通過噴
咀引入所需要的補充水分。所要分離的廢氣中的小液滴，在沉積電極上
凝聚並作為液體除去。應定期地用水膜清洗沉積電極以使電極保持清
潔，從而可防止由於靜電分離過程的反向過程造成氣流的再污染。這種
除塵器可應用在去除率要求極高的情況。

A special design of an *electrostatic separator* for removal of fine mists of process emissions and fine dusts. In a wet electrostatic precipitator the gas stream may

be cold and/or moist. Additional moisture may be introduced through nozzles. The separated droplets of the off-gas are coagulated on the precipitation electrode and are removed as a liquid. The precipitating electrode may be periodically washed with a film of water to keep it clean. This prevents a back-contamination of the gas stream through a reversal of the electrostatic separation process. Applications are in cases where an exceptionally high efficiency is required.

wet flue gas treatment 濕式煙氣處理 *shishi yanqi chuli*

用濕式洗滌法淨化*煙氣*。在*驟冷*和熱回收系統之後，依次組合一些過程步驟來去除酸性氣體氯化氫、溴化氫、二氧化硫和*顆粒物*。可以把此過程設計成由使用不同*洗滌液*的不同*濕式洗滌單元*所組成。此外，還必須包括*微細液滴分離單元*，以減少液滴由上一段攜帶至下一段。為了去除氮氧化物，還需要有*選擇性催化還原過程*。

The purification of *flue gas* by wet scrubbing processes. After heat-quenching (*quench*) and heat recovery systems, a combination of process steps in succession serves to remove acidic gases (HCl, HBr, SO_2) and *particulate matter*. The process may be designed as a combination of different *wet scrubbing units* with different *scrubbing liquors*. In addition, elements for *droplet separation*, to reduce carry-over from stage to stage, are also necessary. To remove oxides of nitrogen (NO_x), techniques such as the *SCR-process* (selective catalytic reduction) are available.

wet sleeve filter 濕式袖式過濾器 *shishi xiushi guolüqi*

一種處理廢氣的*袋式過濾器*，但在氣體入口側通過噴灑洗滌液同時使各個過濾袋增濕。濕式袖式過濾器的分離能力部分基於濕式吸收過程，部分基於真過濾。現在，它只在少數應用中使用。

A *bag filter* for off-gas in which the individual filter bags are at the same time wetted on the gas input side by a spray of a scrubbing liquid. The separation capacity of the wet sleeve filter is based in part on a wet absorption process and in part on true filtration. Today, this design is used only in a few applications.

wetlands 濕地 *shidi*

一種*陸地生態系統*，即沼澤、沼地和淺湖湖岸的特定群落生境。濕地一部分淹沒在水中，一部分是乾地。其動物特徵包括兩棲動物。濕地對人類的影響是非常敏感的。

A *terrestrial ecosystem*, the specific biotope of a swamp, a marsh or the shores of a flat lake. Wetlands are partially submerged in water, partially dry. Characteristic animals of wetlands include amphibians. Wetlands very often are very sensitive to anthropogenic disturbances.

WICE 世界環境工業委員會 *shijie huanjing gongye weiyuanhui*

見：World Industry Council for the Environment。

See *World Industry Council for the Environment.*

winter smog 冬季煙霧 *dongji yanwu*

通常在冬季月份出現的嚴重污染的大氣條件。冬季煙霧最初於1952年在倫敦發現，它主要由濃霧組成，其中含有家庭取暖以及燃煤電廠（煙道氣未充分淨化）排出的二氧化硫、煙灰和煙塵。

A heavily polluted atmospheric condition commonly found in winter months. Winter smog was first recognized as such in about 1952 in London. It is made up mainly of a heavy fog with SO_2, soot and smoke components in emissions from domestic heating and from coal-fired power plants with inadequate flue gas cleaning.

work environment atmosphere 工作環境空氣 *gongzuo huanjing kongqi*

在正常操作條件下工人所接觸的環境空氣。其限值的確定以平均每天接觸8小時為依據。當涉及職業衛生問題時，必須考慮到在一個工作日內往往會出現在不同工作環境中工作一段時間的情況。（參見：職業衛生、最大容許濃度、隨身監測）

The ambient air to which a worker is exposed during normal operations. For the definition of limiting values, it is assumed that the average daily exposure is 8 hours. When dealing with questions of occupational health it has to be taken into account that a work day usually includes several shorter periods in different work environments. (See also *occupational health*; *MAK values*; *man monitoring*)

work hygiene 勞動衛生 *laodong weisheng*

見：occupational health。

See *occupational health*.

workplace measurements 工作場所測定 *gongzuo changsuo ceding*

在實際工作場所對工人的健康具有潛在的不良影響的有害因素進行的測定，重點是測定工作環境空氣中的污染物和噪聲。（參見：作為特殊類型工作場所測定的隨身監測）

Measurements of noxious factors having a potentially negative impact on the health of a worker at his actual workplace. The measurements focus mainly on pollutants in the work atmosphere and on *noise*. (See also *man monitoring* for a special type of workplace measurements)

World Business Council for Sustainable Development (WBCSD) 世界可持續發展工商委員會 *shijie kechixu fazhan gongshang weiyuanhui*

在1995年由*世界環境工業委員會*（WICE）和可持續發展工商委員會（BCSD）合併成立的一個世界範圍的工業組織，它擁有大約120個企業成員，在瑞士日內瓦設有辦公室。其目標是致力於發展中國家和處於過渡時期的國家的可持續發展。具體目的包括：在環境和可持續發展方面推動商業界領導，參與政策的制定，對環境與資源管理方面取得的進展進行示範，以及促進工商界、政府和一切有關組織之間的密切合作。

A world-wide industrial organization which was founded in 1995 through the merger of *WICE* and the Business Council for Sustainable Development, BCSD. The organization has a membership of about 120 enterprises, with offices in Geneva, Switzerland. The WBCSD summarizes its aims as a contribution to a sustainable development of developing nations and nations in transition. Specific objectives encompass the promotion of business leadership, the participation in policy development, the demonstration of progress in environmental and resource management and the development of a close cooperation between business, government and all organizations concerned with the environment and sustainable development.

World Industry Council for the Environment (WICE) 世界環境工業委員會
shijie huanjing gongye weiyuanhui

1993年由*國際商會*發起成立的全世界企業聯合會組織。1995年它與可持續發展工商委員會合併成立了*世界可持續發展工商委員會*。

A global coalition of enterprises initiated in 1993 by the *ICC, International Chamber of Commerce*. In 1995 WICE has merged with the Business Council for Sustainable Development to form the *World Business Council for Sustainable Development*.

WPC 水污染控制 *shuiwuran kongzhi*

見：water pollution control。

See *water pollution control*.

428

X

X-separator for aerosols X型氣溶膠分離器 *aikesixing qirongjiao fenliqi*

　　X型分離器是汽巴－嘉基公司開發的用於空氣污染控制的一種特殊設備的商業名稱。它由許多單個部件構成，用於廢氣流中氣溶膠的分離。

　　它在平壁式部件上並列地安設了具有菱形和三角型截面的兩類塑料棒條，這兩類部件一起開有X型窄縫，在X的中心處窄到1或1.5毫米。詳細地說，這種窄縫截面的形狀形成了*文丘里型氣流系統*，而經預調節的廢氣則從壁的一側通過X型窄縫流到壁的另一側。它基本上是一種*文丘里型系統*。其平壁的尺寸可以根據特殊問題的需要來選定。

'X-separator' is a trade name for a special flat unit in APC-systems developed by CIBA-GEIGY. The X-separator unit is constructed of a number of individual elements and is used for aerosol separation in off-gas streams.

Two types of plastic rods with rhombus-like and triangular cross-section are arranged side-by-side in flat wall-like elements. The two types of elements together leave open X-shaped slits, narrowed to 1 or 1.5 mm at the center of the X. In detail, the shape of the cross-section of the slits forms a *venturi type air flow system*. Preconditioned off-gas passes from one side of the wall to the other through the X-shaped slits. It is basically a *venturi type system*. The dimension of the X-separator wall can be fitted to the specific problem.

xenobiotica 異生化合物 *yisheng huahewu*

　　生物體系中外來的物質。

Substances which are foreign to living systems.

Z

Zahn-Wellens Test 傑恩-魏倫斯測試 *jie'en-weilunsi ceshi*

見：OECD-BOD Test。

See *OECD-BOD Test*.

zeolite 沸石 *fushi*

具有特殊晶體結構的一類天然礦物的集合名稱。其結構能使晶格和溶液之間進行某些陽離子的交換。沸石是天然的離子交換樹脂。（參見：離子交換）

A collective name for a group of natural minerals with a specific crystal structure. This structure allows for the exchange of certain cations between the crystal lattice and a solution. Zeolites are natural ion exchange resins. (See also *ion exchange*)

zero-pollution goal 零污染目標 *lingwuran mubiao*

把某個國家中一切人為活動造成的污染減少到零的政治目標。這一目標是七十年代環境方面討論的政治議題。由於經濟和能源消費方面的原因，零污染是不可能實現的。

The political goal of reducing pollution of all human activities in a country to absolutely zero. This goal was a topic for political discussion in the early environmental area of the 1970s. For economic and energy consumption reasons, zero-pollution is not achievable.

zero-risk goal 零風險目標 *lingfengxian mubiao*

把人類活動造成的一切殘留風險減少到零的目標。實際上，由於經濟和社會的影響，這一目標是不可能實現的。

The goal to reduce all residual risks to man resulting from *anthropogenic activities* to zero. In reality, this goal cannot be reached without unreasonable and unacceptable economic and societal follow-up effects.

zoning laws 區劃法 *quhuafa*

見：municipal zoning laws。

See *municipal zoning laws*.

zoocenose 動物群落 *dongwu qunluo*

某群落生境中各種不同的動物群體。（參見：生物群落、植物群落）

The different populations of animals (fauna) in a given biotope. (See also *biocenose*; *phytocenose*)

zooplankton 浮游動物 *fuyou dongwu*

微小可漂浮的但不能靠自己作長距離移動的*水生異養生物*（動物）。*浮游動物以浮游植物為食物*，因此它是食物鏈中的第二級。浮游動物包括諸如各種水生動物的幼蟲、小蝦、極小的甲殼類動物、海蜇和某些*原生動物*。

Small drifting or floating aquatic *heterotrophic organisms* (animals) which are not able to move around over large distances by themselves. Zooplankton is feeding on *phytoplankton* organisms and is thus the second level of the food chain. Zooplankton includes for example larvae of many different aquatic species, small shrimp, tiny crustaceans, jellyfish and also some forms of *protozoa*.

CHINESE-ENGLISH GLOSSARY

with Chinese entries arranged by the number of strokes of characters

The term in each entry is followed by its pinyin pronunciation in italics and then by its English equivalent. Please refer to the English-Chinese glossary for explanation of the terms.

此中英詞匯中之中文名詞以筆劃排序。每個中文名詞後是其漢語拼音，跟著是其英文名詞。請於英中詞匯中查閱各名詞之解釋。

1972年斯德哥爾摩會議	*yijiuqi'ernian sidege'ermo huiyi*	Stockholm Conference 1972
1989年巴塞爾公約	*yijiubajiunian basai'er gongyue*	Basel Convention of 1989
1992年里約會議	*yijiujiu'ernian liyue huiyi*	RIO Conference 1992
BOD_5/COD比	*BOD_5/COD bi*	BOD_5/COD ratio
DOC/TOC比	*DOC/TOC bi*	DOC/TOC ratio
pH	*pH*	pH
pH計	*pHji*	pH meter
X型氣溶膠分離器	*aikesixing qirongjiao fenliqi*	X-separator for aerosols

一　劃

一次能源	*yici nengyuan*	energy, primary sources
一次澄清池	*yici chengqingchi*	primary clarifier
一般的催化過程	*yibande cuihua guocheng*	catalytic processes in general
一般管理中的控制	*yiban guanlizhongde kongzhi*	controlling in general management
一般管理費用	*yiban guanli feiyong*	overhead costs
一般管理原則	*yiban guanli yuanze*	management principles in general

二　劃

二元參比氣體混合器	*eryuan canbi qiti hunheqi*	binary reference gas mixer
二次澄清池	*erci chengqingchi*	secondary clarifier
二氧化碳	*eryanghuatan*	carbon dioxide
二氧化碳排放收費	*eryanghuatan paifang shoufei*	discharge levy for CO_2

人為排放	*renwei paifang*	anthropogenic emissions
人員參與制度	*renyuan canyu zhidu*	personnel participation schemes
人員激勵制度	*renyuan jili zhidu*	personnel incentive systems
人(類)的活動	*ren (lei) de huodong*	anthropogenic activities
十億分之(幾分)	*shiyifenzhi (jifen)*	ppb

三　劃

下水道系統	*xiaoshuidao xitong*	sewer systems
下游循環	*xiayou xunhuan*	downcycling
千分率	*qianfenlü*	per mille or per mil
土壤	*turang*	soil
土壤生態系統	*turang shengtai xitong*	soil ecosystem
大氣	*daqi*	atmosphere
大腸菌數	*dachangjunshu*	coliform count
小生境	*xiaoshengjing*	ecological niche
工廠的費用中心結構（計劃）	*gongchangde feiyong zhongxin jiegou (jihua)*	cost center structure (plan) of a factory
工廠的指揮機構	*gongchangde zhihui jigou*	line organization of the plant
工廠排放清單	*gongchang paifang qingdan*	inventory of emissions of a factory
工藝廢氣	*gongyi feiqi*	process gases
工藝廢水	*gongyi feishui*	process effluents
工藝措施	*gongyi cuoshi*	in-process measures
工業衛生	*gongye weisheng*	industrial hygiene
工業廢水	*gongye feishui*	industrial effluent
工業廢水處理場	*gongye feishui chulichang*	industrial effluent treatment plant
工作目標或限值	*gongzuo mubiao huo xianzhi*	performance goals or limits
工作場所測定	*gongzuo changsuo ceding*	workplace measurements
工作報告	*gongzuo baogao*	performance report
工作環境空氣	*gongzuo huanjing kongqi*	work environment atmosphere
工作標準	*gongzuo biaozhun*	performance standards
工作職責說明	*gongzuo zhize shuoming*	job description
工程項目評價，環境保護	*gongcheng xiangmu pingjia, huanjing baohu*	project assessment, environmental protection

四 劃

不可生物降解化合物	*bukeshengwu jianjie huahewu*	non-biodegradable compounds
不可再生資源	*bukezaisheng ziyuan*	non-renewable resources
不同形式的能量	*butong xingshide nengliang*	energies
不合格產品	*buhege chanpin*	off-specification products
中空纖維膜	*zhongkong qianweimo*	hollow fiber membrane
中和當量	*zhonghe dangliang*	neutralization equivalents
互聯數據庫	*hulian shujuku*	interconnected databases
五天生物需氧量	*wutian shengwu xuyangliang*	BOD
五天生物需氧量分析	*wutian shengwu xuyangliang fenxi*	BOD5 analysis
內部化的外部費用	*neibuhuade waibu feiyong*	internalizing external costs
內部的污染者付費原則	*neibude wuranzhe fufei yuanze*	internal polluter-pays-principle
內部費用	*neibu feiyong*	internal costs
公司內部收費	*gongsi neibu shoufei*	company internal levy
公司內部的影響評價	*gongsi neibude yingxiang pingjia*	company internal impact assessment
公司的利害關係者	*gongside lihai guanxizhe*	stakeholders of a company
公司環境政策	*gongsi huanjing zhengce*	company environmental policy
公正實施	*gongzheng shishi*	equitable enforcement
公共利益團體	*gonggong liyi tuanti*	public interest groups
分子碎片的重組	*fenzi suipiande chongzu*	recombination of molecular fragments
分貝(A聲級)	*fenbei (A shengji)*	decibel (A)
分光光度測定法	*fenguang guangdu cedingfa*	spectrophotometry
分析中的分離段	*fenxizhongde fenliduan*	separation step in analysis
分析中的偶然誤差	*fenxizhongde ouran wucha*	random error in analysis
分析中的檢測方法	*fenxizhongde jiance fangfa*	detection methods in analysis
分析中的富集步驟	*fenxizhongde fuji buzhou*	enrichment step in analysis
分析中的鑒定段	*fenxizhongde jiandingduan*	identification step in analysis
分析的質量標準	*fenxide zhiliang biaozhun*	quality criteria of analysis
分析的信噪比	*fenxide xinzaobi*	signal-to-noise ratio
分析結果的離散性	*fenxi jieguode lisanxing*	dispersion of analytical results
分析程序的校準	*fenxi chengxude jiaozhun*	calibration of analytical procedures
分流式下水道系統	*fenliushi xiashuidao xitong*	separation sewer systems

分離	*fenli*	separation
化工過程的停留時間	*huagong guochengde tingliu shijian*	residence time in chemical processing
化學吸收	*huaxue xishou*	chemisorption
化學廢物	*huaxue feiwu*	chemical waste
化學濕式分析	*huaxue shishi fenxi*	chemical wet analysis
化學需氧總量	*huaxue xuyang zongliang*	COD load
化學需氧量	*huaxue xuyangliang*	chemical oxygen demand
化學需氧量分析	*huaxue xuyangliang fenxi*	COD analysis
升流式厭氧污泥床處理	*shengliushi yanyang wuni chuang chuli*	upflow anaerobic sludge blanket treatment
反應氣	*fanyingqi*	reaction-gases
反硝化(作用)	*fanxiaohua (zuoyong)*	denitrification
反硝化－硝化過程	*fanxiaohua-xiaohua guocheng*	denitrification-nitrification process
文丘里型洗滌器	*wenqiulixing xidiqi*	venturi type scrubbers
文丘里管道	*wenqiuli guandao*	venturi canal
方法的回收率	*fangfade huishoulü*	recovery rate of a method
方法的檢測極限	*fangfade jiance jixian*	detection limit of a method
比處置費用	*bichuzhi feiyong*	specific disposal costs
比色分析	*bise fenxi*	colorimetric analysis
比吸附表面	*bixifu biaomian*	adsorption surface
比廢物量	*bifeiwuliang*	specific waste quantity
比(單位)費用	*bi (danwei) feiyong*	specific costs
毛細管氣相色譜法	*maoxiguan qixiang sepufa*	capillary gas chromatography
毛細管柱	*maoxiguanzhu*	capillary column
水	*shui*	water
水中溶解氧	*shuizhong rongjieyang*	dissolved oxygen in water
水生生物毒理學	*shuisheng shengwu dulixue*	aquatic toxicology
水污染控制	*shuiwuran kongzhi*	water pollution control
水污染控制觀念	*shuiwuran kongzhi guannian*	water pollution control concept
水污染控制技術	*shuiwuran kongzhi jishu*	water pollution control technology
水體自淨	*shuiti zijing*	self-purification of a waterbody
水體的自動再生	*shuitide zidong zaisheng*	auto regeneration of a waterbody
水質目標	*shuizi mubiao*	water quality goals

水圈	*shuiquan*	hydrosphere
水域管理	*shuiyu guanli*	water management basin
水數據庫	*shui shujuku*	water database
火焰電離檢測器	*huoyan dianli jianceqi*	flame ionization detectors
火焰電離檢測器分析儀	*huoyan dianli jianceqi fenxiyi*	FID analyzer
火焰等離子體	*huoyan dengliziti*	flame plasma

五　劃

世界可持續發展工商委員會	*shijie kechixu fazhan gongshang weiyuanhui*	World Business Council for Sustainable Development
世界環境工業委員會	*shijie huanjing gongye weiyuanhui*	World Industry Council for the Environment
冬季煙霧	*dongji yanwu*	winter smog
包裝廢物	*baozhuang feiwu*	packaging waste
半連續氣相色譜法	*banlianxu qixiang sepufa*	semi-continuous gas chromatography
半抑制濃度	*banyizhi nongdu*	IC-50
半作用濃度	*banzuoyong nongdu*	EC-50
半致死濃度	*banzhisi nongdu*	LC-50
半透膜	*bantoumo*	semi-permeable membrane
去所	*qusuo*	sink
可氯提的有機鹵素化合物	*keqitide youji lusu huahewu*	POX
可生物降解性的連續測定(哈斯曼試驗)	*keshengwu jiangjiexingde lianxu ceding (hasiman shiyan)*	continuous measurement of biodegradability (Husmann Test)
可生物降解的	*keshengwu jiangjiede*	biodegradable
可再生資源	*kezaisheng ziyuan*	renewable resources
可再生能源	*kezaisheng nengyuan*	renewable energy
可吸附的有機鹵素化合物	*kexifude youji lusu huahewu*	absorbable organic halogen compounds
可吸附的有機硫	*kexifude youjiliu*	AOS
可持續發展	*kechixu fazhan*	sustainable development
可獲得而經濟的最佳技術	*kehuode 'er jingjide zuijia jishu*	best available technology economically achievable
可獲得的最佳技術	*kehuodede zuijia jishu*	best available technology

可提取的有機鹵素化合物	*ketiqude youji lusu huahewu*	EOX
可靠性	*kekaoxing*	reliability
外部環境審計	*waibu huanjing shenji*	external environmental audit
外部費用	*waibu feiyong*	external costs
平爐焚燒爐	*pinglu fenshaolu*	open hearth incinerator
平板沉積測試	*pingban chenji ceshi*	plate deposition test
平流層臭氧	*pingliuceng chouyang*	stratospheric ozone
平衡	*pingheng*	equilibrium
末端工藝過程處理	*moduan gongyi guocheng chuli*	end-of-process treatment
末端處理措施	*moduan chuli cuoshi*	end-of-pipe measures
末端改善措施	*moduan gaishan cuoshi*	add-on corrective measures
母液	*muye*	mother liquor
瓦爾德之原則	*wa'erdezhi yuanze*	Valdez Principles
生產工藝文件資料	*shengchan gongyi wenjian ziliao*	process documentation
生產費用	*shengchan feiyong*	manufacturing costs
生命周期分析	*shengming zhouqi fenxi*	life cycle analysis
生命科學	*shengming kexue*	life sciences
生態化學	*shengtai huaxue*	ecochemistry
生態平衡	*shengtai pingheng*	ecobalance
生態平衡	*shengtai pingheng*	ecological equilibrium
生態設計	*shengtai sheji*	eco-design
生態因素	*shengtai yinsu*	ecological factors
生態系統	*shengtai xitong*	ecosystem
生態系統中的生物組成部分	*shengtai xitongzhongde shengwu zucheng bufen*	biotic components of ecosystems
生態系統中的物流循環	*shengtai xitongzhongde wuliu xunhuan*	mass flow cycles in ecosystems
生態系統中的最終去所	*shengtaixitongzhongde zuizhong qusuo*	final sink in the ecosystem
生態系統的非生物組成部分	*shengtai xitongde feishengwu zucheng bufen*	abiotic components of ecosystem
生態評級	*shengtai pingji*	eco-rating
生態學	*shengtaixue*	ecology
生態毒性	*shengtai duxing*	ecotoxicity
生態種群	*shengtai zhongqun*	ecological population

生態圈	*shengtaiquan*	ecosphere
生態基線評價	*shengtai jixian pingjia*	ecological base line assessment
生態群落	*shengtai qunluo*	ecological community
生物反應器(流化床)	*shengwu fanyingqi (liuhuachuang)*	bioreactor, fluidized bed
生物氣(沼氣)	*shengwuqi (zhaoqi)*	biogas
生物處理	*shengwu chuli*	biological treatment
生物轉盤	*shengwu zhuanpan*	rotating biological contactor
生物處理池的隔間	*shengwu chulichide gejian*	compartments in biological basins
生物平衡	*shengwu pingheng*	biotic equilibrium
生物生態因素	*shengwu shengtai yinsu*	biotic ecological factors
生物污泥	*shengwu wuni*	biosludge
生物污泥/活性碳系統	*shengwu wuni/huoxingtan xitong*	biosludge/activated carbon systems
生物多樣性	*shengwu duoyangxing*	biological diversity
生物殺傷劑	*shengwu shashangji*	biocides
生物濾池	*shengwu lüchi*	biofilter
生物技術	*shengwu jishu*	biotechnology
生物學	*shengwuxue*	biology
生物法空氣污染控制系統	*shengwufa kongqi wuran kongzhi xitong*	biological APC-systems
生物法恢復補救	*shengwufa huifu bujiu*	bio-remediation
生物放大	*shengwu fangda*	biomagnification
生物降解	*shengwu jiangjie*	biodegradation
生物降解的抑制	*shengwu jiangjiede yizhi*	inhibition of biodegradation
生物降解速率	*shengwu jiangjie sulü*	biodegradation rate
生物洗滌器(廢氣)	*shengwu xidiqi (feiqi)*	bio scrubber, off-gas
生物活性碳吸附裝置	*shengwu huoxingtan xifu zhuangzhi*	biological activated carbon adsorber
生物測定	*shengwu ceding*	bio-assay
生物科學	*shengwu kexue*	bio sciences
生物適應(馴化)	*shengwu shiying (xunhua)*	bio adaptation
生物監測	*shengwu jiance*	biomonitoring
生物圈	*shengwuquan*	biosphere
生物堆肥	*shengwu duifei*	bio-compost
生物富集	*shengwu fuji*	bioconcentration

生物量	shengwuliang	biomass
生物群落	shengwu qunluo	biocenose
生物需氧量	shengwu xuyangliang	biological oxygen demand
生物膜	shengwumo	biofilm
生活質量	shenghuo zhiliang	quality of life
生境	shengjing	habitat
用於填料塔的 刺猬型填料	yongyu tianliaotade ciweixing tianliao	hedge hog filling for packed columns
用無脊椎動物作為試驗 生物測定的廢水毒性	yongwujizhuidongwu zuowei shiyan shengwu cedingde feishui duxing	effluent toxicity with invertebra as test organisms
由微生物礦質化	you weishengwu kuangzhihua	mineralization by microorganisms
目標管理	mubiao guanli	management by objectives
立法中的排污收費	lifazhongde paiwu shoufei	discharge levy in legislation

六 劃

光化學煙霧	guanghuaxue yanwu	photochemical smog
光電太陽能	guangdian taiyangneng	photovoltaic solar energy
光傳感器	guangchuanganqi	optical sensor
光達	guangda	LIDAR
光度計	guangduji	photometer
光氧化劑	guangyanghuaji	photooxidants
全球大氣的二氧化碳	quanqiu daqide eryanghuatan	global atmospheric carbon dioxide
共生	gongsheng	symbiosis
再生技術	zaisheng jishu	regenerative technologies
再利用	zailiyong	reuse
危險廢物	weixian feiwu	hazardous waste
同化作用	tonghua zuoyong	assimilation
合同	hetong	contracts
合格審計者	hege shenjizhe	certified auditor
合格的環境實驗室	hegede huanjing shiyanshi	certified environmental laboratory
因果關係	yinguo guanxi	cause-effect relationship
回收期	huishouqi	pay back time
回轉窯	huizhuanyao	rotary kiln

回轉窯焚燒爐	*huizhuanyao fenshaolu*	rotary kiln incinerator
地下水	*dixiashui*	groundwater
地下水水質數據	*dixiashui shuizhi shuju*	groundwater quality data
地下水污染	*dixiashui wuran*	groundwater contamination
地下水位計	*dixiashuiweiji*	piezometer
地下水補救恢復措施	*dixiashui bujiuhuifu cuoshi*	groundwater remediation
地下水保護區	*dixiashui baohuqu*	groundwater protection zones
地下水監測	*dixiashui jiance*	groundwater monitoring
地下鹽礦中廢物堆存	*dixia yankuangzhong feiwu duichun*	underground waste deposits in salt mines
在線分析	*zaixian fenxi*	on-line analysis
多功能設備	*duogongneng shebei*	multi-purpose equipment
多爐床焚燒爐	*duoluchuang fenshaolu*	multiple hearth incinerator
多環芳烴	*duohuanfangting*	PAH
多勃遜單位	*duobosun danwei*	Dobson Units
多氯聯苯	*duolülianben*	PCB
多氯聯苯廢物	*duolülianben feiwu*	PCB-wastes
好氧生物降解	*haoyang shengwu jiangjie*	aerobic biodegradation
好氧過程	*haoyang guocheng*	aerobic process
安全審計	*anquan shenji*	safety audit
安全貯存區	*anquan chucunqu*	secure storage area
安全測試	*anquan ceshi*	safety tests
安全填埋	*anquan tianmai*	secure landfill
成本	*chengben*	costs
成本計算方法	*chengben jisuan fangfa*	cost calculation procedure
成本價格	*chengben jiage*	cost covering prices
成問題的廢物	*chengwentide feiwu*	problem wastes
收費	*shoufei*	levy
有關產品和產品使用的法規	*youguan chanpin he chanpin shiyongde fagui*	legislation on products and product use
有關營養的	*youguan yingyangde*	trophic
有規填料	*yougui tianliao*	structured packing
有規填料塔	*yougui tianliaota*	structurally packed column
有毒廢物	*youdu feiwu*	toxic waste
污水	*wushui*	sewage
污水處理廠	*wushui chulichang*	sewage treatment works

污水流量（設計流量）	*wushui liuliang (sheji liuliang)*	sewage flow volumes (design values)
污水管理區	*wushui guanliqu*	sewage management districts
污泥	*wuni*	sludge
污泥衛生處理	*wuni weisheng chuli*	sludge hygienization
污泥乾化床	*wuni ganhuachuang*	sludge drying bed
污泥發酵	*wuni fajiao*	sludge fermentation
污泥處理	*wuni chuli*	sludge treatment
污泥滅菌	*wuni miejun*	sterilization of sludge
污泥厭氧消化	*wuni yanyang xiaohua*	sludge digestion, anaerobic
污泥回流	*wuni huiliu*	sludge recycling
污泥好氧消化	*wuni haoyang xiaohua*	sludge digestion, aerobic
污泥過濾	*wuni guolü*	sludge filtration
污泥體積指數	*wuni tiji zhishu*	sludge volume index
污泥泥齡	*wuni niling*	sludge age
污泥揮發分濃度（混合液揮發性懸浮固體）	*wuni huifafen nongdu (hunheye huifaxing xuanfuguti)*	MLVSS
污泥濃度（混合液懸浮固體）	*wuni nongdu (hunheye xuanfuguti)*	MLSS
污泥濃縮池	*wuni nongsuchi*	sludge thickener
污泥惡臭	*wuni echou*	sludge fouling
污泥離心分離	*wuni lixin fenli*	sludge centrifugation
污泥調節	*wuni tiaojie*	sludge conditioning
污泥脫水	*wuni tuoshui*	sludge dewatering
污染產生者	*wuran changshengzhe*	generator of pollution
污染負荷	*wuran fuhe*	pollution load
污染防治設備	*wuran fangzhi shebei*	anti-pollution equipment
污染物擴散	*wuranwu kuosan*	dispersion of a pollutant
污染物	*wuranwu*	contaminant
污染物	*wuranwu*	pollutant
污染者	*wuranzhe*	polluter
污染者之鏈	*wuranzhezhilian*	chain of polluters
污染者付費原則	*wuranzhe fufei yuanze*	polluter-pays-principle
污染預防	*wuran yufang*	pollution prevention
污染控制中的現代工藝技術	*wuran kongzhizhongde xiandai gongyi jishu*	state-of-the-art in pollution control

污染控制技術	wuran kongzhi jishu	pollution control technologies
污染控制的市場手段	wuran kongzhide shichang shouduan	market instruments in pollution control
污染控制的標準單位成本	wuran kongzhide biaozhun danwei chengben	standard unit costs for pollution control
污染控制費用的分攤	wuran kongzhi feiyongde fentan	allocation of pollution control costs
灰能	huineng	grey energy
百萬分之幾(體積比)	baiwanfenzhiji (tijibi)	ppm by volume
百萬分之(幾分)	biawanfenzhi (jifen)	ppm
百分比	baifenbi	percent
自動取樣器	zidong quyangqi	autosampler
自養生物	ziyang shengwu	autotrophic organisms
色譜法	sepufa	chromatography
行業合同	hangye hetong	branch contract
行業契約	hangye qiyue	branch covenant

七　劃

作用濃度	zuoyong nongdu	effect concentration
克氏氮分析	keshidan fenxi	Kjeldahl nitrogen analysis
冷卻水	lengqueshui	cooling water
冷卻和液化	lengque he yehua	cooling and liquefaction
冷等離子體電極	lengdengliziti dianji	cold plasma electrode
冷壁式焚燒爐	lengbishi fenshaolu	cold wall incinerator
冷壁式焚燒爐	lengbishi fenshaolu	incinerator, cold wall
社會費用原則	shehui feiyong yuanze	social cost principle
利用植物恢復補救	liyong zhiwu huifu bujiu	phytoremediation
助濾劑	zhulüji	filter aid
吸收(作用)	xishou (zuoyong)	absorption
吸附(作用)	xifu (zuoyong)	adsorption
吸附劑	xifuji	adsorbent
吸附塔	xifuta	adsorption columns
吸著過程	xizhao guocheng	sorption processes
含鹽量	hanyanliang	salinity
均衡	junheng	equalization
均衡池	junhengchi	equalization tanks

完整的工藝過程開發	wanzhengde gongyi guocheng kaifa	integral process development
宏觀經濟問題	hongguan jingji wenti	macroeconomic aspects
尾氣氧化處理	weiqi yanghua chuli	oxidative off-gas treatment
技術的管理	jishude guanli	management of technology
折板	zheban	deflector elements
投資收益率	touzi shouyilü	return on investment
投資費用	touzi feiyong	investment costs
抑制濃度 (活性污泥)	yizhi nongdu (huoxing wuni)	inhibitory concentration (activated sludge)
沉降	chenjiang	sedimentation
沉降除塵箱	chenjiang chuchenxiang	settling box dust eliminator
沉降速度	chenjiang sudu	sedimentation velocity
沉降箱	chenjiangxiang	drop out box
沉積電極	chenji dianji	precipitation electrode
沉澱劑	chendianji	precipitating agent
決策方法 (技術)	juece fangfa (jishu)	decision techniques
決策因子	juece yinzi	decision factors
汽車尾氣排放	qiche weiqi paifang	traffic emissions
汽車催化轉化器	qiche cuihua zhuanhuaqi	automotive catalytic converters
系統邊界	xitong bianjie	system boundary
系統誤差	xitong wucha	systematic error
辛醇-水系數	xinchun-shui xishu	octanol-water coefficient

八　劃

並流	bingliu	co-current
乳液	ruyi	emulsion
乳液分離	ruye fenli	separation of emulsions
免疫測定法分析	mianyi cedingfa fenxi	immunoassay analysis
免費物品	mianfei wupin	free goods
事故報告渠道	shigu baogao qudao	reporting channels in accidents
事故貯水池	shigu chushuichi	retention basins
事故控制計劃	shigu kongzhi jihua	accident control plan
事故遏制措施	shigu ezhi cuoshi	containment measures
事故釋放	shigu shifang	accidental releases
事故影響	shigu yingxiang	impact of an accident
亞硝化桿菌	yaxiaohua ganjun	nitrosomonas bacteria

具有生態效益的活動或行動	juyou shengtai xiaoyide huodong huo xingdong	ecoefficient activities or actions
刺激(獎勵)制度	ciji (jiangli) zhidu	incentive systems
取樣和測量口	quyang he celiangkou	sampling and measuring ports
取樣程序	quyang chengxu	sampling procedure
受納水體	shouna shuiti	receiving waterbody
受控處置場	shoukong chuzhichang	controlled disposal site
受控填埋場	shoukong tianmaichang	controlled landfill
呼吸測定法	huxi cedingfa	respirometry
周轉率	zhouzhuanlü	turnover rate
固體廢物	guti feiwu	solid waste
固體特殊廢物	guti teshu feiwu	solid special waste
固定化廢物	gudinghua feiwu	immobilized waste
固定床生物反應器	gudingchuang shengwu fanyingqi	fixed bed bioreactor
固定床系統	gudingchuang xitong	fixed bed systems
固定膜生物系統	gudingmo shengwu xitong	fixed film biological systems
固相萃取	guxiang cuiqu	solid phase extraction
垃圾袋處置費	lajidai chuzhifei	garbage bag disposal fee
岩石圈	yanshiquan	lithosphere
帕克斯系統	pakesi xitong	Paques System
底棲生態系統	diqi shengtai xitong	benthnic ecosystem
底棲生物	diqi shengwu	benthon, benthos
延時曝氣系統	yanshi baoqi xitong	extended aeration systems
拆建廢物	chaijian feiwu	demolition waste
林多克斯系統	linduokesi xitong	LINDOX system
板式塔	banshita	plate columns
泥漿	nijiang	slurry
河口生態系統	hekou shengtai xitong	estuarine ecosystem
河流生態系統中的指示生物	heliu shengtai xitongzhongde zhishi shengwu	bioindicators in river ecosystems
河流質量	heliu zhiliang	river quality
河流質量目標	heliu zhiliang mubiao	river quality goal
河流流域管理部門	heliu liuyu guanli bumen	river basin authority
法規	fagui	legislation
法規符合審計	fagui fuhe shenji	compliance audit
法律手段	falü shouduan	legal instruments

法律責任	*falü zeren*	legal responsibility
沸石	*fushi*	zeolite
泄漏	*xielou*	spills
泡沫控制	*paomo kongzhi*	foam control
泡罩原理	*paozhao yuanli*	bubble principle
泡罩塔	*paozhaota*	bubble column
泡罩塔洗滌器	*paozhaota xidiqi*	bubble column scrubber
物質守恆定律	*wuzhi shouheng dinglü*	law on the conservation of material
物料平衡	*wuliao pingheng*	material balance
物流	*wuliu*	mass flow
直接污染控制費用	*zhijie wuran kongzhi feiyong*	direct pollution control costs
直接費用	*zhijie feiyong*	direct costs
空氣	*kongqi*	air
空氣氣提	*kongqiqiti*	air stripping
空氣污染	*kongqi wuran*	air pollution
空氣污染的指示生物	*kongqi wurande zhishi shengwu*	air pollution bioindicators
空氣污染造成的森林破壞	*kongqi wuran zaochengde senlin pohuai*	forest damages through air pollution
空氣污染控制	*kongqi wuran kongzhi*	air pollution control
空氣污染控制中的催化過程	*kongqi wuran kongzhizhongde cuihua guocheng*	catalytic processes in APC
空氣污染控制觀念	*kongqi wuran kongzhi guannian*	APC-concept
空氣污染控制設備	*kongqi wuran kongzhi shebei*	air pollution control equipment
空氣污染控制技術	*kongqi wuran kongzhi jishu*	air pollution control technology
空氣污染控制裝置的壓差	*kongqi wuran kongzhi zhuangzhi de yacha*	pressure difference in APC units
空氣和氧噴射系統	*kongqiheyang penshe xitong*	air and oxygen injection systems
空氣質量目標	*kongqi zhiliang mubiao*	air quality goals
表面再生	*biaomian zaisheng*	surface regeneration
表面曝氣機	*biaomian baoqiji*	surface aerator
雨水	*yushui*	meteor water
雨水溢流	*yushui yiliu*	rainwater overflow
非反應性廢物	*feifanyingxing feiwu*	non-reactive wastes
非生物的	*feishengwude*	abiotic
非生物的生態因子	*feishengwude shengtai yinzi*	abiotic ecological factors
非安全填埋	*feianquan tianmai*	non-secure landfill

非色散紅外分析	*feisesan hongwai fenxi*	infrared analysis, non-dispersive
非經濟的法律手段	*feijingjide falü shouduan*	non-economic legal instruments
非政府組織	*feizhengfu zuzhi*	NGO
非敏感性催化劑	*feiminganxing cuihuaji*	non-sensitive catalysts

九　劃

前澄清池	*qianchengqingchi*	preclarifier
哈龍	*halong*	Halon
哈斯曼試驗	*hasiman shiyan*	Husmann Test
城市區劃法	*chengshi quhuafa*	municipal zoning laws
城市污水	*chengshi wushui*	municipal sewage
城市污水處理場	*chengshi wushui chulichang*	municipal effluent treatment plant
城市垃圾	*chengshi laji*	municipal trash
城市廢物處置費	*chengshi feiwu chuzhifei*	municipal waste disposal fees
城市廢物管理	*chengshi feiwu guanli*	municipal waste management
城市(廢物)焚燒爐	*chengshi (feiwu) fenshaolu*	municipal incinerator
室內空氣	*shineikongqi*	room air
建議制度(人事管理)	*jianyi zhidu (renshi guanli)*	suggestion system (personnel management)
建議箱	*jianyixiang*	suggestion box
建設工程的生態評價	*jianshe gongchengde shengtai pingjia*	ecological assessment (of a project)
建築廢物	*jianzhu feiwu*	construction waste
後燃燒室	*houranshaoshi*	post-combustion chamber
急性效應	*jixing xiaoying*	acute effect
恢復	*huifu*	rehabilitation
恢復/補救	*huifu/bujiu*	remediation
恢復/補救目標	*huifu/bujiu mubiao*	remediation goals
恢復補救費用	*huifu bujiu feiyong*	costs of remediation
恢復補救，項目管理	*huifu bujiu, xiangmu guanli*	remediation, project management
恢復措施	*huifu cuoshi*	rehabilitation measures
持有證書的檢查者	*chiyou zhengshude jianchazhe*	certified verifier
指示生物	*zhishi shengwu*	bioindicators
指揮職責	*zhihui zhize*	line responsibility
指揮職能	*zhihui zhineng*	line function

447

指標(參數)	zhibiao (canshu)	parameter
染色廢水	ranse feishui	dye house effluents
柱上注射器(氣相色譜法中)	zhushang zhusheqi (qixiang sepufazhong)	on-column injector (in GC)
毒物	duwu	toxicant
毒理學	dulixue	toxicology
氟里昂	fuli'ang	FREON
洗滌液	xidiye	scrubbing liquid
洗滌器	xidiqi	scrubber
洗脱液測試	xiduoye ceshi	eluate test
活化池(塔)	huohuachi (ta)	activation basin (tower)
活性污泥	huoxing wuni	activated sludge
活性污泥池	huoxingwunichi	activated sludge basin
活性污泥BOD₅負荷	huoxingwuni BOD₅ fuhe	BOD₅ loading of biosludge
活性污泥技術	huoxingwuni jishu	activated sludge technology
活性污泥的適應(馴化)	huoxing wunide shiyin (xunhua)	adaptation of activated sludge
活性污泥的熱衛生處理	huoxing wunide reweisheng chuli	thermal hygienization of activated sludge
活性表面	huoxing biaomian	active surface
活性物質	huoxingwuzhi	active substance
活性碳	huoxingtan	activated carbon
活性碳再生	huoxingtan zaisheng	regeneration of activated carbon
活性碳吸附	huoxingtan xifu	adsorption on activated carbon
活性碳緩衝系統	huoxingtan huanchong xitong	activated carbon buffer system
活性碳解吸	huoxingtan jiexi	desorption from activated carbon
洛普羅克斯技術(拜耳公司)	luopuluokesi jishu (bai'er gongsi)	LOPROX technology (Bayer)
相分離	xiangfenli	phase separation
相關系數	xiangguan xishu	correlation coefficient
相界面	xiangjiemian	phase boundary
砂礫	shali	grit
砂濾池	shalüchi	sandfilter
科特雷爾靜電分離器	ketelei'er jingdian fenliqi	Cottrell separator, electrostatic
穿透點	chuantoudian	breakthrough point
紅外分析	hongwai fenxi	IR
紅外光譜分析	hongwai guangpu fenxi	infrared spectral analysis

美國材料試驗學會標準	*meiguo cailiao shiyan xuehui biaozhun*	ASTM Standards
美國政府職業安全與衛生署	*meiguo zhengfu zhiye anquan yu weisheng shu*	OSHA
耐火襯	*naihuochen*	refractory lining
背景噪聲	*beijing zaosheng*	background noise
負離子化電極（靜電除塵器）	*fulizihua dianji (jingdian chuchenqi)*	negative ionizing electrode (electrostatic precipitator)
負荷限值	*fuhe xianzhi*	load limits
負載限值	*fuzai xianzhi*	freight limits
重力沉降箱	*zhongli chenjiangxiang*	gravity settling box
重組過程	*chongzu guocheng*	recombination processes
重金屬	*zhongjinshu*	heavy metals
降水水	*jiangshui shui*	precipitation water
風險	*fengxian*	risk
風險分析	*fengxian fenxi*	risk analysis
風速計	*fengsuji*	anemometer
食肉動物	*shirou dongwu*	carnivore
食物／生物量比	*shiwu/shengwuliang bi*	food to biomass ratio
食物鏈	*shiwulian*	food chain
食物鏈中的積累	*shiwulianzhongde jilei*	accumulation in the food chain
食草動物	*shicao dongwu*	herbivore

十 劃

致死效應	*zhisi xiaoying*	lethal effect
致癌多環芳烴	*zhi'ai duohuanfangting*	cPAH
借貸的基建費用	*jiedaide jijian feiyong*	costs of borrowed capital
個人監測	*geren jiance*	personal monitoring
個體生態學	*geti shengtaixue*	autecology
修正措施	*xiuzheng cuoshi*	corrective measures
原子吸收分析	*yuanzi xishou fenxi*	atomic absorption analysis
原子吸收火焰光譜法	*yuanzi xishou huoyan guangpufa*	atomic absorption flame spectrometry
原生動物	*yuansheng dongwu*	protozoa
原生物	*yuanshengwu*	protists
原料收費	*yuanliao shoufei*	levy on raw materials

原核細胞	yuanhe xibao	procaryotic cells
夏季煙霧	xiaji yanwu	summer smog
夏季煙霧中的臭氧	xiaji yanwuzhongde chouyang	ozone in summer smog
容許日攝入量	rongxu risheruliang	acceptable daily intake
射流曝氣系統	sheliu baoqi xitong	jet aeration system
捕食者	bushizhe	predator
捕食者與被捕食者的 關係	bushizhe yu beibushizhe de guanxi	predator-prey relationship
效益－費用分析	xiaoyi-feiyong fenxi	benefit-cost analysis
時間平均樣	shijian pingjunyang	time average sample
時間範疇上的難題	shijian fanchoushangde nanti	time horizon dilemma
校準，氣相色譜法	jiaozhun, qixiang sepufa	calibration, gas chromatography
校準標準	jiaozhun biaozhun	calibration standards
校準標準氣	jiaozhun biaozhunqi	gas standards for calibration
校準混合氣	jiaozhun hunheqi	calibration gas mixtures
柴油發動機	chaiyou fadongji	diesel engines
柴油發動機的催化 轉化器	chaiyou fadongjide cuihua zhuanhuaqi	catalytic converters for diesel engines
氣體平衡管	qiti pingheng guan	gas balancing pipes
氣體監測器	qiti jianceqi	gas monitor
氣體檢測管	qiti jianceguan	gas detector tubes
氣味	qiwei	odor
氣相色譜法	qixiang sepufa	gas chromatography
氣相色譜法使用的 滲透管	qixiang sepufa shiyongde shentouguan	permeation tube in gas chromatography
氣相色譜法保留時間	qixiang sepufa baoliu shijian	retention time in gas chromatography
氣相色譜－質譜分析法	qixiang sepu-zhipu fenxifa	gas chromatography-mass spectroscopy
氣候變化	qihou bianhua	climatic changes
氣提	qiti	stripping
氣提塔	qitita	stripping column
氣溶膠	qirongjiao	aerosol
氣溶膠分離	qirongjiao fenli	aerosol separation
氧化溝	yanghuagou	oxidation ditch
氧化鈣當量	yanghuagai dangliang	calcium oxide equivalents
氧循環	yangxunhuan	oxygen cycle

流化床生物處理	*liuhuachuang shengwu chuli*	fluidized bed biology
流化床系統	*liuhuachuang xitong*	fluidized bed systems
流化床焚燒	*liuhuachuang fenshao*	fluidized bed incineration
流動相	*liudongxiang*	fluid phase
流量平均樣	*liuliang pingjunyang*	flow average sample
消泡劑	*xiaopaoji*	antifoaming agent
消毒（一般的）	*xiaodu (yibande)*	disinfection, general
消毒（滅菌）	*xiaodu (miejun)*	sterilization
消費品的預付處置費	*xiaofeipinde yufu chuzhifei*	prepaid disposal fee for consumer goods
浸出液	*jinchuye*	leachate
浸出液測試	*jinchuye ceshi*	leachate test
海洋生態系統	*haiyang shengtai xitong*	marine ecosystems
浮選	*fuxuan*	flotation
浮選澄清池	*fuxuan chengqingchi*	flotation clarifier
浮游生物	*fuyou shengwu*	plankton
浮游動物	*fuyou dongwu*	zooplankton
浮游植物	*fuyou zhiwu*	phytoplankton
特殊廢物	*teshu feiwu*	special wastes
特殊廢物分析	*teshu feiwu fenxi*	analysis of special wastes
特殊廢物分析	*teshu feiwu fenxi*	special waste analysis
特殊廢物處置的管理	*teshu feiwu chuzhide guanli*	management of special waste disposal
特殊廢物的管理	*teshu feiwude guanli*	management of special wastes
特殊廢物焚燒爐	*teshu feiwu fenshaolu*	special waste incinerator
特殊廢物管理	*teshu feiwu guanli*	special waste management
特殊廢物，定義	*teshu feiwu, dingyi*	special wastes, definitions
真核細胞	*zhenhe xibao*	eucaryotic cell
真菌	*zhenjun*	fungi
破乳	*poru*	emulsion splitting
破乳	*poru*	splitting of emulsions
粉末活性碳處理	*fenmo huoxingtan chuli*	powdered activated carbon treatment
粉塵	*fenchen*	dust
粉塵監測	*fenchen jiance*	dust monitoring
粉碎	*fensui*	milling

紡織廢水	*fangzhi feishui*	textile effluents
純氧廢水生物處理	*chunyang feishui shengwu chuli*	pure oxygen biological effluent treatment
純氧曝氣系統	*chunyang baoqi xitong*	aeration systems with oxygen
純粹過濾器	*chuncui guolüqi*	absolute filter
缺氧生物降解	*queyang shengwu jiangjie*	anoxic biodegradation
脂肪組織	*zhifang zuzhi*	fatty tissue
脂溶性的	*zhirongxingde*	fat soluble
能量	*nengliang*	energy
能量共生	*nengliang gongsheng*	co-generation of energy
能量回收	*nengliang huishou*	energy recuperation
能量轉換	*nengliang zhuanhuan*	energy transformation
能源，礦物資源	*nengyuan, kuangwu ziyuan*	energy, fossil resources
臭閾濃度	*chouyu nongdu*	threshold odor concentration
臭味指數	*chouwei zhishu*	odor index
臭氧	*chouyang*	ozone
臭氧空洞	*chouyang kongdong*	ozone hole
臭氧氧化技術	*chouyang yanghua jishu*	ozone oxidation technology
臭氧耗竭(破壞)	*chouyang haojie (pohuai)*	ozone depletion
逆相高壓液相色譜法	*nixiang gaoya yexiang sepufa*	reversed phase HPLC
逆流	*niliu*	counter-current
逆滲透過程	*nishentou guoceng*	reversed osmosis process
逆滲透過程的持留液	*nishentou guochengde chiliuye*	retentate in reversed osmosis
逆滲透技術	*nishentou jishu*	reversed osmosis technology
除砂礫	*chushali*	grit removal
除氮氧化物轉化器	*chudanyanghuawu zhuanhuaqi*	DENOX converter
除霧器	*chuwuqi*	demister
高壓氧化室	*gaoya yanghuashi*	high pressure oxidation chamber
高壓液相色譜法	*gaoya yexiang sepufa*	high pressure liquid chromatography
高壓濕式空氣氧化	*gaoya shishi kongqi yanghua*	high pressure wet air oxidation
高負荷生物處理系統	*gaofuhe shengwu chuli xitong*	high load biology
高溫焚燒	*gaowen fenshao*	high temperature incineration
高溫燃燒	*gaowen ranshao*	high temperature combustion
高錳酸鹽值	*gaomengsuanyanzhi*	permanganate value

十一劃

乾式空氣污染控制系統	*ganshi kongqi wuran kongzhi xitong*	APC-systems, dry
乾式除塵器	*ganshi chuchenqi*	dry dust eliminators
乾式煙氣處理	*ganshi yanqi chuli*	dry flue gas treatment
乾式靜電除塵器	*ganshi jingdian chuchenqi*	dry electrostatic precipitator
乾預值	*ganyuzhi*	intervention values
做功的能	*zuogongdeneng*	exergy
偶然誤差	*ouran wucha*	random errors
動態平衡	*dongtai pingheng*	dynamic equilibrium
動物群落	dongwu qunluo	zoocenose
區劃法	*quhuafa*	zoning laws
參謀職責	*canmou zhize*	staff responsibility
參謀職能	*canmou zhineng*	staff function
國際水域管理	*guoji shuiyu guanli*	international water management
國際標準組織	*guoji biaozhun zuzhi*	International Standards Organization
國際標準組織14000標準系列	*guoji biaozhun zuzhi 14000 biaozhun xilie*	*ISO 14000 Norm series*
國際商會	*guoji shanghui*	ICC
國際商會鹿特丹憲章	*guoji shanghui lutedan xianzhang*	Rotterdam Charter of the ICC
堆放場	*duifangchang*	dumps
堆肥	*duifei*	compost
基建投資	*jijian touzi*	capital investments
基建項目的環境影響	*jijian xiangmude huanjing yingxiang*	environmental impact of capital projects
"從搖籃到墳墓"控制	*cong yaolan dao fenmu kongzhi*	cradle-to-grave control
控制論	*kongzhilun*	cybernetics
控制論管理方法	*kongzhilun guanli fangfa*	cybernetic management approach
控制過濾器	*kongzhi guolüqi*	police filter
控制和報告體系	*kongzhi he baogao tixi*	controlling and reporting systems
接種	*jiezhong*	inoculation
接觸層(帶)	*jiechucheng (dai)*	contact zone

排水收費	paishui shoufei	effluent levy
排水收費法	paishui shoufeifa	law on effluent levies
排污費	paiwufei	discharge fees
排序方法	paixu fangfa	ranking methods
排序判據	paixu panju	ranking criteria
排放	paifang	emission
排放	paifang	immission
排放因子	paifang yinzi	emission factor
排放權	paifangquan	emission right
排放權交易	paifangquan jiaoyi	trading of emission rights
排放權證的市場價格	paifangquanzhengde, shicang jiage	emission right certificates, market price
排放權證的初始價格	paifangquanzhengde, chushi jiage	emission right certificates, initial price
排放限值	paifang xianzhi	discharge limit
排放限值	paifang xianzhi	emission limit
排放限值	paifang xianzhi	immission limits
排放穿透	paifang chuantou	emission breakthrough
排放清單	paifang qingdan	emission inventory
排放數據庫	paifang shujuku	emission database
排放源	paifangyuan	emission sources
旋風分離器	xuanfeng fenliqi	cyclone separators
梯度洗脫	tidu xiduo	gradient elution
毫克	haoke	milligram
毫微米	haoweimi	nanometer
毫微米級過濾	haoweimiji guolü	nanofiltration
毫微克	haoweike	nanogram
毫微微克	haoweiweike	femtogram
液上氣體分析	yeshang qiti fenxi	headspace analysis
液化	yehua	liquefaction
液相色譜法	yexiang sepufa	liquid chromatography
液液萃取	yeye cuiqu	liquid/liquid extraction
液滴分離器	yedi fenliqi	droplet separator
清潔工藝	qingjie gongyi	clean technologies
清潔化學	qingjie huaxue	clean chemistry
清潔生產過程	qingjie shengchan guocheng	clean production processes
清除淨化費用	qingchu jinghua feiyong	clean up costs

混合液	hunheye	mixed liquor
混凝	hunning	coagulation
混凝劑	hunningji	coagulating agents
深井曝氣技術	shenjing baoqi jishu	deep shaft technology
理論板	lilunban	theoretical plate
理論需氧量	lilun xuyangliang	theoretical oxygen demand
現代工藝水平	xiandai gongyi shuiping	state-of-the-art
現場處理	xianchang chuli	on-site treatment
現場恢復(補救)	xianchang huifu (bujiu)	in-site remediation
現場恢復補救處理	xianchang huifu bujiu chuli	on-site treatment in remediation
產物攜帶	chanwu xiedai	product carry-over
產品生產總成本	chanping shengchan zongchengben	total product manufacturing costs
產品生態平衡	chanpin shengtai pingheng	product ecobalances
產品生命周期分析	chanpin shengming zhouqi fenxi	product life cycle analysis
產品評價	chanpin pingjia	product assessment
產品的生命周期	chanpinde shengming zhouqi	life cycle of a product
產品的真實總成本	chanpingde zhenshi zongchengben	true total product costs
產品線生命周期分析	chanpinxian shengming zhouqi fenxi	product-line analysis
產品管理	chanpin guanli	product stewardship
異生化合物	yisheng huahewu	xenobiotica
異養生物	yiyang shengwu	heterotrophic organisms
移動床吸附塔	yidongchuang xifuta	moving bed adsorption column
移動爐箅式焚燒爐	yidong lubishi fenshaolu	moving grate incinerator
符合公司內部標準	fuhe gongsi neibu biaozhun	compliance with company internal norms
符合法律要求	fuhe falü yaoqiu	compliance with legal requirements
符合法律要求	fuhe falü yaoqiu	legal compliance
粒狀活性碳處理	lizhuang huoxingtan chuli	granular activated carbon treatment
粒狀活性碳過濾器	lizhuang huoxingtan guolüqi	granular activated carbon filter
統計	tongji	statistics
細菌	xijun	bacteria

組合空氣污染控制系統	*zuhe kongqi wuran kongzhi xitong*	combined APC-systems
脫氣區	*tuoqiqu*	degassing zone
脫臭方法	*tuochou fangfa*	deodorizing processes
荷蘭限值表	*helan xianzhibiao*	Holland List
處置場	*chuzhichang*	disposal site
被捕食者	*beibushizhe*	prey
袖式過濾器	*xiushiguolüqi*	sleeve filter
袋式過濾器	*daishi guolüqi*	bag filter
袋濾室	*dailüshi*	baghouse
設備廢氣	*shebei feiqi*	object off-gas
責任	*zeren*	liability
責任關懷計劃	*zeren guanhuai jihua*	Responsible Care Program
貧營養的	*pinyingyangde*	oligotrophic
連續分析	*lianxu fenxi*	continuous analysis
逕流水	*jingliushui*	runoff water
閉合物流系統	*bihe wuliu xitong*	closed mass flow system
閉環活動	*bihuan huodong*	closed cycle activity
陸地生態系統	*ludi shengtai xitong*	terrestrial ecosystems
陸圈	*luquan*	geosphere
陰離子交換	*yinlizi jiaohuan*	anion exchange
魚試驗	*yu shiyan*	fish tests

十二劃

傑欣-魏倫斯測試	*jie'en-weilunsi ceshi*	Zahn-Wellens Test
最大容許濃度	*zuida rongxu nongdu*	MAK
最終處理塘	*zuizhong chulitang*	finishing pond
最終澄清池	*zuizhong chengqingchi*	final clarifier
最佳實用技術	*zuijia shiyong jishu*	best practicable technology
勞動衛生	*laodong weisheng*	work hygiene
喀斯特水	*kasite shui*	karst water
單一物質的水生生物毒性	*danyi wuzhide shuisheng shengwu duxing*	aquatic, toxicity, of single substances
單一要素指標(參數)	*danyi yaosu zhibiao (canshu)*	single element parameter
場外恢復補救處理	*changwai huifu bujiu chuli*	off-site treatment in remediation
富營養的，富營養化	*fuyingyangde, fuyingyanghua*	eutrophic, eutrophication
循環利用	*xunhuan liyong*	recycling

循環利用技術	xunhuan liyong jishu	recycling technologies
惡臭	echou	malodors
惡臭閾限值	echou yuxianzhi	odor recognition threshold limit
惡臭測定法	echou cedingfa	olfactometry
揮發性有機化合物	huifaxing youji huahewu	volatile organic compounds
揮發性有機化合物的收費	huifaxing youji huahewude shoufei	levy on VOC input
揮發性有機鹵素化合物	huifaxing youji lusu huahewu	VOX
景觀保護	jingguan baohu	landscape protection
植物毒性	zhiwu duxing	phytotoxicity
植物群落	zhiwu qunluo	phytocenose
殘存影響	cancun yingxiang	residual impact
氮氧化物	danyanghuawu	NO_X
氮循環	danxunhuan	nitrogen cycle
氯化	lühua	chlorination
氯化烴	lühuating	chlorinated hydrocarbons
氯氟烴	lüfuting	chloro fluoro carbons
減壓段	jianyaduan	decompression stage
湖沼生態系統	huzhao shengtai xitong	limnic ecosystems
測量口	celiang kou	measurement ports
焚燒	fenshao	incineration
焚燒爐	fenshaolu	incinerator
焚燒室	fenshaoshi	incineration chamber
無可見作用水平	wukejian zuoyong shuiping	no observable effect level
無可見作用濃度	wukejian zuoyong nongdu	no observable effect concentration
無形價值	wuxing jiazhi	immaterial values
無形商品	wuxing shangpin	immaterial goods
無規填料塔	wugui tianliaota	randomly packed column
發光法廢水毒性	faguangfa feishui duxing	luminescence effluent toxicity
發酵	fajiao	fermentation
短期接觸限值	duanqi jiechu xianzhi	STEL
硝化桿菌	xiaohua ganjun	nitrobacter
硝化（作用）	xiaohua (zuoyong)	nitrification
硝化細菌	xiaohua xijun	nitrifying bacteria
硝化–反硝化過程	xiaohua-fanxiaohua guocheng	nitrification-denitrification process

稅	*shui*	tax
等離子體焚燒技術	*dengliziti fenshao jishu*	plasma incineration technique
結論的有效性	*jielunde youxiaoxing*	validity of a conclusion
結果的重複性	*jieguode chongfuxing*	repeatability of results
紫外輻射	*ziwai fushe*	UV radiation
絮凝(作用)	*xuning (zuoyong)*	flocculation
絮凝劑	*xuningji*	flocculating agent
萃取技術	*cuiqu jishu*	extraction technology
萃取溶劑	*cuiqu rongji*	extraction solvent
費用	*feiyong*	fee
費用中心	*feiyong zhongxin*	cost center
費用-效益分析	*feiyong-xiaoyi fenxi*	cost-benefit analysis
超級基金計劃	*chaoji jijin jihua*	SUPERFUND program
超聲波質量流量計	*chaoshengbo zhiliang liuliangji*	ultrasonic mass flow meter
超臨界流體色譜法	*chaolinjie liuti sepufa*	supercritical fluid chromatography
超臨界流體萃取	*chaolinjie liuti cuiqu*	supercritical fluid extraction
超濾	*chaolü*	ultrafiltration
開式物流系統	*kaishi wuliu xitong*	open mass flow system
間接的污染控制費用	*jianjiede wuran kongzhi feiyong*	indirect pollution control costs
間接費用	*jianjie feiyong*	indirect costs
陽離子交換	*yanglizi jiaohuan*	cation exchange
集水池	*jishuichi*	catch basin

十三劃

傳輸現象	*chuanshu xianxiang*	transport phenomena
催化還原去除氮氧化物	*cuihua huanyuan quchu danyanghuawu*	catalytic reductive removal of NO_x
催化劑	*cuihuaji*	catalysts
催化劑中毒	*cuihuaji zhongdu*	poisoning of catalysts
催化劑床	*cuihuajichuang*	catalyst bed
催化劑毒物	*cuihuaji duwu*	catalyst poisons
催化轉化器	*cuihua zhuanhuaqi*	catalytic converters
催化氧化	*cuihua yanghua*	catalytic oxidation
催化燃燒/焚燒	*cuihua ranshao/fenshao*	catalytic combustion/ incineration

塑料廢物	suliao feiwu	plastic wastes
塑料廢物標記	suliao feiwu biaoji	codes for plastic wastes
塑料廢物標記	suliao feiwu biaoji	plastic waste codes
塔式生物反應器	tashi shengwu fanyingqi	tower bioreactor
塔式滴濾池（廢水）	tashi delüchi (feishui)	trickling towers (effluents)
塔填料	tatianliao	column packing
填充段（區）	tianchongduan (qu)	packing zone
填埋氣體	tianmai qiti	landfill gas
填埋場中浸出液的 控制和處理	tianmaichangzhong jinchuyede kongzhi he chuli	leachate control and treatment in landfill
填料洗滌塔	tianliao xidita	packed column scrubber
填料塔	tianliaota	packed column
微生物處理	weishengwu chuli	microbiological treatment
微生物學	weishengwuxue	microbiology
微生物礦質化	weishengwu kuangzhihua	microbiological mineralization
微生物群落	weishengwu qunluo	microcenose
微米	weimi	micrometer
微觀經濟問題	weiguan jingji wenti	microeconomic aspects
微過濾	weiguolü	microfiltration
微克	weike	microgram
微粒沉降	weili chenjiang	precipitation of particulates
微微克	weiweike	picogram
感應耦合等離子體 質譜分析	ganying ouhe dengliziti zhipu fenxi	inductively coupled plasma mass spectrometry
損害恢復費用	sunhai huifu feiyong	damage repair costs
損害預防費用	sunhai yufang feiyong	damage prevention costs
新陳代謝	xinchendaixie	metabolism
溶解性有機碳	rongjiexing youjitan	dissolved organic carbon
溶解氧	rongjieyang	DO
源頭污染預防	yuantou wuran yufang	PPS
源頭空氣	yuantou kongqi	source air
源頭削減	yuantouxuejian	source reduction
溫室氣體	wenshi qiti	greenhouse gases
溫室效應	wenshi xiaoying	greenhouse effect
煙氣處理	yanqi chuli	flue gas treatment
煙囪排放	yancong paifang	stack emissions
煙道氣	yandaoqi	flue gas

煙霧	*yanwu*	smog
煙霧	*yanwu*	smoke
煅燒殘渣	*duanshao canzha*	calcinated residue
當期費用	*dangqi feiyong*	period costs
經濟合作與發展組織	*jingji hezuo yu fazhan zuzhi*	OECD
經濟合作與發展組織- 生物需氧量測定	*jingji hezuo yu fazhan zuzhi- shengwu xuyangliang ceding*	OECD-BOD Test
經濟刺激	*jingji ciji*	economic incentives
置信界限	*zhixin jiexian*	confidence limits
群落生態學	*qunluo shengtaixue*	synecology
群落生境	*qunluo shengjing*	biotope
補貼或減稅	*butie huo jianshui*	subsidies or tax reductions
裝置廢氣	*zhuangzhi feiqi*	apparatus off-air
試樣母體	*shiyang muti*	matrix of a sample
試驗方法的準確度	*shiyan fangfade zhunquedu*	accuracy of a test method
試驗方法的精密度	*shiyang fangfede jingmidu*	precision of a test method
試驗魚	*shiyanyu*	test fish
試驗數據總體	*shiyang shuju zongti*	population of test data
資源	*ziyuan*	resources
載氣	*zaiqi*	carrier gas
載體上的生物	*zaitishangde shengwu*	biologies on supports
載體上的活性污泥	*zaitishangde huoxing wuni*	activated sludge on supports
運用市場機制的法律 手段	*yunyong shichang jizhide falü shouduan*	legal instruments using market mechanisms
運行成本(費用)	*yunxing chengben (feiyong)*	running costs
運行費用	*yunxing feiyong*	operating costs
過程專用設備	*guocheng zhuanyong shebei*	process-dedicated equipment
過程開發	*guocheng kaifa*	process development
過程優化	*guocheng youhua*	process optimization
過濾	*guolü*	filtration
過濾介質	*guolü jiezhi*	filter medium
過濾纖維	*guolü xianwei*	filter fibers
過濾器穿透	*guolüqi chuantou*	filter breakthrough
電子捕獲檢測器	*dianzi buhuo jianceqi*	electron capture detector
電導率	*diandaolü*	conductivity
電煙霧	*dianyanwu*	electrosmog
電滲析	*dianshenxi*	electrodialysis

電鍍廢物	*diandu feiwu*	galvanic waste
零風險目標	*lingfengxian mubiao*	zero-risk goal
零污染目標	*lingwuran mubiao*	zero-pollution goal
靶生物	*bashengwu*	target organism
預防性安全措施	*yufangxing anquan cuoshi*	preventive safety measures
預防性維修	*yufangxing weixiu*	preventive maintenance
預防性維修制度	*yufangxing weixiu zhidu*	preventive maintenance system
預防措施	*yufang cuoshi*	preventive measures
預濃縮	*yunongsuo*	preconcentration

十 四 劃

厭氧水體	*yanyang shuiti*	anaerobic waterbody
厭氧生物反應器系統	*yanyang shengwu fanyingqi xitong*	biothane system
厭氧生物過程	*yanyang shengwu guocheng*	anaerobic biological processes
厭氧生物降解	*yanyang shengwu jiangjie*	anaerobic biodegradation
厭氧廢水處理	*yanyang feishui chuli*	anaerobic effluent treatment
厭氧微生物	*yanyang weishengwu*	anaerobic microorganisms
實時數據	*shishi shuju*	real-time data
實驗室間的相互測試	*shiyanshijiande xianghu ceshi*	inter-laboratory tests
實驗室廢水	*shiyanshi feishui*	laboratory effluents
對生態系統的影響	*duishengtai xitongde yingxiang*	impact on ecosystems
對魚有毒	*duiyuyoudu*	fish-toxic
對魚的毒性	*duiyude duxing*	toxicity versus fish
對蝦的毒性	*duixiade duxing*	toxicity versus shrimp
慢性效應	*manxing xiaoying*	chronic effect
滯後損害費用	*zhihou sunhai feiyong*	delayed damage costs
滯後效應	*zhihou xiaoying*	delayed effects
滲流水	*shenliushui*	seepage water
滲透水	*shentoushui*	permeating water
滲透壓	*shentouya*	osmotic pressure
滲透(作用)	*shentou (zuoyong)*	osmosis
滲透液	*shentouye*	permeate
碳循環	*tanxunhuan*	carbon cycle
管制法	*guanzhifa*	police laws
管理手段	*guanli shouduan*	management instruments

管理體系	guanli tixi	management system
管理規程	guanli guicheng	management procedure
管理控制手段	guanli kongzhi shouduan	management control instruments
精密氣體標準	jingmi qiti biaozhun	precision gas standards
綜合環境管理	zonghe huanjing guanli	integral environmental management
綜合指標	zonghe zhibiao	sum parameters
綜合指標(參數)	zonghe zhibiao (canshu)	collective parameter
綜合指標(參數)	zonghe zhibiao (canshu)	aggregate parameter
綜合指標(參數)	zonghe zhibiao (canshu)	group parameter
綠色產品	lüse chanpin	green products
緊急事故管理	jinji shigu guanli	emergency management
緊急響應管理	jinji xiangying guanli	emergency response management
維護維修費用	weihu weixiu feiyong	maintenance and repair costs
維修與維護	weixiu yu weihu	repair and maintenance
罰款	fakuan	fine
聚電解質	judianjiezhi	polyelectrolyte
聚結	jujie	coalescence
聚結器	jujieqi	coalescer unit
聚醯胺膜	juxian'anmo	polyamide membranes
蒙特利爾議定書	mengteli'er yidingshu	Montreal Protocol
蒸汽汽提	zhengqi qiti	stripping with steam
蒸餾分離	zhengliu fenli	distillation separation
輔助操作	fuzhucaozuo	auxiliary operations
輔助燃料燒咀	fuzhu ranliao shaozui	support fuel burner
遙感	yaogan	remote sensing
酸性水	suanxingshui	sour water
酸雨	suanyu	acid rain

十 五 劃

億萬分之(幾分)	yiwanfenzhi (jifen)	ppt
儀器分析方法	yiqi fenxi fangfa	instrumental methods of analysis
價值	jiazhi	values
價值創造鏈	jiazhi chuangzaolian	chain of value creation
噴射洗滌器	penshe xidiqi	jet scrubber
噴射器	pensheqi	injector

噴淋折板	penlin zheban	spray deflector
噴淋塔(洗滌器)	penlinta (xidiqi)	spray tower (scrubber)
審計	shenji	audit
審計小組	shenji xiaozu	audit team
審計者	shenjizhe	auditor
審計標準	shenji biaozhun	audit standards
審計結果	shenji jieguo	audit findings
審計程序	shenji chengxu	audit procedure
層次分析	cengci fenxi	tiers analysis
層次分析計劃	cengci fenxi jihua	analysis plan in tiers
廢氣	feiqi	off-gas
廢氣中的顆粒物，連續測定	feiqizhongde keliwu, lianxu ceding	particulates in off-gas, continuous measurement
廢氣分別處理概念	feiqi fenbie chuli gainian	off-gas separation concept
廢氣分析	feiqi fenxi	off-gas analysis
廢氣分類	feiqi fenlei	classes of off-gases
廢氣處理	feiqi chuli	off-gas treatment
廢氣處理生物滴濾池	feiqi chuli shengwu dilüchi	bio trickling filter for off-gas
廢氣生物處理	feiqi shengwu chuli	biological off-gas treatment
廢氣過濾	feiqi guolü	filtration of off-gas
廢氣洗滌器	feiqi xidiqi	off-gas scrubbers
廢氣類別	feiqi leibie	off-gas classes
廢氣流速	feiqi liusu	off-gas velocity
廢氣預處理(調節)	feiqi yuchuli (tiaojie)	preconditioning of off-gases
廢氣排放的管理	feiqi paifangde guanli	management of off-gas emissions
廢氣數據卡	feiqi shujuka	off-gas data sheet
廢氣燃燒	feiqi ranshao	off-gas combustion
廢水	feishui	effluent
廢水三級處理	feishui sanji chuli	tertiary treatment of effluents
廢水中的揮發性有機碳	feishuizhongde huifaxing youjitan	volatile organic carbon in effluents, VOC
廢水分析	feishui fenxi	effluent analysis
廢水處理	feishui chuli	effluent treatment
廢水處理技術	feishui chuli jishu	effluent treatment technologies
廢水處理的營養物	feishui chulide yingyangwu	nutrients in effluent treatment
廢水處理費用	feishui chuli feiyong	fees for effluent treatment

廢水處理場	*feishui chulichang*	effluent treatment plant
廢水生物處理的停留時間	*feishui shengwu chulide tingliu shijian*	residence time in biological effluent treatment
廢水生物處理場	*feishui shengwu chulichang*	biological effluent plant
廢水自然處理系統	*feishui ziran chuli xitong*	natural treatment systems for effluents
廢水低壓氧化	*feishui diya yanghua*	low pressure effluent oxidation
廢水沉澱技術	*feishui chendian jishu*	precipitation technology for effluents
廢水取樣	*feishui quyang*	sampling of effluents
廢水的毒性	*feishuide duxing*	toxicity of effluents
廢水毒性	*feishui duxing*	effluent toxicity
廢水毒性，發光法	*feishui duxing, faguangfa*	effluent toxicity, luminescence method
廢水消毒	*feishui xiaodu*	effluent disinfection
廢水預處理	*feishui yuchuli*	pretreatment of effluents
廢水預處理過濾器	*feishui yuchuli guolüqi*	effluent pretreatment filter
廢水懸浮物的測定	*feishui xuanfuwude ceding*	suspended solids in effluents, measurement
廢水清單	*feishui qingdan*	effluent inventory
廢水脫色	*feishui tuose*	decolorization of effluent
廢水數據卡	*feishui shujuka*	effluent data sheet
廢物	*feiwu*	waste
廢物分類	*feiwu fenlei*	waste classification
廢物去毒	*feifu qudu*	detoxification of waste
廢物處理的分離方法	*feiwu chulide fenli fangfa*	separation processes in waste treatment
廢物處理的穩定化方法	*feiwu chulide wendinghua fangfa*	stabilization processes in waste treatment
廢物處置許可證	*feiwu chuzhi xukezheng*	disposal permit for wastes
廢物處置	*feiwu chuzhi*	waste disposal
廢物處置費	*feiwu chuzhifei*	fee, waste disposal
廢物交換	*feiwu jiaohuan*	waste exchange
廢物多聯單	*feiwu duoliandan*	waste manifest
廢物多聯單管理方法	*feiwu duoliangdan guanli fangfa*	waste manifest procedure
廢物固化	*feiwu guhua*	solidification of wastes

廢物固定化	feiwu gudinghua	immobilization of wastes
廢物的地下堆存	feiwude dixia duicun	underground waste deposits
廢物，浸出性	feiwu, jinchuxing	wastes, leaching properties
廢物熱解	feiwu rejie	pyrolysis of wastes
廢物預處理	feiwu yuchuli	pretreatment of wastes
廢物堆存	feiwu duicun	waste deposits
廢物堆肥	feiwu duifei	composting of wastes
廢物最少量化	feiwu zushaolianghua	waste minimization
廢物焚燒爐	feiwu fenshaolu	waste incinerators
廢物焚燒預濃縮過程	feiwu fenshao yunongsuo guocheng	preconcentration processes in waste incineration
廢物數據卡	feiwu shujuka	waste data sheet
廢物數據庫	feiwu shujuku	wastes database
廢物源頭削減	feiwu yuantou xuejian	waste reduction at the source
廢物跟蹤程序	feiwu genzhong chengxu	waste tracking procedure
廢物管理	feiwu guanli	waste management
廠內措施	changnei cuoshi	in-plant measures
廠外處理	changwai chuli	off-site treatment
影響評價和分析	yingxiang pingjia he fenxi	impact assessment and analysis
德拉格管	delageguan	Drager tubes
撞擊式分離器	zhuangjishi fenliqi	impingement separator
數字記錄儀	shuzi jiluyi	digital recorder
數字信號	shuzi xinhao	digital signal
數學離散模式	shuxue lisan moshi	mathematical dispersion models
數據記錄儀	shuju jiluyi	data logger
數據類別，空氣質量	shuju leibie, kongqi zhiliang	data types, air quality
數據管理系統	shuju guanli xitong	data management systems
暴雨水分流排管	baoyushui fenliu paiguan	stormwater overflow
樣品富集	yangpin fuji	sample enrichment
標準分析方法	biaozhun fenxi fangfa	standard methods in analysis
標準成本計算	biaozhun chengben jisuan	standard cost calculation
標準單位成本	biaozhun danwei chengben	standard unit costs
標準偏差	biaozhun piancha	standard deviation
模擬記錄器	moni jiluqi	analog recorder
模擬信號	moni xinhao	analog signal
歐洲化學工業聯合會中心	ouzhou huaxue gongye lianhehui zhongxin	CEFIC

歐洲管理和審計指令	ouzhou guanli he shenji zhiling	European Management and Auditing Directive
歐盟《生態管理和審計體系》指令	oumeng shengtai guanli he shenji tixi zhiling	EMAS Directive of the European Union
澄清池	chengqingchi	clarifier
潛在風險	qianzai fengxian	potential risk
熱交換器	rejiaohuanqi	heat exchanger
熱氧化系統	reyanghua xitong	thermal oxidation systems
熱解	rejie	pyrolysis
熱解吸	rejiexi	thermodesorption
線性活動	xianxing huodong	linear activity
緩衝池	huanchongchi	buffer tanks
膜	mo	membrane
衛生化處理	weishenghua chuli	hygienization
複合技術焚燒爐	fuhe jishu fenshaolu	hybrid technology incinerator
質量目標	zhiliang mubiao	quality goals
質譜分析	zhipu fenxi	mass spectrometry, MS
鞍形填料	anxing tianliao	saddle column packing

十六劃

凝集	ningji	agglomeration
噪聲	zaosheng	noise
噪聲，干擾限值	zaosheng, ganrao xianzhi	noise, disturbance limits
噪聲污染	zaosheng wuran	noise pollution
噪聲防護區	zaosheng fanghuqu	noise protection zones
噪聲防護措施	zaosheng fanghu cuoshi	noise protection measures
噪聲監測	zaosheng jiance	noise monitoring
噪聲散發限值	zaosheng sanfa xianzhi	noise emission limits
操作規程	caozuo guicheng	operating procedure
濃度限值	nongdu xianzhi	concentration limits
濃度限值與總量限值	nongdu xianzhi yu zongliang xianzhi	concentration limits versus mass flow limits
濃縮液	nongsuye	concentrate
激光在分析中的應用	jiguan zaifenxizhongde yingyong	laser in analysis
燃燒	ranshao	combustion
燃燒室	ranshaoshi	combustion chamber

螢光分光光度法	yingguang fenguang guangdufa	fluorescence spectrophotometry
螢光法	yingguangfa	fluorometry
親油(脂)的	qingyou (zhi) de	lipophilic
選擇性非催化還原法	xuanzexing feicuihua huanyuanfa	selective non-catalytic reduction process
選擇性非催化還原法	xuanzexing feicuihua huanyuanfa	SNCR process
選擇催化還原法	xuanze cuihua huanyanfa	SCR process
隨身監測	suishen jiance	man monitoring
靜電分離器	jingdianfenliqi	electrostatic separator
靜電除塵器(乾式)	jingdian chuchenqi (ganshi)	electrostatic precipitators, dry
靜電除塵器(濕式)	jingdian chuchenqi (shishi)	electrostatic precipitators, wet
靜態平衡	jingtai pingheng	static equilibrium
靜態燃燒室焚燒爐	jingtai ranshaoshi fenshaolu	static chamber incinerator
鮑爾環填料	bao'erhuan tianliao	pall ring column packing
閾限值	yuxianzhi	threshold limit values

十七劃

優諾克斯系統	younuokesi xitong	UNOX System
儲罐(無組織)排放	chuguan (wuzuzhi) paifang	tank storage emissions
壓力法五天生物需氧量測定	yalifa wutian shengwu xuyangliang ceding	barometric BOD_5 measurement
壓降	yajiang	pressure drop
壓濾機	yalüji	filterpress
應急貯存池	yingji chucunchi	emergency holding basins
檢測極限	jiance jixian	limit of detection
檢測器信號	jianceqi xinhao	detector signal
濕地	shidi	wetlands
濕式空氣污染控制系統	shishi kongqi wuran kongzhi xitong	wet APC-systems
濕式煙氣處理	shishi yanqi chuli	wet flue gas treatment
濕式煙氣處理系統	shishi yanqi chuli xitong	flue gas treatment, wet systems
濕式袖式過濾器	shishi xiushi guolüqi	wet sleeve filter
濕式靜電除塵器	shishi jingdian chuchenqi	wet electrostatic precipitator
營養級	yingyangji	trophic level
環形噴射器	huanxing pensheqi	ring-jet unit

環(實驗室)測試	*huan (shiyanshi) ceshi*	ring-tests
環境	*huanjing*	environment
環境人員	*huanjing renyuan*	environmental officer
環境工作報告	*huanjing gongzuo baogao*	environmental performance report
環境分析	*huanjing fenxi*	environmental analysis
環境可接受的	*huanjing kejieshoude*	environmentally acceptable
環境可接受的產品	*huanjing kejieshoude chanpin*	environmentally acceptable products
環境立法	*huanjing lifa*	environmental legislation
環境立法中的經濟手段	*huanjing lifa zhongde jingji shouduan*	economic instruments in *environmental legislation*
環境行為評級	*huanjing xingwei pingji*	environmental performance rating
環境行為指標	*huanjing xingwei zhibiao*	environmental performance indicators
環境負責組織聯盟	*huanjing fuze zuzhi liangmeng*	Coalition for Environmentally Responsible Economies
環境負責組織聯盟原則	*huanjing fuze zuzhi lianmeng yuanze*	CERES Principles
環境技術	*huanjing jishu*	environmental technology
環境投資，回收期	*huanjing touzi, huishouqi*	environmental investments, pay-back time
環境報告	*huanjing baogao*	environmental report
環境狀況報告	*huanjing zhuangkuang baogao*	environmental statement
環境審計	*huanjing shenji*	environmental audit
環境的組成部分	*huanjingde zucheng bufen*	environmental compartments
環境的組成部分	*huanjingde zucheng bufen*	environmental segment
環境空氣	*huanjing kongqi*	ambient air
環境空氣目標	*huanjing kongqi mubiao*	ambient air goals
環境空氣濃度	*huanjing kongqi nongdu*	ambient air concentrations
環境經濟學	*huanjing jingjixue*	environmental economics
環境責任	*huanjing zeren*	environmental liability
環境質量目標	*huanjing zhiliang mubiao*	environmental quality goals
環境保護	*huanjing baohu*	environmental protection
環境保護投資費用	*huanjing baohu touzi feiyong*	investment costs, environmental protection

環境保護的管理手段	huanjing baohude guanli shouduan	management instruments in environmental protection
環境保護政策報告	huanjing baohu zhengce baogao	policy statement on environmental protection
環境保護費用	huanjing baohu feiyong	environmental protection costs
環境政策	huanjing zhengce	environmental policy
環境政策報告	huanjing zhengce baogao	environmental policy statement
環境科技人員	huanjing keji renyuan	environmental scientists
環境費用的管理	huanjing feiyongde guanli	management of environmental costs
環境損害的社會費用	huanjing sunhaide shehui feiyong	social costs from environmental damages
環境損害費用	huanjing sunhai feiyong	damage costs, environmental
環境框架法	huanjing kuangjiafa	environmental framework laws
環境控制	huanjing kongzhi	environmental controlling
環境清單	huanjing qingdan	environmental inventory
環境數據	huanjing shuju	environmental data
環境管理中的刺激手段	huanjin guanlizhongde ciji shouduan	incentives in environmental management
環境管理體系	huanjing guanli tixi	environmental management systems
環境影響報告	huanjing yingxiang baogao	environmental impact report
環境影響評價	huanjing yingxiang pingjia	environmental impact assessment
環境噪聲	huanjing zaosheng	ambient noise
環境噪聲(散發)限值	huanjing zaosheng (sanfa) xianzhi	ambient noise (immission) limits
瞬時樣	shunshiyang	grab sample
磷酸鹽去除	linsuanyan quchu	phosphate removal
總有機鹵素化合物	zhongyoujilusu huahewu	TOX
總有機碳	zongyoujitan	total organic carbon
總有機碳分析	zongyoujitan fenxi	TOC analysis
總有機碳負荷	zongyoujitan fuhe	TOC load
總有機碳/總碳分析	zongyoujitan/zongtan fenxi	TOC/TC analysis
總懸浮固體	zongxuanfu guti	TSS
總量限值	zongliang xianzhi	mass flow limits

總碳	*zongtan*	TC
總需氧量	*zongxuyangliang*	total oxygen demand
聲功率級	*shenggonglüji*	sound power level
聲壓級	*shengyaji*	sound pressure level
聲音	*shengyin*	sound
聲音發散	*shengyin fasan*	sound emission
聲音發散限值	*shengyin fasan xianzhi*	sound emission limits
聲音的傳播	*shengyinde chuanbo*	sound transmission
聲音的影響	*shengyinde yingxiang*	impact of sound
聯合國環境規劃署	*lianheguo huanjing guihuashu*	UNEP
聯合國環境規劃署 廢物分類系統	*liangheguo huanjing guihuashu feiwu fenlei xitong*	UNEP waste classification system
聯合國環境規劃署 廢物轉運公約	*lianheguo huanjing guihuashu feiwu zhuanyun gongyue*	UNEP Convention on waste transports
薄層色譜法	*boceng sepufa*	thin layer chromatography
購買獲得前審計	*goumai huodeqian shenji*	preacquisition audit
避免損害的費用	*bimian sunhaide feiyong*	costs of avoided damage
鍋爐房中廢氣的燃燒	*guolufangzhong feiqide ranshao*	combustion of off-gases in a boiler house
顆粒物	*keliwu*	particulates
點排放源	*dianpai fangyuan*	point emission source

十八劃

擴散排放	*kuosan paifang*	diffuse emissions
擴散排放源	*kuosan paifang yuan*	diffuse emission source
擴散模式	*kuosan moshi*	dispersion models
擴散器	*kuosanqi*	diffusors
職工的人員目標	*zhigongde renyuan mubiao*	personal objectives for individual employee
職業衛生	*zhiye weisheng*	occupational health
轉盤式反應器	*zhuanpanshi fanyingqi*	disc reactor, rotating
離子分析	*lizi fenxi*	ion analysis
離子交換	*lizi jiaohuan*	ion exchange
離子交換分離	*lizi jiaohuan fenli*	ion exchange separation
離子交換柱	*lizi jiaohuanzhu*	ion exchange columns
離子交換樹脂	*lizi jiaohuan shuzhi*	ion exchange resins
離子色譜法	*lizi sepufa*	ion chromatography

離子色譜法中的抑制 作用	*lizi sepufazhongde yizhi zuoyong*	suppression in ion chromatography
離子色譜法，化學抑制	*lizi sepufa, huaxue yizhi*	ion chromatography, chemical suppression
離子選擇(特效性)電極	*lizi xuanze (texiaoxing) dianji*	ion specific electrode
離心分離力	*lixin fenlili*	centrifugal separation forces
雙介質過濾	*shuangjiezhi guolü*	dual media filtration
雙重包封	*shuangchong baofeng*	double containment

十九劃

曝氣池(淺平式)	*baoqichi (qianpingshi)*	aeration basin, flat
曝氣池(塔式)	*baoqichi (tashi)*	aeration basin, tall/tower
曝氣設備	*baoqi shebei*	aeration equipment
曝氣停留時間	*baoqi tingliu shijian*	aeration residence time
曝氣器	*baoqiqi*	aerators
曝氣	*baoqi*	aeration
羅馬俱樂部	*luoma julebu*	Club of Rome
邊界層	*bianjieceng*	boundary layer
難降解化合物	*nanjiangjie huahewu*	recalcitrant compounds
難降解污染物	*nanjiangjie wuranwu*	refractory pollutants
難降解的離子化合物	*nanjiangjiede lizi huahewu*	refractory ionic compounds

二十劃

懸浮固體(懸浮物)	*xuanfu guti (xuanfuwu)*	suspended solids
爐渣	*luzha*	slag
爐箅式焚燒爐	*lubishi fenshaolu*	grate incinerator
礦質化	*kuangzhihua*	mineralization
藻類	*zaolei*	algae
蘊涵標準	*yunhan biaozhun*	standards by implication
警報計劃	*jingbao jihua*	alarm plan

二十一劃

蘭達傳感器	*landa chuanganqi*	lambda sensor

二十三劃

纖毛蟲	*qianmaochong*	ciliata

471

顯著性(百分率水平)	*xianzhuxing (baifenlü shuiping)*	significance (percentage level)
體積/體積比	*tiji/tiji bi*	volume/volume ratio

二 十 四 劃

驟冷	*zhouleng*	quench

CHINESE-ENGLISH GLOSSARY

with Chinese entries arranged by their pinyin pronunciation

The term in each entry is followed by its pinyin pronunciation in italics and then by its English equivalent. Please refer to the English-Chinese glossary for explanation of the terms.

此中英詞匯中之中文名詞以漢語拼音排序。每個中文名詞後是其漢語拼音，跟著是其英文名詞。請於英中詞匯中查閱各名詞之解釋。

A

X型氣溶膠分離器	*aikesixing qirongjiao fenliqi*	X-separator for aerosols
安全測試	*anquan ceshi*	safety tests
安全貯存區	*anquan chuncunqu*	secure storage area
安全審計	*anquan shenji*	safety audit
安全填埋	*anquan tianmai*	secure landfill
鞍形填料	*anxing tianliao*	saddle column packing

B

BOD_3/COD比	*BOD_5/COD bi*	BOD_3/COD ratio
百分比	*baifenbi*	percent
百萬分之幾(體積比)	*baiwanfenzhiji (tijibi)*	ppm by volume
半連續氣相色譜法	*banlianxu qixiang sepufa*	semi-continuous gas chromatography
板式塔	*banshita*	plate columns
半透膜	*bantoumo*	semi-permeable membrane
半抑制濃度	*banyizhi nongdu*	IC-50
半致死濃度	*banzhisi nongdu*	LC-50
半作用濃度	*banzuoyong nongdu*	EC-50
鮑爾環填料	*bao'erhuan tianliao*	pall ring column packing
薄層色譜法	*baoceng sepufa*	chromatography, thin layer, TLC
曝氣	*baoqi*	aeration
曝氣設備	*baoqi shebei*	aeration equipment
曝氣停留時間	*baoqi tingliu shijian*	aeration residence time

曝氣池(淺平式)	*baoqichi (qianpingshi)*	aeration basin, flat
曝氣池(塔式)	*baoqichi (tashi)*	aeration basin, tall/tower
曝氣器	*baoqiqi*	aerators
暴雨水分流排管	*baoyushui fenliu paiguan*	stormwater overflow
包裝廢物	*baozhuang feiwu*	packaging waste
靶生物	*bashengwu*	target organism
被捕食者	*beibushizhe*	prey
背景噪聲	*beijing zaosheng*	background noise
比(單位)費用	*bi (danwei) feiyong*	specific costs
邊界層	*bianjieceng*	boundary layer
表面曝氣機	*biaomian baoqiji*	surface aerator
表面再生	*biaomian zaisheng*	surface regeneration
標準成本計算	*biaozhun chengben jisuan*	standard cost calculation
標準單位成本	*biaozhun danwei chengben*	standard unit costs
標準分析方法	*biaozhun fenxi fangfa*	standard methods in analysis
標準偏差	*biaozhun piancha*	standard deviation
百萬分之(幾分)	*biawanfenzhi (jifen)*	ppm
比處置費用	*bichuzhi feiyong*	specific disposal costs
比廢物量	*bifeiwuliang*	specific waste quantity
閉合物流系統	*bihe wuliu xitong*	closed mass flow system
閉環活動	*bihuan huodong*	closed cycle activity
避免損害的費用	*bimian sunhaide feiyong*	costs of avoided damage
並流	*bingliu*	co-current
比色分析	*bise fenxi*	colorimetric analysis
比吸附表面	*bixifu biaomian*	adsorption surface
薄層色譜法	*boceng sepufa*	thin layer chromatography
不合格產品	*buhege chanpin*	off-specification products
不可生物降解化合物	*bukeshengwu jianjie huahewu*	non-biodegradable compounds
不可再生資源	*bukezaisheng ziyuan*	non-renewable resources
捕食者	*bushizhe*	predator
捕食者與被捕食者的關係	*bushizhe yu beibushizhe de guanxi*	predator-prey relationship
補貼或減税	*butie huo jianshui*	subsidies or tax reductions
不同形式的能量	*butong xingshide nengliang*	energies

C

殘存影響	*cancun yingxiang*	residual impact

參謀職能	canmou zhineng	staff function
參謀職責	canmou zhize	staff responsibility
操作規程	caozuo guicheng	operating procedure
測量口	celiang kou	measurement ports
層次分析	cengci fenxi	tiers analysis
層次分析計劃	cengci fenxi jihua	analysis plan in tiers
拆建廢物	chaijian feiwu	demolition waste
柴油發動機	chaiyou fadongji	diesel engines
柴油發動機的催化轉化器	chaiyou fadongjide cuihua zhuanhuaqi	catalytic converters for diesel engines
廠內措施	changnei cuoshi	in-plant measures
廠外處理	changwai chuli	off-site treatment
場外恢復補救處理	changwai huifu bujiu chuli	off-site treatment in remediation
產品管理	chanpin guanli	product stewardship
產品評價	chanpin pingjia	product assessment
產品生命周期分析	chanpin shengming zhouqi fenxi	product life cycle analysis
產品生態平衡	chanpin shengtai pingheng	product ecobalances
產品的生命周期	chanpinde shengming zhouqi	life cycle of a product
產品生產總成本	chanping shengchan zongchengben	total product manufacturing costs
產品的真實總成本	chanpingde zhenshi zongchengben	true total product costs
產品線生命周期分析	chanpinxian shengming zhouqi fenxi	product-line analysis
產物攜帶	chanwu xiedai	product carry-over
超級基金計劃	chaoji jijin jihua	SUPERFUND program
超臨界流體萃取	chaolinjie liuti cuiqu	supercritical fluid extraction
超臨界流體色譜法	chaolinjie liuti sepufa	supercritical fluid chromatography
超濾	chaolü	ultrafiltration
超聲波質量流量計	chaoshengbo zhiliang liuliangji	ultrasonic mass flow meter
沉澱劑	chendianji	precipitating agent
成本	chengben	costs
成本價格	chengben jiage	cost covering prices
成本計算方法	chengben jisuan fangfa	cost calculation procedure
澄清池	chengqingchi	clarifier

城市(廢物)焚燒爐	chengshi (feiwu) fenshaolu	municipal incinerator
城市廢物處置費	chengshi feiwu chuzhifei	municipal waste disposal fees
城市廢物管理	chengshi feiwu guanli	municipal waste management
城市垃圾	chengshi laji	municipal trash
城市區劃法	chengshi quhuafa	municipal zoning laws
城市污水	chengshi wushui	municipal sewage
城市污水處理場	chengshi wushui chulichang	municipal effluent treatment plant
成問題的廢物	chengwentide feiwu	problem wastes
沉積電極	chenji dianji	precipitation electrode
沉降	chenjiang	sedimentation
沉降除塵箱	chenjiang chuchenxiang	settling box dust eliminator
沉降速度	chenjiang sudu	sedimentation velocity
沉降箱	chenjiangxiang	drop out box
持有證書的檢查者	chiyou zhengshude jianchazhe	certified verifier
重組過程	chongzu guocheng	recombination processes
臭味指數	chouwei zhishu	odor index
臭氧	chouyang	ozone
臭氧耗竭(破壞)	chouyang haojie (pohuai)	ozone depletion
臭氧空洞	chouyang kongdong	ozone hole
臭氧氧化技術	chouyang yanghua jishu	ozone oxidation technology
臭閾濃度	chouyu nongdu	threshold odor concentration
傳輸現象	chuanshu xianxiang	transport phenomena
穿透點	chuantoudian	breakthrough point
除氮氧化物轉化器	chudanyanghuawu zhuanhuaqi	DENOX converter
儲罐(無組織)排放	chuguan (wuzuzhi) paifang	tank storage emissions
純粹過濾器	chuncui guolüqi	absolute filter
純氧曝氣系統	chunyang baoqi xitong	aeration systems with oxygen
純氧廢水生物處理	chunyang feishui shengwu chuli	pure oxygen biological effluent treatment
除砂礫	chushali	grit removal
除霧器	chuwuqi	demister
處置場	chuzhichang	disposal site
刺激(獎勵)制度	ciji (jiangli) zhidu	incentive systems
"從搖籃到墳墓"控制	cong yaolan dao fenmu kongzhi	cradle-to-grave control
催化還原去除氮氧化物	cuihua huanyuan quchu danyanghuawu	catalytic reductive removal of NO$_x$

催化燃燒/焚燒	cuihua ranshao/fenshao	catalytic combustion/incineration
催化氧化	cuihua yanghua	catalytic oxidation
催化轉化器	cuihua zhuanhuaqi	catalytic converters
催化劑	cuihuaji	catalysts
催化劑毒物	cuihuaji duwu	catalyst poisons
催化劑中毒	cuihuaji zhongdu	poisoning of catalysts
催化劑床	cuihuajichuang	catalyst bed
萃取技術	cuiqu jishu	extraction technology
萃取溶劑	cuiqu rongji	extraction solvent

D

DOC/TOC比	DOC/TOC bi	DOC/TOC ratio
大腸菌數	dachangjunshu	coliform count
袋濾室	dailüshi	baghouse
袋式過濾器	daishi guolüqi	bag filter
當期費用	dangqi feiyong	period costs
氮循環	danxunhuan	nitrogen cycle
氮氧化物	danyanghuawu	NO_X
單一物質的水生 生物毒性	danyi wuzhide shuisheng shengwu duxing	aquatic, toxicity, of single substances
單一要素指標(參數)	danyi yaosu zhibiao (canshu)	single element parameter
大氣	daqi	atmosphere
德拉格管	delageguan	Drager tubes
等離子體焚燒技術	dengliziti fenshao jishu	plasma incineration technique
電導率	diandaolü	conductivity
電鍍廢物	diandu feiwu	galvanic waste
點排放源	dianpai fangyuan	point emission source
電滲析	dianshenxi	electrodialysis
電煙霧	dianyanwu	electrosmog
電子捕獲檢測器	dianzi buhuo jianceqi	electron capture detector
底棲生態系統	diqi shengtai xitong	benthnic ecosystem
底棲生物	diqi shengwu	benthon, benthos
地下鹽礦中廢物堆存	dixia yankuangzhong feiwu duichun	underground waste deposits in salt mines
地下水	dixiashui	groundwater
地下水保護區	dixiashui baohuqu	groundwater protection zones
地下水補救恢復措施	dixiashui bujiuhuifu cuoshi	groundwater remediation

地下水監測	*dixiashui jiance*	groundwater monitoring
地下水水質數據	*dixiashui shuizhi shuju*	groundwater quality data
地下水污染	*dixiashui wuran*	groundwater contamination
地下水位計	*dixiashuiweiji*	piezometer
冬季煙霧	*dongji yanwu*	winter smog
動態平衡	*dongtai pingheng*	dynamic equilibrium
動物群落	*dongwu qunluo*	zoocenose
短期接觸限值	*duanqi jiechu xianzhi*	STEL
燒殘渣	*duanshao canzha*	calcinated residue
堆放場	*duifangchang*	dumps
堆肥	*duifei*	compost
對生態系統的影響	*duishengtai xitongde yingxiang*	impact on ecosystems
對蝦的毒性	*duixiade duxing*	toxicity versus shrimp
對魚的毒性	*duiyude duxing*	toxicity versus fish
對魚有毒	*duiyuyoudu*	fish-toxic
毒理學	*dulixue*	toxicology
多勃遜單位	*duobosun danwei*	Dobson Units
多功能設備	*duogongneng shebei*	multi-purpose equipment
多環芳烴	*duohuanfangting*	PAH
多爐床焚燒爐	*duoluchuang fenshaolu*	multiple hearth incinerator
多氯聯苯	*duolülianben*	PCBs
多氯聯苯廢物	*duolülianben feiwu*	PCB-wastes
毒物	*duwu*	toxicant

E

惡臭	*echou*	malodors
惡臭測定法	*echou cedingfa*	olfactometry
惡臭閾限值	*echou yuxianzhi*	odor recognition threshold limit
二次澄清池	*erci chengqingchi*	secondary clarifier
二氧化碳	*eryanghuatan*	carbon dioxide
二氧化碳排放收費	*eryanghuatan paifang shoufei*	discharge levy for CO_2
二元參比氣體混合器	*eryuan canbi qiti hunheqi*	binary reference gas mixer

F

| 發光法廢水毒性 | *faguangfa feishui duxing* | luminescence effluent toxicity |
| 法規 | *fagui* | legislation |

法規符合審計	*fagui fuhe shenji*	compliance audit
發酵	*fajiao*	fermentation
罰款	*fakuan*	fine
法律手段	*falü shouduan*	legal instruments
法律責任	*falü zeren*	legal responsibility
方法的回收率	*fangfade huisoulw*	recovery rate of a method
方法的檢測極限	*fangfade jiance jixian*	detection limit of a method
紡織廢水	*fangzhi feishui*	textile effluents
反硝化(作用)	*fanxiaohua (zuoyong)*	denitrification
反硝化-硝化過程	*fanxiaohua-xiaohua guocheng*	denitrification-nitrification process
反應氣	*fanyingqi*	reaction-gases
非安全填埋	*feianquan tianmai*	non-secure landfill
非反應性廢物	*feifanyingxing feiwu*	non-reactive wastes
廢物去毒	*feifu qudu*	detoxification of waste
非經濟的法律手段	*feijingjide falü shouduan*	non-economic legal instruments
非敏感性催化劑	*feiminganxing cuihuaji*	non-sensitive catalysts
廢氣	*feiqi*	off-gas
廢氣處理	*feiqi chuli*	off-gas treatment
廢氣處理生物滴濾池	*feiqi chuli shengwu dilüchi*	bio trickling filter for off-gas
廢氣分別處理概念	*feiqi fenbie chuli gainian*	off-gas separation concept
廢氣分類	*feiqi fenlei*	classes of off-gases
廢氣分析	*feiqi fenxi*	off-gas analysis
廢氣過濾	*feiqi guolü*	filtration of off-gas
廢氣類別	*feiqi leibie*	off-gas classes
廢氣流速	*feiqi liusu*	off-gas velocity
廢氣排放的管理	*feiqi paifangde guanli*	management of off-gas emissions
廢氣燃燒	*feiqi ranshao*	off-gas combustion
廢氣生物處理	*feiqi shengwu chuli*	biological off-gas treatment
廢氣數據卡	*feiqi shujuka*	off-gas data sheet
廢氣洗滌器	*feiqi xidiqi*	off-gas scrubbers
廢氣預處理(調節)	*feiqi yuchuli (tiaojie)*	preconditioning of off-gases
廢氣中的顆粒物，連續測定	*feiqizhongde keliwu, lianxu ceding*	particulates in off-gas, continuous measurement
非色散紅外分析	*feisesan hongwai fenxi*	infrared analysis, non-dispersive
非生物的	*feishengwude*	abiotic
非生物的生態因子	*feishengwude shengtai yinzi*	abiotic ecological factors

廢水	*feishui*	effluent
廢水沉澱技術	*feishui chendian jishu*	precipitation technology for effluents
廢水處理	*feishui chuli*	effluent treatment
廢水處理費用	*feishui chuli feiyong*	fees for effluent treatment
廢水處理技術	*feishui chuli jishu*	effluent treatment technologies
廢水處理場	*feishui chulichang*	effluent treatment plant
廢水處理的營養物	*feishui chulide yingyangwu*	nutrients in effluent treatment
廢水低壓氧化	*feishui diya yanghua*	low pressure effluent oxidation
廢水毒性	*feishui duxing*	effluent toxicity
廢水毒性，發光法	*feishui duxing, faguangfa*	effluent toxicity, luminescence method
廢水分析	*feishui fenxi*	effluent analysis
廢水清單	*feishui qingdan*	effluent inventory
廢水取樣	*feishui quyang*	sampling of effluents
廢水三級處理	*feishui sanji chuli*	tertiary treatment of effluents
廢水生物處理場	*feishui shengwu chulichang*	biological effluent plant
廢水生物處理的停留時間	*feishui shengwu chulide tingliu shijian*	residence time in biological effluent treatment
廢水數據卡	*feishui shujuka*	effluent data sheet
廢水脫色	*feishui tuose*	decolorization of effluent
廢水消毒	*feishui xiaodu*	effluent disinfection
廢水懸浮物的測定	*feishui xuanfuwude ceding*	suspended solids in effluents, measurement
廢水預處理	*feishui yuchuli*	pretreatment of effluents
廢水預處理過濾器	*feishui yuchuli guolüqi*	effluent pretreatment filter
廢水自然處理系統	*feishui ziran chuli xitong*	natural treatment systems for effluents
廢水的毒性	*feishuide duxing*	toxicity of effluents
廢水中的揮發性有機碳	*feishuizhongde huifaxing youjitan*	volatile organic carbon in effluents, VOC
廢物	*feiwu*	waste
廢物處理的分離方法	*feiwu chulide fenli fangfa*	separation processes in waste treatment
廢物處理的穩定化方法	*feiwu chulide wendinghua fangfa*	stabilization processes in waste treatment
廢物處置	*feiwu chuzhi*	waste disposal

廢物處置許可證	*feiwu chuzhi xukezheng*	disposal permit for wastes
廢物處置費	*feiwu chuzhifei*	fee, waste disposal
廢物堆存	*feiwu duicun*	waste deposits
廢物堆肥	*feiwu duifei*	composting of wastes
廢物多聯單	*feiwu duoliandan*	waste manifest
廢物多聯單管理方法	*feiwu duoliangdan guanli fangfa*	waste manifest procedure
廢物分類	*feiwu fenlei*	waste classification
廢物焚燒預濃縮過程	*feiwu fenshao yunongsuo guocheng*	preconcentration processes in waste incineration
廢物焚燒爐	*feiwu fenshaolu*	waste incinerators
廢物跟蹤程序	*feiwu genzhong chengxu*	waste tracking procedure
廢物管理	*feiwu guanli*	waste management
廢物固定化	*feiwu gudinghua*	immobilization of wastes
廢物固化	*feiwu guhua*	solidification of wastes
廢物交換	*feiwu jiaohuan*	waste exchange
廢物熱解	*feiwu rejie*	pyrolysis of wastes
廢物數據卡	*feiwu shujuka*	waste data sheet
廢物數據庫	*feiwu shujuku*	wastes database
廢物源頭削減	*feiwu yuantou xuejian*	waste reduction at the source
廢物預處理	*feiwu yuchuli*	pretreatment of wastes
廢物最少量化	*feiwu zushaolianghua*	waste minimization
廢物，浸出性	*feiwu, jinchuxing*	wastes, leaching properties
廢物的地下堆存	*feiwude dixia duicun*	underground waste deposits
費用	*feiyong*	fee
費用中心	*feiyong zhongxin*	cost center
費用-效益分析	*feiyong-xiaoyi fenxi*	cost-benefit analysis
非政府組織	*feizhengfu zuzhi*	NGO
分貝（A聲級）	*fenbei (A shengji)*	decibel (A)
粉塵	*fenchen*	dust
粉塵監測	*fenchen jiance*	dust monitoring
風速計	*fengsuji*	anemometer
分光光度測定法	*fenguang guangdu cedingfa*	spectrophotometry
風險	*fengxian*	risk
風險分析	*fengxian fenxi*	risk analysis
分離	*fenli*	separation
分流式下水道系統	*fenliushi xiashuidao xitong*	separation sewer systems

粉末活性碳處理	*fenmo huoxingtan chuli*	powdered activated carbon treatment
焚燒	*fenshao*	incineration
焚燒爐	*fenshaolu*	incinerator
焚燒室	*fenshaoshi*	incineration chamber
粉碎	*fensui*	milling
分析程序的校準	*fenxi chengxude jiaozhun*	calibration of analytical procedures
分析結果的離散性	*fenxi jieguode lisanxing*	dispersion of analytical results
分析的信噪比	*fenxide xinzaobi*	signal-to-noise ratio
分析的質量標準	*fenxide zhiliang biaozhun*	quality criteria of analysis
分析中的分離段	*fenxizhongde fenliduan*	separation step in analysis
分析中的富集步驟	*fenxizhongde fuji buzhou*	enrichment step in analysis
分析中的檢測方法	*fenxizhongde jiance fangfa*	detection methods in analysis
分析中的鑒定段	*fenxizhongde jiandingduan*	identification step in analysis
分析中的偶然誤差	*fenxizhongde ouran wucha*	random error in analysis
分子碎片的重組	*fenzi suipiande chongzu*	recombination of molecular fragments
符合法律要求	*fuhe falü yaoqiu*	compliance with legal requirements
符合法律要求	*fuhe falü yaoqiu*	legal compliance
符合公司内部標準	*fuhe gongsi neibu biaozhun*	compliance with company internal norms
复合技術焚燒爐	*fuhe jishu fenshaolu*	hybrid technology incinerator
負荷限值	*fuhe xianzhi*	load limits
氟裡昂	*fuli'ang*	FREON
負離子化電極 （靜電除塵器）	*fulizihua dianji (jingdian chuchenqi)*	negative ionizing electrode (electrostatic precipitator)
沸石	*fushi*	zeolite
浮選	*fuxuan*	flotation
浮選澄清池	*fuxuan chengqingchi*	flotation clarifier
富營養的，富營養化	*fuyingyangde, fuyingyanghua*	eutrophic, eutrophication
浮游動物	*fuyou dongwu*	zooplankton
浮游生物	*fuyou shengwu*	plankton
浮游植物	*fuyou zhiwu*	phytoplankton
負載限值	*fuzai xianzhi*	freight limits
輔助燃料燒咀	*fuzhu ranliao shaozui*	support fuel burner

輔助操作	*fuzhucaozuo*	auxiliary operations

G

乾式除塵器	*ganshi chuchenqi*	dry dust eliminators
乾式靜電除塵器	*ganshi jingdian chuchenqi*	dry electrostatic precipitator
乾式空氣污染控制系統	*ganshi kongqi wuran kongzhi xitong*	APC-systems, dry
乾式煙氣處理	*ganshi yanqi chuli*	dry flue gas treatment
感應耦合等離子體質譜分析	*ganying ouhe dengliziti zhipu fenxi*	inductively coupled plasma mass spectrometry
干預值	*ganyuzhi*	intervention values
高負荷生物處理系統	*gaofuhe shengwu chuli xitong*	high load biology
高錳酸鹽值	*gaomengsuanyanzhi*	permanganate value
高溫焚燒	*gaowen fenshao*	high temperature incineration
高溫燃燒	*gaowen ranshao*	high temperature combustion
高壓濕式空氣氧化	*gaoya shishi kongqi yanghua*	high pressure wet air oxidation
高壓氧化室	*gaoya yanghuashi*	high pressure oxidation chamber
個人監測	*geren jiance*	personal monitoring
個體生態學	*geti shengtaixue*	autecology
工廠排放清單	*gongchang paifang qingdan*	inventory of emissions of a factory
工廠的費用中心結構(計劃)	*gongchangde feiyong zhongxin jiegou (jihua)*	cost center structure (plan) of a factory
工廠的指揮機構	*gongchangde zhihui jigou*	line organization of the plant
工程項目評價，環境保護	*gongcheng xiangmu pingjia, huanjing baohu*	project assessment, environmental protection
公共利益團體	*gonggong liyi tuanti*	public interest groups
共生	*gongsheng*	symbiosis
公司環境政策	*gongsi huanjing zhengce*	company environmental policy
公司內部收費	*gongsi neibu shoufei*	company internal levy
公司內部的影響評價	*gongsi neibude yingxiang pingjia*	company internal impact assessment
公司的利害關係者	*gongside lihai guanxizhe*	stakeholders of a company
工業廢水	*gongye feishui*	industrial effluent
工業廢水處理場	*gongye feishui chulichang*	industrial effluent treatment plant
工業衛生	*gongye weisheng*	industrial hygiene
工藝措施	*gongyi cuoshi*	in-process measures
工藝廢氣	*gongyi feiqi*	process gases

工藝廢水	gongyi feishui	process effluents
公正實施	gongzheng shishi	equitable enforcement
工作報告	gongzuo baogao	performance report
工作標準	gongzuo biaozhun	performance standards
工作場所測定	gongzuo changsuo ceding	workplace measurements
工作環境空氣	gongzuo huanjing kongqi	work environment atmosphere
工作目標或限值	gongzuo mubiao huo xianzhi	performance goals or limits
工作職責說明	gongzuo zhize shuoming	job description
購買獲得前審計	goumai huodeqian shenji	preacquisition audit
光傳感器	guangchuanganqi	optical sensor
光達	guangda	LIDAR
光電太陽能	guangdian taiyangneng	photovoltaic solar energy
光度計	guangduji	photometer
光化學煙霧	guanghuaxue yanwu	photochemical smog
光氧化劑	guangyanghuaji	photooxidants
管理規程	guanli guicheng	management procedure
管理控制手段	guanli kongzhi shouduan	management control instruments
管理手段	guanli shouduan	management instruments
管理體系	guanli tixi	management system
管制法	guanzhifa	police laws
固定床生物反應器	gudingchuang shengwu fanyingqi	fixed bed bioreactor
固定床系統	gudingchuang xitong	fixed bed systems
固定化廢物	gudinghua feiwu	immobilized wastes
固定膜生物系統	gudingmo shengwu xitong	fixed film biological systems
過程開發	guocheng kaifa	process development
過程優化	guocheng youhua	process optimization
過程專用設備	guocheng zhuanyong shebei	process-dedicated equipment
國際標準組織	guoji biaozhun zuzhi	International Standards Organization
國際標準組織14000標準系列	guoji biaozhun zuzhi 14000 biaozhun xilie	ISO 14000 Norm series
國際商會	guoji shanghui	ICC
國際商會鹿特丹憲章	guoji shanghui lutedan xianzhang	Rotterdam Charter of the ICC
國際水域管理	guoji shuiyu guanli	international water management
過濾	guolü	filtration

過濾介質	*guolü jiezhi*	filter medium
過濾纖維	*guolü xianwei*	filter fibers
鍋爐房中廢氣的燃燒	*guolufangzhong feiqide ranshao*	combustion of off-gases in a boiler house
過濾器穿透	*guolüqi chuantou*	filter breakthrough
固體廢物	*guti feiwu*	solid waste
固體特殊廢物	*guti teshu feiwu*	solid special waste
固相萃取	*guxiang cuiqu*	solid phase extraction

H

海洋生態系統	*haiyang shengtai xitong*	marine ecosystems
哈龍	*halong*	Halon
行業合同	*hangye hetong*	branch contract
行業契約	*hangye qiyue*	branch covenant
含鹽量	*hanyanliang*	salinity
毫克	*haoke*	milligram
毫微克	*haoweike*	nanogram
毫微米	*haoweimi*	nanometer
毫微米級過濾	*haoweimiji guolw*	nanofiltration
毫微微克	*haoweiweike*	femtogram
好氧過程	*haoyang guocheng*	aerobic process
好氧生物降解	*haoyang shengwu jiangjie*	aerobic biodegradation
哈斯曼試驗	*hasiman shiyan*	Husmann Test
合格審計者	*hege shenjizhe*	certified auditor
合格的環境實驗室	*hegede huanjing shiyanshi*	certified environmental laboratory
河口生態系統	*hekou shengtai xitong*	estuarine ecosystem
荷蘭限值表	*helan xianzhibiao*	Holland List
河流流域管理部門	*heliu liuyu guanli bumen*	river basin authority
河流生態系統中的 指示生物	*heliu shengtai xitongzhongde zhishi shengwu*	bioindicators in river ecosystems
河流質量	*heliu zhiliang*	river quality
河流質量目標	*heliu zhiliang mubiao*	river quality goal
合同	*hetong*	contracts
宏觀經濟問題	*hongguan jingji wenti*	macroeconomic aspects
紅外分析	*hongwai fenxi*	IR
紅外光譜分析	*hongwai guangpu fenxi*	infrared spectral analysis
後燃燒室	*houranshaoshi*	post-combustion chamber

化工過程的停留時間	*huagong guochengde tingliu shijian*	residence time in chemical processing
環(實驗室)測試	*huan (shiyanshi) ceshi*	ring-tests
緩衝池	*huanchongchi*	buffer tanks
環境管理中的刺激手段	*huanjin guanlizhongde ciji shouduan*	incentives in environmental management
環境	*huanjing*	environment
環境報告	*huanjing baogao*	environmental report
環境保護	*huanjing baohu*	environmental protection
環境保護費用	*huanjing baohu feiyong*	environmental protection costs
環境保護投資費用	*huanjing baohu touzi feiyong*	investment costs, environmental protection
環境保護政策報告	*huanjing baohu zhengce baogao*	policy statement on environmental protection
環境保護的管理手段	*huanjing baohude guanli shouduan*	management instruments in environmental protection
環境費用的管理	*huanjing feiyongde guanli*	management of environmental costs
環境分析	*huanjing fenxi*	environmental analysis
環境負責組織聯盟	*huanjing fuze zuzhi liangmeng*	Coalition for Environmentally Responsible Economies
環境負責組織聯盟原則	*huanjing fuze zuzhi lianmeng yuanze*	CERES Principles
環境工作報告	*huanjing gongzuo baogao*	environmental performance report
環境管理體系	*huanjing guanli tixi*	environmental management systems
環境經濟學	*huanjing jingjixue*	environmental economics
環境技術	*huanjing jishu*	environmental technology
環境科技人員	*huanjing keji renyuan*	environmental scientists
環境可接受的	*huanjing kejieshoude*	environmentally acceptable
環境可接受的產品	*huanjing kejieshoude chanpin*	environmentally acceptable products
環境空氣	*huanjing kongqi*	ambient air
環境空氣目標	*huanjing kongqi mubiao*	ambient air goals
環境空氣濃度	*huanjing kongqi nongdu*	ambient air concentrations
環境控制	*huanjing kongzhi*	environmental controlling
環境框架法	huanjing kuangjiafa	environmental framework laws

環境立法	huanjing lifa	environmental legislation
環境立法中的經濟手段	huanjing lifa zhongde jingji shouduan	economic instruments in environmental legislation
環境清單	huanjing qingdan	environmental inventory
環境人員	huanjing renyuan	environmental officer
環境審計	huanjing shenji	environmental audit
環境數據	huanjing shuju	environmental data
環境損害費用	huanjing sunhai feiyong	damage costs, environmental
環境損害的社會費用	huanjing sunhaide shehui feiyong	social costs from environmental damages
環境投資，回收期	huanjing touzi, huishouqi	environmental investments, pay-back time
環境行為評級	huanjing xingwei pingji	environmental performance rating
環境行為指標	huanjing xingwei zhibiao	environmental performance indicators
環境影響報告	huanjing yingxiang baogao	environmental impact report
環境影響評價	huanjing yingxiang pingjia	environmental impact assessment
環境噪聲	huanjing zaosheng	ambient noise
環境噪聲(散發)限值	huanjing zaosheng (sanfa) xianzhi	ambient noise (immission) limits
環境責任	huanjing zeren	environmental liability
環境政策	huanjing zhengce	environmental policy
環境政策報告	huanjing zhengce baogao	environmental policy statement
環境質量目標	huanjing zhiliang mubiao	environmental quality goals
環境狀況報告	huanjing zhuangkuang baogao	environmental statement
環境的組成部分	huanjingde zucheng bufen	environmental compartments
環境的組成部分	huanjingde zucheng bufen	environmental segment
環形噴射器	huanxing pensheqi	ring-jet unit
化學廢物	huaxue feiwu	chemical waste
化學濕式分析	huaxue shishi fenxi	chemical wet analysis
化學吸收	huaxue xishou	chemisorption
化學需氧總量	huaxue xuyang zongliang	COD load
化學需氧量	huaxue xuyangliang	chemical oxygen demand
化學需氧量分析	huaxue xuyangliang fenxi	COD analysis
揮發性有機化合物	huifaxing youji huahewu	volatile organic compounds
揮發性有機化合物的收費	huifaxing youji huahewude shoufei	levy on VOC input

揮發性有機鹵素 化合物	*huifaxing youji lusu huahewu*	VOX
恢復	*huifu*	rehabilitation
恢復補救費用	*huifu bujiu feiyong*	costs of remediation
恢復補救，項目管理	*huifu bujiu, xiangmu guanli*	remediation, project management
恢復措施	*huifu cuoshi*	rehabilitation measures
恢復／補救	*huifu/bujiu*	remediation
恢復／補救目標	*huifu/bujiu mubiao*	remediation goals
灰能	*huineng*	grey energy
回收期	*huishouqi*	pay back time
回轉窯	*huizhuanyao*	rotary kiln
回轉窯焚燒爐	*huizhuanyao fenshaolu*	rotary kiln incinerator
互聯數據庫	*hulian shujuku*	interconnected databases
混合液	*hunheye*	mixed liquor
混凝	*hunning*	coagulation
混凝劑	*hunningji*	coagulating agents
活化池(塔)	*huohuachi (ta)*	activation basin (tower)
活性表面	*huoxing biaomian*	active surface
活性污泥	*huoxing wuni*	activated sludge
活性污泥的熱衛生 處理	*huoxing wunide reweisheng chuli*	thermal hygienization of activated sludge
活性污泥的適應 （馴化）	*huoxing wunide shiyin (xunhua)*	adaptation of activated sludge
活性碳	*huoxingtan*	activated carbon
活性碳緩衝系統	*huoxingtan huanchong xitong*	activated carbon buffer system
活性碳解吸	*huoxingtan jiexi*	desorption from activated carbon
活性碳吸附	*huoxingtan xifu*	adsorption on activated carbon
活性碳再生	*huoxingtan zaisheng*	regeneration of activated carbon
活性污泥BOD_5負荷	*huoxingwuni BOD_5 fuhe*	BOD_5 loading of biosludge
活性污泥技術	*huoxingwuni jishu*	activated sludge technology
活性污泥池	*huoxingwunichi*	activated sludge basin
活性物質	*huoxingwuzhi*	active substance
火焰等離子體	*huoyan dengliziti*	flame plasma
火焰電離檢測器	*huoyan dianli jianceqi*	flame ionization detectors
火焰電離檢測器 分析儀	*huoyan dianli jianceqi fenxiyi*	FID analyzer
呼吸測定法	*huxi cedingfa*	respirometry

| 湖沼生態系統 | huzhao shengtai xitong | limnic ecosystems |

J

檢測極限	jiance jixian	limit of detection
檢測器信號	jianceqi xinhao	detector signal
降水水	jiangshui shui	precipitation water
間接費用	jianjie feiyong	indirect costs
間接的污染控制費用	jianjiede wuran kongzhi feiyong	indirect pollution control costs
建設工程的生態評價	jianshe gongchengde shengtai pingjia	ecological assessment (of a project)
減壓段	jianyaduan	decompression stage
建議制度（人事管理）	jianyi zhidu (renshi guanli)	suggestion system (personnel management)
建議箱	jianyixiang	suggestion box
建築廢物	jianzhu feiwu	construction waste
校準標準	jiaozhun biaozhun	calibration standards
校準標準氣	jiaozhun biaozhunqi	gas standards for calibration
校準混合氣	jiaozhun hunheqi	calibration gas mixtures
校準，氣相色譜法	jiaozhun, qixiang sepufa	calibration, gas chromatography
價值	jiazhi	values
價值創造鏈	jiazhi chuangzaolian	chain of value creation
傑恩－魏倫斯測試	jie'en-weilunsi ceshi	Zahn-Wellens Test
接觸層（帶）	jiechucheng (dai)	contact zone
借貸的基建費用	jiedaide jijian feiyong	costs of borrowed capital
結果的重復性	jieguode chongfuxing	repeatability of results
結論的有效性	jielunde youxiaoxing	validity of a conclusion
接種	jiezhong	inoculation
激光在分析中的應用	jiguan zaifenxizhongde yingyong	laser in analysis
基建投資	jijian touzi	capital investments
基建項目的環境影響	jijian xiangmude huanjing yingxiang	environmental impact of capital projects
浸出液	jinchuye	leachate
浸出液測試	jinchuye ceshi	leachate test
警報計劃	jingbao jihua	alarm plan
靜電除塵器（乾式）	jingdian chuchenqi (ganshi)	electrostatic precipitators, dry

靜電除塵器（濕式）	*jingdian chuchenqi (shishi)*	electrostatic precipitators, wet
靜電分離器	*jingdianfenliqi*	electrostatic separator
景觀保護	*jingguan baohu*	landscape protection
經濟刺激	*jingji ciji*	economic incentives
經濟合作與發展組織	*jingji hezuo yu fazhan zuzhi*	OECD
經濟合作與發展 組織-生物需氧量 測定	*jingji hezuo yu fazhan zuzhi- shengwu xuyangliang ceding*	OECD-BOD Test
逕流水	*jingliushui*	runoff water
精密氣體標準	*jingmi qiti biaozhun*	precision gas standards
靜態平衡	*jingtai pingheng*	static equilibrium
靜態燃燒室焚燒爐	*jingtai ranshaoshi fenshaolu*	static chamber incinerator
緊急事故管理	*jinji shigu guanli*	emergency management
緊急響應管理	*jinji xiangying guanli*	emergency response management
技術的管理	*jishude guanli*	management of technology
集水池	*jishuichi*	catch basin
急性效應	*jixing xiaoying*	acute effect
聚電解質	*judianjiezhi*	polyelectrolyte
決策方法（技術）	*juece fangfa (jishu)*	decision techniques
決策因子	*juece yinzi*	decision factors
聚結	*jujie*	coalescence
聚結器	*jujieqi*	coalescer unit
均衡	*junheng*	equalization
均衡池	*junhengchi*	equalization tanks
聚醯胺膜	*juxian'anmo*	polyamide membranes
具有生態效益的活動 或行動	*juyou shengtai xiaoyide huodong huo xingdong*	ecoefficient activities or actions

K

開式物流系統	*kaishi wuliu xitong*	open mass flow system
喀斯特水	*kasite shui*	karst water
可持續發展	*kechixu fazhan*	sustainable development
可獲得而經濟的 最佳技術	*kehuode 'er jingjide zuijia jishu*	best available technology economically achievable
可獲得的最佳技術	*kehuodede zuijia jishu*	best available technology
可靠性	*kekaoxing*	reliability
顆粒物	*keliwu*	particulates

可氯提的有機鹵素化合物	*keqitide youji lusu huahewu*	POX
可生物降解的	*keshengwu jiangjiede*	biodegradable
可生物降解性的連續測定(哈斯曼試驗)	*keshengwu jiangjiexingde lianxu ceding (hasiman shiyan)*	continuous measurement of biodegradability (Husmann Test)
克氏氮分析	*keshidan fenxi*	Kjeldahl nitrogen analysis
科特雷爾靜電分離器	*ketele sier jingdian fenliqi*	Cottrell separator, electrostatic
可提取的有機鹵素化合物	*ketiqude youji lusu huahewu*	EOX
可吸附的有機鹵素化合物	*kexifude youji lusu huahewu*	absorbable organic halogen compounds
可吸附的有機硫	*kexifude youjiliu*	AOS
可再生能源	*kezaisheng nengyuan*	renewable energy
可再生資源	*kezaisheng ziyuan*	renewable resources
空氣	*kongqi*	air
空氣污染	*kongqi wuran*	air pollution
空氣污染控制	*kongqi wuran kongzhi*	air pollution control
空氣污染控制觀念	*kongqi wuran kongzhi guannian*	APC-concept
空氣污染控制技術	*kongqi wuran kongzhi jishu*	air pollution control technology
空氣污染控制設備	*kongqi wuran kongzhi shebei*	air pollution control equipment
空氣污染控制裝置的壓差	*kongqi wuran kongzhi zhuangzhi de yacha*	pressure difference in APC units
空氣污染控制中的催化過程	*kongqi wuran kongzhizhongde cuihua guocheng*	catalytic processes in APC
空氣污染造成的森林破壞	*kongqi wuran zaochengde senlin pohuai*	forest damages through air pollution
空氣污染的指示生物	*kongqi wurande zhishi shengwu*	air pollution bioindicators
空氣質量目標	*kongqi zhiliang mubiao*	air quality goals
空氣和氧噴射系統	*kongqiheyang penshe xitong*	air and oxygen injection systems
空氣氣提	*kongqiqiti*	air stripping
控制過濾器	*kongzhi guolüqi*	police filter
控制和報告體系	*kongzhi he baogao tixi*	controlling and reporting systems
控制論	*kongzhilun*	cybernetics
控制論管理方法	*kongzhilun guanli fangfa*	cybernetic management approach

礦質化	*kuangzhihua*	mineralization
擴散模式	*kuosan moshi*	dispersion models
擴散排放	*kuosan paifang*	diffuse emissions
擴散排放源	*kuosanpaifang yuan*	diffuse emission source
擴散器	*kuosanqi*	diffusors

L

垃圾袋處置費	*lajidai chuzhifei*	garbage bag disposal fee
蘭達傳感器	*landa chuanganqi*	lambda sensor
勞動衛生	*laodong weisheng*	work hygiene
冷壁式焚燒爐	*lengbishi fenshaolu*	cold wall incinerator
冷壁式焚燒爐	*lengbishi fenshaolu*	incinerator, cold wall
冷等離子體電極	*lengdengl(izi)ti dianji*	cold plasma electrode
冷卻和液化	*lengque he yehua*	cooling and liquefaction
冷卻水	*lengqueshui*	cooling water
聯合國環境規劃署 廢物分類系統	*liangheguo huanjing guihuashu feiwu fenlei xitong*	UNEP waste classification system
聯合國環境規劃署	*lianheguo huanjing guihuashu*	UNEP
聯合國環境規劃署 廢物轉運公約	*lianheguo huanjing guihuashu feiwu zhuanyun gongyue*	UNEP Convention on waste transports
連續分析	*lianxu fenxi*	continuous analysis
立法中的排污收費	*lifazhongde paiwu shoufei*	discharge levy in legislation
理論需氧量	*lilun xuyangliang*	theoretical oxygen demand
理論板	*lilunban*	theoretical plate
林多克斯系統	*linduokesi xitong*	LINDOX system
零風險目標	*lingfengxian mubiao*	zero-risk goal
零污染目標	*lingwuran mubiao*	zero-pollution goal
磷酸鹽去除	*linsuanyan quchu*	phosphate removal
流動相	*liudongxiang*	fluid phase
流化床焚燒	*liuhuachuang fenshao*	fluidized bed incineration
流化床生物處理	*liuhuachuang shengwu chuli*	fluidized bed biology
流化床系統	*liuhuachuang xitong*	fluidized bed systems
流量平均樣	*liuliang pingjunyang*	flow average sample
離心分離力	*lixin fenlili*	centrifugal separation forces
利用植物恢復補救	*liyong zhiwu huifu bujiu*	phytoremediation
粒狀活性碳處理	*lizhuang huoxingtan chuli*	granular activated carbon treatment

粒狀活性碳過濾器	lizhuang huoxingtan guolüqi	granular activated carbon filter
離子分析	lizi fenxi	ion analysis
離子交換	lizi jiaohuan	ion exchange
離子交換分離	lizi jiaohuan fenli	ion exchange separation
離子交換樹脂	lizi jiaohuan shuzhi	ion exchange resins
離子交換柱	lizi jiaohuanzhu	ion exchange columns
離子色譜法	lizi sepufa	ion chromatography
離子色譜法，化學抑制	lizi sepufa, huaxue yizhi	ion chromatography, chemical suppression
離子色譜法中的抑制作用	lizi sepufazhongde yizhi zuoyong	suppression in ion chromatography
離子選擇(特效性)電極	lizi xuanze (texiaoxing) dianji	ion specific electrode
爐箅式焚燒爐	lubishi fenshaolu	grate incinerator
陸地生態系統	ludi shengtai xitong	terrestrial ecosystems
氯氟烴	lüfuting	chloro fluoro carbons
氯化	lühua	chlorination
氯化烴	lühuating	chlorinated hydrocarbons
羅馬俱樂部	luoma julebu	Club of Rome
洛普羅克斯技術（拜耳公司）	luopuluokesi jishu (bai'er gongsi)	LOPROX technology (Bayer)
陸圈	luquan	geosphere
綠色產品	lüse chanpin	green products
爐渣	luzha	slag

M

慢性效應	manxing xiaoying	chronic effect
毛細管氣相色譜法	maoxiguan qixiang sepufa	capillary gas chromatography
毛細管柱	maoxiguanzhu	capillary column
美國材料試驗學會標準	meiguo cailiao shiyan xuehui biaozhun	ASTM Standards
美國政府職業安全與衛生署	meiguo zhengfu zhiye anquan yu weisheng shu	OSHA
蒙特利爾議定書	mengteli'er yidingshu	Montreal Protocol
免費物品	mianfei wupin	free goods
免疫測定法分析	mianyi cedingfa fenxi	immunoassay analysis
膜	mo	membrane

末端處理措施	moduan chuli cuoshi	end-of-pipe measures
末端改善措施	moduan gaishan cuoshi	add-on corrective measures
末端工藝過程處理	moduan gongyi guocheng chuli	end-of-process treatment
模擬記錄器	moni jiluqi	analog recorder
模擬信號	moni xinhao	analog signal
目標管理	mubiao guanli	management by objectives
母液	muye	mother liquor

N

耐火襯	naihuochen	refractory lining
難降解化合物	nanjiangjie huahewu	recalcitrant compounds
難降解污染物	nanjiangjie wuranwu	refractory pollutants
難降解的離子化合物	nanjiangjiede lizi huahewu	refractory ionic compounds
內部費用	neibu feiyong	internal costs
內部的污染者付費原則	neibude wuranzhe fufei yuanze	internal polluter-pays-principle
內部化的外部費用	neibuhuade waibu feiyong	internalizing external costs
能量	nengliang	energy
能量共生	nengliang gongsheng	co-generation of energy
能量回收	nengliang huishou	energy recuperation
能量轉換	nengliang zhuanhuan	energy transformation
能源，礦物資源	nengyuan, kuangwu ziyuan	energy, fossil resources
泥漿	nijiang	slurry
逆流	niliu	counter-current
凝集	ningji	agglomeration
逆滲透過程	nishentou guoceng	reversed osmosis process
逆滲透過程的持留液	nishentou guochengde chiliuye	retentate in reversed osmosis
逆滲透技術	nishentou jishu	reversed osmosis technology
逆相高壓液相色譜法	nixiang gaoya yexiang sepufa	reversed phase HPLC
濃度限值	nongdu xianzhi	concentration limits
濃度限值與總量限值	nongdu xianzhi yu zongliang xianzhi	concentration limits versus mass flow limits
濃縮液	nongsuye	concentrate

O

| 歐盟《生態管理和審計體系》指令 | oumeng shengtai guanli he shenji tixi zhiling | EMAS Directive of the European Union |

偶然誤差	*ouran wucha*	random errors
歐洲管理和審計指令	*ouzhou guanli he shenji zhiling*	European Management and Auditing Directive
歐洲化學工業聯合會中心	*ouzhou huaxue gongye lianhehui zhongxin*	CEFIC

P

pH	*pH*	pH
pH計	*pHji*	pH meter
排放	*paifang*	emission
排放	*paifang*	immission
排放穿透	*paifang chuantou*	emission breakthrough
排放清單	*paifang qingdan*	emission inventory
排放數據庫	*paifang shujuku*	emission database
排放限值	*paifang xianzhi*	discharge limit
排放限值	*paifang xianzhi*	emission limit
排放限值	*paifang xianzhi*	immission limits
排放因子	*paifang yinzi*	emission factor
排放權	*paifangquan*	emission right
排放權交易	*paifangquan jiaoyi*	trading of emission rights
排放權證的初始價格	*paifangquanzheng, chushi jiage*	emission right certificates, initial price
排放權證的市場價格	*paifangquanzheng, shicang jiage*	emission right certificates, market price
排放源	*paifangyuan*	emission sources
排水收費	*paishui shoufei*	effluent levy
排水收費法	*paishui shoufeifa*	law on effluent levies
排污費	*paiwufei*	discharge fees
排序方法	*paixu fangfa*	ranking methods
排序判據	*paixu panju*	ranking criteria
帕克斯系統	*pakesi xitong*	Paques System
泡沫控制	*paomo kongzhi*	foam control
泡罩原理	*paozhao yuanli*	bubble principle
泡罩塔	*paozhaota*	bubble column
泡罩塔洗滌器	*paozhaota xidiqi*	bubble column scrubber
噴淋折板	*penlin zheban*	spray deflector
噴淋塔(洗滌器)	*penlinta (xidiqi)*	spray tower (scrubber)

噴射洗滌器	*penshe xidiqi*	jet scrubber
噴射器	*pensheqi*	injector
平板沉積測試	*pingban chenji ceshi*	plate deposition test
平衡	*pingheng*	equilibrium
平流層臭氧	*pingliuceng chouyang*	stratospheric ozone
平爐焚燒爐	*pinglu fenshaolu*	open hearth incinerator
貧營養的	*pinyingyangde*	oligotrophic
破乳	*poru*	splitting of emulsions

Q

前澄清池	*qianchengqingchi*	preclarifier
千分率	*qianfenlw*	per mille or per mil
纖毛蟲	*qianmaochong*	ciliata
潛在風險	*qianzai fengxian*	potential risk
汽車催化轉化器	*qiche cuihua zhuanhuaqi*	automotive catalytic converters
汽車尾氣排放	*qiche weiqi paifang*	traffic emissions
氣候變化	*qihou bianhua*	climatic changes
清除淨化費用	*qingchu jinghua feiyong*	clean up costs
清潔工藝	*qingjie gongyi*	clean technologies
清潔化學	*qingjie huaxue*	clean chemistry
清潔生產過程	*qingjie shengchan guocheng*	clean production processes
親油(脂)的	*qingyou(zhi)de*	lipophilic
氣溶膠	*qirongjiao*	aerosol
氣溶膠分離	*qirongjiao fenli*	aerosol separation
氣提	*qiti*	stripping
氣體檢測管	*qiti jianceguan*	gas detector tubes
氣體監測器	*qiti jianceqi*	gas monitor
氣體平衡管	*qiti pingheng guan*	gas balancing pipes
氣提塔	*qitita*	stripping column
氣味	*qiwei*	odor
氣相色譜－質譜分析法	*qixiang sepu-zhipu fenxifa*	gas chromatography-mass spectroscopy
氣相色譜法	*qixiang sepufa*	gas chromatography
氣相色譜法保留時間	*qixiang sepufa baoliu shijian*	retention time in gas chromatograph
氣相色譜法使用的滲透管	*qixiang sepufa shiyongde shentouguan*	permeation tube in gas chromatography

全球大氣的二氧化碳	quanqiu daqide eryanghuatan	global atmospheric carbon dioxide
缺氧生物降解	queyang shengwu jiangjie	anoxic biodegradation
區劃法	quhuafa	zoning laws
群落生境	qunluo shengjing	biotope
群落生態學	qunluo shengtaixue	synecology
去所	qusuo	sink
取樣程序	quyang chengxu	sampling procedure
取樣和測量口	quyang he celiangkou	sampling and measuring ports

R

染色廢水	ranse feishui	dye house effluents
燃燒	ranshao	combustion
燃燒室	ranshaoshi	combustion chamber
熱交換器	rejiaohuanqi	heat exchanger
熱解	rejie	pyrolysis
熱解吸	rejiexi	thermodesorption
人（類）的活動	ren (lei) de huodong	anthropogenic activities
人為排放	renwei paifang	anthropogenic emissions
人員參與制度	renyuan canyu zhidu	personnel participation schemes
人員激勵制度	renyuan jili zhidu	personnel incentive systems
熱氧化系統	reyanghua xitong	thermal oxidation systems
溶解性有機碳	rongjiexing youjitan	dissolved organic carbon
溶解氧	rongjieyang	DO
容許日攝入量	rongxu risheruliang	acceptable daily intake
乳液分離	ruye fenli	separation of emulsions
乳液	ruyi	emulsion

S

色譜法	sepufa	chromatography
砂礫	shali	grit
砂濾池	shalüchi	sandfilter
設備廢氣	shebei feiqi	object off-gas
社會費用原則	shehui feiyong yuanze	social cost principle
射流曝氣系統	sheliu baoqi xitong	jet aeration system
生產費用	shengchan feiyong	manufacturing costs
生產工藝文件資料	shengchan gongyi wenjian ziliao	process documentation

497

聲功率級	shenggonglüji	sound power level
生活質量	shenghuo zhiliang	quality of life
生境	shengjing	habitat
升流式厭氧污泥床處理	shengliushi yanyang wuni chuang chuli	upflow anaerobic sludge blanket treatment
生命科學	shengming kexue	life sciences
生命周期分析	shengming zhouqi fenxi	life cycle analysis
生態毒性	shengtai duxing	ecotoxicity
生態化學	shengtai huaxue	ecochemistry
生態基線評價	shengtai jixian pingjia	ecological base line assessment
生態平衡	shengtai pingheng	ecobalance
生態平衡	shengtai pingheng	ecological equilibrium
生態評級	shengtai pingji	eco-rating
生態群落	shengtai qunluo	ecological community
生態設計	shengtai sheji	eco-design
生態系統	shengtai xitong	ecosystem
生態系統的非生物組成部分	shengtai xitongde feishengwu zucheng bufen	abiotic components of an ecosystem
生態系統中的生物組成部分	shengtai xitongzhongde shengwu zucheng bufen	biotic components of ecosystems
生態系統中的物流循環	shengtai xitongzhongde wuliu xunhuan	mass flow cycles in ecosystems
生態因素	shengtai yinsu	ecological factors
生態種群	shengtai zhongqun	ecological population
生態圈	shengtaiquan	ecosphere
生態系統中的最終去所	shengtaixitongzhongde zuizhong qusuo	final sink in the ecosystem
生態學	shengtaixue	ecology
生物測定	shengwu ceding	bio-assay
生物處理	shengwu chuli	biological treatment
生物處理池的隔間	shengwu chulichide gejian	compartments in biological basins
生物堆肥	shengwu duifei	bio-compost
生物多樣性	shengwu duoyangxing	biological diversity
生物放大	shengwu fangda	biomagnification
生物反應器(流化床)	shengwu fanyingqi (liuhuachuang)	bioreactor, fluidized bed

生物富集	*shengwu fuji*	bioconcentration
生物活性碳吸附裝置	*shengwu huoxingtan xifu zhuangzhi*	biological activated carbon adsorber
生物監測	*shengwu jiance*	biomonitoring
生物降解	*shengwu jiangjie*	biodegradation
生物降解速率	*shengwu jiangjie sulü*	biodegradation rate
生物降解的抑制	*shengwu jiangjiede yizhi*	inhibition of biodegradation
生物技術	*shengwu jishu*	biotechnology
生物科學	*shengwu kexue*	bio sciences
生物濾池	*shengwu lüchi*	biofilter
生物平衡	*shengwu pingheng*	biotic equilibrium
生物群落	*shengwu qunluo*	biocenose
生物殺傷劑	*shengwu shashangji*	biocides
生物生態因素	*shengwu shengtai yinsu*	biotic ecological factors
生物適應(馴化)	*shengwu shiying (xunhua)*	bio adaptation
生物污泥	*shengwu wuni*	biosludge
生物污泥/活性碳 系統	*shengwu wuni/ huoxingtan xitong*	biosludge/activated carbon systems
生物洗滌器(廢氣)	*shengwu xidiqi (feiqi)*	bio scrubber, off-gas
生物需氧量	*shengwu xuyangliang*	biological oxygen demand
生物轉盤	*shengwu zhuanpan*	rotating biological contactor
生物法恢復補救	*shengwufa huifu bujiu*	bio-remediation
生物法空氣污染 控制系統	*shengwufa kongqi wuran kongzhi xitong*	biological APC-systems
生物量	*shengwuliang*	biomass
生物膜	*shengwumo*	biofilm
生物氣(沼氣)	*shengwuqi (zhaoqi)*	biogas
生物圈	*shengwuquan*	biosphere
生物學	*shengwuxue*	biology
聲壓級	*shengyaji*	sound pressure level
聲音	*shengyin*	sound
聲音發散	*shengyin fasan*	sound emission
聲音發散限值	*shengyin fasan xianzhi*	sound emission limits
聲音的傳播	*shengyinde chuanbo*	sound transmission
聲音的影響	*shengyinde yingxiang*	impact of sound
審計	*shenji*	audit
審計標準	*shenji biaozhun*	audit standards

審計程序	shenji chengxu	audit procedure
審計結果	shenji jieguo	audit findings
審計小組	shenji xiaozu	audit team
深井曝氣技術	shenjing baoqi jishu	deep shaft technology
審計者	shenjizhe	auditor
滲流水	shenliushui	seepage water
滲透(作用)	shentou (zuoyong)	osmosis
滲透水	shentoushui	permeating water
滲透壓	shentouya	osmotic pressure
滲透液	shentouye	permeate
食草動物	shicao dongwu	herbivore
濕地	shidi	wetlands
事故報告渠道	shigu baogao qudao	reporting channels in accidents
事故貯水池	shigu chushuichi	retention basins
事故遏制措施	shigu ezhi cuoshi	containment measures
事故控制計劃	shigu kongzhi jihua	accident control plan
事故釋放	shigu shifang	accidental releases
事故影響	shigu yingxiang	impact of an accident
時間範疇上的難題	shijian fanchoushangde nanti	time horizon dilemma
時間平均樣	shijian pingjunyang	time average sample
世界環境工業委員會	shijie huanjing gongye weiyuanhui	World Industry Council for the Environment
世界可持續發展工商委員會	shijie kechixu fazhan gongshang weiyuanhui	World Business Council for Sustainable Development
室內空氣	shineikongqi	room air
食肉動物	shirou dongwu	carnivore
濕式靜電除塵器	shishi jingdian chuchenqi	wet electrostatic precipitator
濕式空氣污染控制系統	shishi kongqi wuran kongzhi xitong	wet APC-systems
實時數據	shishi shuju	real-time data
濕式袖式過濾器	shishi xiushi guolüqi	wet sleeve filter
濕式煙氣處理	shishi yanqi chuli	wet flue gas treatment
濕式煙氣處理系統	shishi yanqi chuli xitong	flue gas treatment, wet systems
食物/生物量比	shiwu/shengwuliang bi	food to biomass ratio
食物鏈	shiwulian	food chain
食物鏈中的積累	shiwulianzhongde jilei	accumulation in the food chain
試驗方法的準確度	shiyan fangfade zhunquedu	accuracy of a test method

試驗方法的精密度	*shiyang fangfede jingmidu*	precision of a test method
試樣母體	*shiyang muti*	matrix of a sample
試驗數據總體	*shiyang shuju zongti*	population of test data
實驗室廢水	*shiyanshi feishui*	laboratory effluents
實驗室間的相互測試	*shiyanshijiande xianghu ceshi*	inter-laboratory tests
試驗魚	*shiyanyu*	test fish
十億分之(幾分)	*shiyifenzhi (jifen)*	ppb
收費	*shoufei*	levy
受控處置場	*shoukong chuzhichang*	controlled disposal site
受控填埋場	*shoukong tianmaichang*	controlled landfill
受納水體	*shouna shuiti*	receiving waterbody
雙重包封	*shuangchong baofeng*	double containment
雙介質過濾	*shuangjiezhi guolü*	dual media filtration
水	*shui*	water
稅	*shui*	tax
水數據庫	*shui shujuku*	water database
水圈	*shuiquan*	hydrosphere
水生生物毒理學	*shuisheng shengwu dulixue*	aquatic toxicology
水體自淨	*shuiti zijing*	self-purification of a waterbody
水體的自動再生	*shuitide zidong zaisheng*	auto regeneration of a waterbody
水污染控制	*shuiwuran kongzhi*	water pollution control
水污染控制觀念	*shuiwuran kongzhi guannian*	water pollution control concept
水污染控制技術	*shuiwuran kongzhi jishu*	water pollution control technology
水域管理	*shuiyu guanli*	water management basin
水中溶解氧	*shuizhong rongjieyang*	dissolved oxygen in water
水質目標	*shuizi mubiao*	water quality goals
數據管理系統	*shuju guanli xitong*	data management systems
數據記錄儀	*shuju jiluyi*	data logger
數據類別，空氣質量	*shuju leibie, kongqi zhiliang*	data types, air quality
瞬時樣	*shunshiyang*	grab sample
數學離散模式	*shuxue lisan moshi*	mathematical dispersion models
數字記錄儀	*shuzi jiluyi*	digital recorder
數字信號	*shuzi xinhao*	digital signal
酸性水	*suanxingshui*	sour water
酸雨	*suanyu*	acid rain

隨身監測	*suishen jiance*	man monitoring
塑料廢物	*suliao feiwu*	plastic wastes
塑料廢物標記	*suliao feiwu biaoji*	codes for plastic wastes
塑料廢物標記	*suliao feiwu biaoji*	plastic waste codes
損害恢復費用	*sunhai huifu feiyong*	damage repair costs
損害預防費用	*sunhai yufang feiyong*	damage prevention costs

T

碳循環	*tanxunhuan*	carbon cycle
塔式滴濾池(廢水)	*tashi delüchi (feishui)*	trickling towers (effluents)
塔式生物反應器	*tashi shengwu fanyingqi*	tower bioreactor
塔填料	*tatianliao*	column packing
特殊廢物	*teshu feiwu*	special wastes
特殊廢物處置的管理	*teshu feiwu chuzhide guanli*	management of special waste disposal
特殊廢物焚燒爐	*teshu feiwu fenshaolu*	special waste incinerator
特殊廢物分析	*teshu feiwu fenxi*	analysis of special wastes
特殊廢物分析	*teshu feiwu fenxi*	special waste analysis
特殊廢物管理	*teshu feiwu guanli*	special waste management
特殊廢物，定義	*teshu feiwu, dingyi*	special wastes, definitions
特殊廢物的管理	*teshu feiwude guanli*	management of special wastes
填充段(區)	*tianchongduan (qu)*	packing zone
填料洗滌塔	*tianliao xidita*	packed column scrubber
填料塔	*tianliaota*	packed column
填埋氣體	*tianmai qiti*	landfill gas
填埋場中浸出液的控制和處理	*tianmaichangzhong jinchuyede kongzhi he chuli*	leachate control and treatment in landfill
梯度洗脫	*tidu xiduo*	gradient elution
體積/體積比	*tiji/tiji bi*	volume/volume ratio
同化作用	*tonghua zuoyong*	assimilation
統計	*tongji*	statistics
投資費用	*touzi feiyong*	investment costs
投資收益率	*touzi shouyilw*	return on investment
脫臭方法	*tuochou fangfa*	deodorizing processes
脫氣區	*tuoqiqu*	degassing zone
土壤	*turang*	soil
土壤生態系統	*turang shengtai xitong*	soil ecosystem

W

瓦爾德之原則	wa'erdezhi yuanze	Valdez Principles
外部費用	waibu feiyong	external costs
外部環境審計	waibu huanjing shenji	external environmental audit
完整的工藝過程開發	wanzhengde gongyi guocheng kaifa	integral process development
微觀經濟問題	weiguan jingji wenti	microeconomic aspects
微過濾	weiguolü	microfiltration
維護維修費用	weihu weixiu feiyong	maintenance and repair costs
微克	weike	microgram
微粒沉降	weili chenjiang	precipitation of particulates
微米	weimi	micrometer
尾氣氧化處理	weiqi yanghua chuli	oxidative off-gas treatment
衛生化處理	weishenghua chuli	hygienization
微生物處理	weishengwu chuli	microbiological treatment
微生物礦質化	weishengwu kuangzhihua	microbiological mineralization
微生物群落	weishengwu qunluo	microcenose
微生物學	weishengwuxue	microbiology
微微克	weiweike	picogram
危險廢物	weixian feiwu	hazardous waste
維修與維護	weixiu yu weihu	repair and maintenance
文丘里管道	wenqiuli guandao	venturi canal
文丘里型洗滌器	wenqiulixing xidiqi	venturi type scrubbers
溫室氣體	wenshi qiti	greenhouse gases
溫室效應	wenshi xiaoying	greenhouse effect
無規填料塔	wugui tianliaota	randomly packed column
無可見作用濃度	wukejian zuoyong nongdu	no observable effect concentration
無可見作用水平	wukejian zuoyong shuiping	no observable effect level
物料平衡	wuliao pingheng	material balance
物流	wuliu	mass flow
污泥	wuni	sludge
污泥處理	wuni chuli	sludge treatment
污泥惡臭	wuni echou	sludge fouling
污泥發酵	wuni fajiao	sludge fermentation
污泥乾化床	wuni ganhuachuang	sludge drying bed

污泥過濾	wuni guolü	sludge filtration
污泥好氧消化	wuni haoyang xiaohua	sludge digestion, aerobic
污泥揮發分濃度	wuni huifafen nongdu	MLVSS
（混合液揮發性懸	(hunheye huifaxing	
浮固體）	xuanfuguti)	
污泥回流	wuni huiliu	sludge recycling
污泥離心分離	wuni lixin fenli	sludge centrifugation
污泥滅菌	wuni miejun	sterilization of sludge
污泥泥齡	wuni niling	sludge age
污泥濃度（混合液	wuni nongdu (hunheye	MLSS
懸浮固體）	xuanfuguti)	
污泥濃縮池	wuni nongsuchi	sludge thickener
污泥調節	wuni tiaojie	sludge conditioning
污泥體積指數	wuni tiji zhishu	sludge volume index
污泥脫水	wuni tuoshui	sludge dewatering
污泥衛生處理	wuni weisheng chuli	sludge hygienization
污泥厭氧消化	wuni yanyang xiaohua	sludge digestion, anaerobic
污染產生者	wuran changshengzhe	generator of pollution
污染防治設備	wuran fangzhi shebei	anti-pollution equipment
污染負荷	wuran fuhe	pollution load
污染控制費用的分攤	wuran kongzhi feiyongde	allocation of pollution control
	fentan	costs
污染控制技術	wuran kongzhi jishu	pollution control technologies
污染控制的標準	wuran kongzhide biaozhun	standard unit costs for pollution
單位成本	danwei chengben	control
污染控制的市場手段	wuran kongzhide shichang	market instruments in pollution
	shouduan	control
污染控制中的現代	wuran kongzhizhongde	state-of-the-art in pollution control
工藝技術	xiandai gongyi jishu	
污染預防	wuran yufang	pollution prevention
污染物	wuranwu	contaminant
污染物	wuranwu	pollutant
污染物擴散	wuranwu kuosan	dispersion of a pollutant
污染者	wuranzhe	polluter
污染者付費原則	wuranzhe fufei yuanze	polluter-pays-principle
污染者之鏈	wuranzhezhilian	chain of polluters
污水	wushui	sewage

污水處理廠	*wushui chulichang*	sewage treatment works
污水管理區	*wushui guanliqu*	sewage management districts
污水流量(設計流量)	*wushui liuliang (sheji liuliang)*	sewage flow volumes (design values)
五天生物需氧量	*wutian shengwu xuyangliang*	BOD
五天生物需氧量分析	*wutian shengwu xuyangliang fenxi*	BOD$_5$ analysis
無形價值	*wuxing jiazhi*	immaterial values
無形商品	*wuxing shangpin*	immaterial goods
物質守恆定律	*wuzhi shouheng dinglü*	law on the conservation of material

X

夏季煙霧	*xiaji yanwu*	summer smog
夏季煙霧中的臭氧	*xiaji yanwuzhongde chouyang*	ozone in summer smog
現場處理	*xianchang chuli*	on-site treatment
現場恢復(補救)	*xianchang huifu (bujiu)*	in-site remediation
現場恢復補救處理	*xianchang huifu bujiu chuli*	on-site treatment in remediation
現代工藝水平	*xiandai gongyi shuiping*	state-of-the-art
相分離	*xiangfenli*	phase separation
相關系數	*xiangguan xishu*	correlation coefficient
相界面	*xiangjiemian*	phase boundary
線性活動	*xianxing huodong*	linear activity
顯著性(百分率水平)	*xianzhuxing (baifenlü shuiping)*	significance (percentage level)
消毒(滅菌)	*xiaodu (miejun)*	sterilization
消毒(一般的)	*xiaodu (yibande)*	disinfection, general
消費品的預付處置費	*xiaofeipinde yufu chuzhifei*	prepaid disposal fee for consumer goods
硝化(作用)	*xiaohua (zuoyong)*	nitrification
硝化桿菌	*xiaohua ganjun*	nitrobacter
硝化細菌	*xiaohua xijun*	nitrifying bacteria
硝化-反硝化過程	*xiaohua-fanxiaohua guocheng*	nitrification-denitrification process
消泡劑	*xiaopaoji*	antifoaming agent
小生境	*xiaoshengjing*	ecological niche
下水道系統	*xiaoshuidao xitong*	sewer systems
效益-費用分析	*xiaoyi-feiyong fenxi*	benefit-cost analysis

下游循環	xiayou xunhuan	downcycling
洗滌器	xidiqi	scrubber
洗滌液	xidiye	scrubbing liquid
洗脫液測試	xiduoye ceshi	eluate test
泄漏	xielou	spills
吸附(作用)	xifu (zuoyong)	adsorption
吸附劑	xifuji	adsorbent
吸附塔	xifuta	adsorption columns
細菌	xijun	bacteria
新陳代謝	xinchendaixie	metabolism
辛醇－水系數	xinchun-shui xishu	octanol-water coefficient
吸收(作用)	xishou (zuoyong)	absorption
系統邊界	xitong bianjie	system boundary
系統誤差	xitong wucha	systematic error
袖式過濾器	xiushiguolüqi	sleeve filter
修正措施	xiuzheng cuoshi	corrective measures
吸著過程	xizhao guocheng	sorption processes
旋風分離器	xuanfeng fenliqi	cyclone separators
懸浮固體(懸浮物)	xuanfu guti (xuanfuwu)	suspended solids
選擇催化還原法	xuanze cuihua huanyanfa	SCR process
選擇性非催化還原法	xuanzexing feicuihua huanyuanfa	selective non-catalytic reduction process, SNCR
循環利用	xunhuan liyong	recycling
循環利用技術	xunhuan liyong jishu	recycling technologies
絮凝(作用)	xuning (zuoyong)	flocculation
絮凝劑	xuningji	flocculating agent

Y

壓降	yajiang	pressure drop
壓力法五天生物需氧量測定	yalifa wutian shengwu xuyangliang ceding	barometric BOD_5 measurement
壓濾機	yalwji	filterpress
煙囪排放	yancong paifang	stack emissions
煙道氣	yandaoqi	flue gas
氧化鈣當量	yanghuagai dangliang	calcium oxide equivalents
氧化溝	yanghuagou	oxidation ditch
陽離子交換	yanglizi jiaohuan	cation exchange

樣品富集	*yangpin fuji*	sample enrichment
氧循環	*yangxunhuan*	oxygen cycle
煙氣處理	*yanqi chuli*	flue gas treatment
延時曝氣系統	*yanshi baoqi xitong*	extended aeration systems
岩石圈	*yanshiquan*	lithosphere
煙霧	*yanwu*	smog
煙霧	*yanwu*	smoke
厭氧廢水處理	*yanyang feishui chuli*	anaerobic effluent treatment
厭氧生物反應器 系統	*yanyang shengwu fanyingqi xitong*	biothane system
厭氧生物過程	*yanyang shengwu guocheng*	anaerobic biological processes
厭氧生物降解	*yanyang shengwu jiangjie*	anaerobic biodegradation
厭氧水體	*yanyang shuiti*	anaerobic waterbody
厭氧微生物	*yanyang weishengwu*	anaerobic microorganisms
遙感	*yaogan*	remote sensing
亞硝化桿菌	*yaxiaohua ganjun*	nitrosomonas bacteria
液滴分離器	*yedi fenliqi*	droplet separator
液化	*yehua*	liquefaction
液上氣體分析	*yeshang qiti fenxi*	headspace analysis
液相色譜法	*yexiang sepufa*	liquid chromatography
液液萃取	*yeye cuiqu*	liquid/liquid extraction
一般管理費用	*yiban guanli feiyong*	overhead costs
一般管理原則	*yiban guanli yuanze*	management principles in general
一般管理中的控制	*yiban guanlizhongde kongzhi*	controlling in general management
一般的催化過程	*yibande cuihua guocheng*	catalytic processes in general
一次澄清池	*yici chengqingchi*	primary clarifier
一次能源	*yici nengyuan*	energy, primary sources
移動爐箄式焚燒爐	*yidong lubishi fenshaolu*	moving grate incinerator
移動床吸附塔	*yidongchuang xifuta*	moving bed adsorption column
1'989年巴塞爾公約	*yijiubajiunian basai'er gongyue*	Basel Convention of 1989
1992年里約會議	*yijiujiu'ernian liyue huiyi*	RIO Conference 1992
1972年斯德哥爾摩 會議	*yijiuqi'ernian sidege'ermo huiyi*	Stockholm Conference 1972
螢光分光光度法	*yingguang fenguang guangdufa*	fluorescence spectrophotometry

熒光法	*yingguangfa*	fluorometry
應急貯存池	*yingji chucunchi*	emergency holding basins
因果關係	*yinguo guanxi*	cause-effect relationship
影響評價和分析	*yingxiang pingjia he fenxi*	impact assessment and analysis
營養級	*yingyangji*	trophic level
陰離子交換	*yinlizi jiaohuan*	anion exchange
儀器分析方法	*yiqi fenxi fangfa*	instrumental methods of analysis
異生化合物	*yisheng huahewu*	xenobiotica
億萬分之(幾分)	*yiwanfenzhi (jifen)*	ppt
異養生物	*yiyang shengwu*	heterotrophic organisms
抑制濃度(活性污泥)	*yizhi nongdu (huoxing wuni)*	inhibitory concentration (activated sludge)
用無脊椎動物作為試驗生物測定的廢水毒性	*yongwujizhuidongwu zuowei shiyan shengwu cedingde feishui duxing*	effluent toxicity with invertebra as test organisms
用於填料塔的刺猬型填料	*yongyu tianliaotade ciweixing tianliao*	hedge hog filling for packed columns
由微生物礦質化	*you weishengwu kuangzhihua*	mineralization by microorganisms
有毒廢物	*youdu feiwu*	toxic waste
有關產品和產品使用的法規	*youguan chanpin he chanpin shiyongde fagui*	legislation on products and product use
有關營養的	*youguan yingyangde*	trophic
有規填料	*yougui tianliao*	structured packing
有規填料塔	*yougui tianliaota*	structurally packed column
優諾克斯系統	*younuokesi xitong*	UNOX System
魚試驗	*yu shiyan*	fish tests
原核細胞	*yuanhe xibao*	procaryotic cells
原料收費	*yuanliao shoufei*	levy on raw materials
原生動物	*yuansheng dongwu*	protozoa
原生物	*yuanshengwu*	protists
源頭空氣	*yuantou kongqi*	source air
源頭污染預防	*yuantou wuran yufang*	PPS
源頭削減	*yuantouxuejian*	source reduction
原子吸收分析	*yuanzi xishou fenxi*	atomic absorption analysis
原子吸收火焰光譜法	*yuanzi xishou huoyan guangpufa*	atomic absorption flame spectrometry
預防措施	*yufang cuoshi*	preventive measures

預防性安全措施	*yufangxing anquan cuoshi*	preventive safety measures
預防性維修	*yufangxing weixiu*	preventive maintenance
預防性維修制度	*yufangxing weixiu zhidu*	preventive maintenance system
蘊涵標準	*yunhan biaozhun*	standards by implication
預濃縮	*yunongsuo*	preconcentration
運行成本(費用)	*yunxing chengben (feiyong)*	running costs
運行費用	*yunxing feiyong*	operating costs
運用市場機制的 法律手段	*yunyong shichang jizhide falü shouduan*	legal instruments using market mechanisms
雨水	*yushui*	meteor water
雨水溢流	*yushui yiliu*	rainwater overflow
固體廢物	*yuti feiwu*	solid waste
閾限值	*yuxianzhi*	threshold limit values

Z

再利用	*zailiyong*	reuse
載氣	*zaiqi*	carrier gas
再生技術	*zaisheng jishu*	regenerative technologies
載體上的活性污泥	*zaitishangde huoxing wuni*	activated sludge on supports
載體上的生物	*zaitishangde shengwu*	biologies on supports
在線分析	*zaixian fenxi*	on-line analysis
藻類	*zaolei*	algae
噪聲	*zaosheng*	noise
噪聲防護措施	*zaosheng fanghu cuoshi*	noise protection measures
噪聲防護區	*zaosheng fanghuqu*	noise protection zones
噪聲監測	*zaosheng jiance*	noise monitoring
噪聲散發限值	*zaosheng sanfa xianzhi*	noise emission limits
噪聲污染	*zaosheng wuran*	noise pollution
噪聲,干擾限值	*zaosheng, ganrao xianzhi*	noise, disturbance limits
責任	*zeren*	liability
責任關懷計劃	*zeren guanhuai jihua*	Responsible Care Program
折板	*zheban*	deflector elements
蒸餾分離	*zhengliu fenli*	distillation separation
蒸汽汽提	*zhengqi qiti*	stripping with steam
真核細胞	*zhenhe xibao*	eucaryotic cell
真菌	*zhenjun*	fungi
致癌多環芳烴	*zhi'ai duohuanfangting*	cPAH

指標（參數）	*zhibiao (canshu)*	parameter
脂肪組織	*zhifang zuzhi*	fatty tissue
職工的人員目標	*zhigongde renyuan mubiao*	personal objectives for individual employee
滯後損害費用	*zhihou sunhai feiyong*	delayed damage costs
滯後效應	*zhihou xiaoying*	delayed effects
指揮職能	*zhihui zhineng*	line function
指揮職責	*zhihui zhize*	line responsibility
直接費用	*zhijie feiyong*	direct costs
直接污染控制費用	*zhijie wuran kongzhi feiyong*	direct pollution control costs
質量目標	*zhiliang mubiao*	quality goals
質譜分析	*zhipu fenxi*	mass spectrometry, MS
脂溶性的	*zhirongxingde*	fat soluble
指示生物	*zhishi shengwu*	bioindicators
致死效應	*zhisi xiaoying*	lethal effect
植物毒性	*zhiwu duxing*	phytotoxicity
植物群落	*zhiwu qunluo*	phytocenose
置信界限	*zhixin jiexian*	confidence limits
職業衛生	*zhiye weisheng*	occupational health
中和當量	*zhonghe dangliang*	neutralization equivalents
重金屬	*zhongjinshu*	heavy metals
中空纖維膜	*zhongkong qianweimo*	hollow fiber membrane
重力沉降箱	*zhongli chenjiangxiang*	gravity settling box
總有機鹵素化合物	*zhongyoujilusu huahewu*	TOX
驟冷	*zhouleng*	quench
周轉率	*zhouzhuanlü*	turnover rate
撞擊式分離器	*zhuangjishi fenliqi*	impingement separator
裝置廢氣	*zhuangzhi feiqi*	apparatus off-air
轉盤式反應器	*zhuanpanshi fanyingqi*	disc reactor, rotating
助濾劑	*zhulüji*	filter aid
柱上注射器（氣相色譜法中）	*zhushang zhusheqi (qixiang sepufazhong)*	on-column injector (in GC)
自動取樣器	*zidong quyangqi*	autosampler
紫外輻射	*ziwai fushe*	UV radiation
自養生物	*ziyang shengwu*	autotrophic organisms
資源	*ziyuan*	resources

510

綜合環境管理	*zonghe huanjing guanli*	integral environmental management
綜合指標	*zonghe zhibiao*	sum parameters
綜合指標 (參數)	*zonghe zhibiao (canshu)*	collective parameter
綜合指標 (參數)	*zonghe zhibiao (canshu)*	aggregate parameter
綜合指標 (參數)	*zonghe zhibiao (canshu)*	group parameter
總量限值	*zongliang xianzhi*	mass flow limits
總碳	*zongtan*	TC
總懸浮固體	*zongxuanfu guti*	TSS
總需氧量	*zongxuyangliang*	total oxygen demand
總有機碳	*zongyoujitan*	total organic carbon
總有機碳分析	*zongyoujitan fenxi*	TOC analysis
總有機碳負荷	*zongyoujitan fuhe*	TOC load
總有機碳/總碳分析	*zongyoujitan/zongtan fenxi*	TOC/TC analysis
組合空氣污染控制 系統	*zuhe kongqi wuran kongzhi xitong*	combined APC-systems
最大容許濃度	*zuida rongxu nongdu*	MAK
最佳實用技術	*zuijia shiyong jishu*	best practicable technology
最佳實用技術	*zuijia shiyong jishu*	BPT
最終澄清池	*zuizhong chengqingchi*	final clarifier
最終處理塘	*zuizhong chulitang*	finishing pond
做功的能	*zuogongdeneng*	exergy
作用濃度	*zuoyong nongdu*	effect concentration